Student's Solutions Manual

Milton Loyer
Penn State University

Essentials of Statistics

Fourth Edition

Mario F. Triola
Dutchess Community College

Addison-Wesley
is an imprint of

Reproduced by Pearson Addison-Wesley from electronic files supplied by the author.

Publishing as Pearson Addison-Wesley, 75 Arlington Street, Boston, MA 02116.

ISBN-13: 978-0-321-64151-9
ISBN-10: 0-321-64151-5

1 2 3 4 5 6 BRR 12 11 10 09

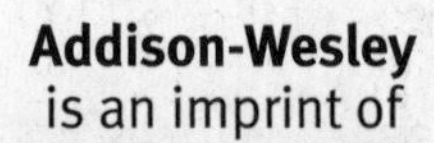

www.pearsonhighered.com

TABLE OF CONTENTS

INTRODUCTION

by Milton Loyer

This *Student's Solutions Manual* contains detailed solutions to all the the odd-numbered exercises for each section in the text *Essentials of Statistics*, Fourth Edition, by Mario Triola. It also contains the solutions to all of the end-of-chapter exercises: the Statistical Literacy and Critical Thinking exercises, the Chapter Quick Quiz, the Review Exercises, and the Cumulative Review Exercises. To aid in the comprehension of the calculations, worked problems typically include intermediate steps of algebraic or computer/calculator notation. When appropriate, additional hints and comments are included and prefaced by NOTE.

Many statistical problems are best solved using particular formats. Because recognizing and following these patterns promote understanding, this manual identifies and employs such formats whenever practicable.

I would like to thank Mario Triola for writing an excellent elementary statistics textbook and for inviting me to prepare this solutions manual.

Chapter 1

Introduction to Statistics

1-2 Statistical Thinking

1. A voluntary response sample, also called a self-selected selected, is one in which the subjects themselves decide whether or not to participate in the study.

3. Statistical significance occurs when the sample data indicate that a particular conclusion is mathematically justified. Practical significance occurs when the sample data lead to a conclusion that is meaningful and useful. A conclusion may be mathematically justified (i.e., statistically significant) but of no meaningful or useful value (i.e., of no practical significance). For example, the data may lead to the conclusion that a gasoline additive improves mileage – but if the estimated improvement is only 0.1 mpg, common sense dictates that using the additive is not worth the time or the money. That would be an example of a case where there is statistical significance but not practical significance.

5. Determining whether or not the weight loss is statistically significant properly requires applying the techniques presented in future chapters. But even if one can conclude that the program is effective (i.e., that the 3.0 lb weight loss is statistically significant), such a weight loss does not have practical significance. Most people would not be willing to subject themselves for one full year to a regimen that produced such a small weight loss.

7. (b) Possible, but very unlikely. Possible because there are no physical constraints or rules of the game that would prohibit such scores.

9. (c) Possible and likely. Likely in the sense of "not unusual enough to raise suspicion," not likely in the sense of "more probable to happen than not to happen."

11. (b) Possible, but very unlikely. This would be either an amazing coincidence or the result of specific planning.

13. (a) Impossible. The highest possible total on a normal pair of dice is 6+6 = 12.

15. There is nothing in the presentation of the data to suggest that the x and y values are matched. If the x and y values are not matched as presented, it does not make sense to calculate the differences between the x and y values – moving the x values around, for example, would yield a different set of differences.

17. The data can address the issue of whether or not the two types of cigarette (menthol and king-size) contain the same amounts of nicotine.

19. The presentation of the data suggests that the weight and mpg values are matched, each pair of values referring to a particular car. Even though the x (lbs) and y (mpg) values are matched, it does not make sense to calculate their differences. In general, quantities must have the same units in order to be added or subtracted – and then the resulting sum or difference will also have those same units. No meaningful units or interpretation can be assigned to "pounds minus mpg."

21. If car manufacturers supplied the information, there is reason to suspect that the data might be biased. Since the industry profits from increased sales, and since consumers tend to look for cars with high mpg ratings, the individual manufacturers might tend to supply information only for their better models (i.e., the one with the highest mpg values).

23. The weight loss program has statistical significance (i.e., one can conclude that the program produces a weight loss) because it is very unlikely (only 3 chances in 1000) that such a weight loss could have occurred by chance alone. The program does not have practical significance because the amount of lost weigh (3.3 lbs) is so small.

25. Determining whether or not the difference between the survey results (85%) and the statement of the industry representative (50%) is statistically significant properly requires applying the techniques presented in future chapters. But common sense suggests that this is statistically significant evidence against the representative's claim because (1) the observed result of 85% is so much greater than the stated claim of the 50% and (2) the sample of 1038 randomly selected adults appears to be properly chosen and large enough to provide reliable data.

27. a. Yes. In every case the highway amount is substantially greater than the city amount. If there were no difference, we would expect the highway and city amounts to be approximately equal – and when they were not, we would expect the highway amount to be larger about half the time and the city amount to be larger about half the time.
    b. Yes. The differences appear to be substantial – generally ranging from 6 to 10 mpg higher for the highway amounts.
    c. One practical implication of the difference is the increased cost of city driving compared to highway driving. Given a choice between two otherwise equal routes, financial considerations would suggest choosing the highway route over the city one.

1-3 Types of Data

1. A parameter is a numerical value that describes a population, while a statistic is a numerical value that describes a sample.

3. Discrete data are limited to certain finite or countable number of values, while continuous data can assume any value within a specified range.

5. Statistic, since it was determined from a sample of households.

7. Parameter, since it was determined from the population of all 2223 passengers.

9. Parameter, since it was determined from the population of all 50 states.

11. Statistic, since it was determined from a sample of 40 days.

13. Discrete, since the value must be one of the integers 0,1,2,...3249,3250.

15. Continuous, since the amount of nicotine could be any value on a continuum – even though it is reported to the nearest 0.1 mg.

17. Discrete, since the number of baby girls must be one of the integers 0,1,2,...,725,726.

19. Continuous, since the weight could be any value on a continuum – even though it is reported to the nearest 0.1 kg.

21. Ratio, since differences are meaningful and there is a meaningful zero.

23. Ordinal, since the categories have a natural ordering but the differences between the categories are not necessarily uniform.

25. Nominal, since the data give category names only and there is no natural ordering.

27. Interval, since differences between years are meaningful but there is no meaningful zero.

IMPORTANT NOTE for exercises 29-32: The population and sample are determined by the intent of the researcher, which must be clearly defined at the outset of any project. Unfortunately these exercises state only what the researcher did, and do not specifically identify the intent of the researcher. Consequently there may be differences of interpretation in some of the exercises, but some general principles apply.
(1) The sample is a subset of the population of interest and must have the same units as the population. If the population of interest is all households, for example, then the sample must be a selection of households and not a selection of adults – as households
with more adults would have a higher chance of being included in the study, thus creating a bias.
(2) The problem of nonresponse must be addressed. If 500 persons are randomly selected and asked a personal question, for example, but only 400 choose to answer the question, what is the sample? Depending on the situation, the sample could be either the 500 people randomly selected (and the 100 "no response" answers be reported as part of the sample data) or just the 400 people who actually gave data in the form of a specific answer.

29. a. The sample is the readers who completed and returned the survey.
b. The population is all people who read *USA Today*.
c. No, since the sample is self-selected it is not likely to be representative of the population. Only those with special interest in health matters are likely to return the survey.

31. a. The sample is the people who responded to the request.
b. The intended population is likely all persons over 18. The actual population is actually all persons over 18 who have opportunity to receive the request. If the request went out over the Internet, for example, the population would be all Internet users over 18. If the request went out over several radio and/or TV stations, the population would be all persons over 18 who tune in to those stations. And so on.
c. No, since the sample is self-selected it is not likely to be representative of the population. Only those with special interest in abortion issues are likely to spend the time and money to respond to the survey.

33. Temperature ratios are not meaningful because a temperature of 0° does not represent the absence of temperature in the same sense that \$0 represents the absence of money. The zero temperature in the exercise (whether Fahrenheit or Centigrade) was determined by a criterion other than "the absence of temperature."

35. This is an example of ordinal data. It is not interval data because differences are not meaningful – i.e., the difference between the ratings +4 and +5 does not necessarily represent the same differential in the quality of food as the difference between 0 and +1.

1-4 Critical Thinking

1. A voluntary response sample is one in which the subjects themselves choose whether or not to participate. This is generally unsuitable for statistical purposes because it is persons with strong feelings and/or a personal interest in the topic that tend to respond – and the opinions of such people are not necessarily representative of the entire population.

3. a. No. An association between two variables does not imply that there is a cause and effect. Even if there is a cause and effect between two variables, the statistics alone cannot identify which is the cause and which is the effect – saying that an increase in the number of registered weapons caused an increase in the murder rate is no more justified than saying it was an increase in the murder rate that caused the increase in the number of registered weapons.
   b. No. Since there is not necessarily a cause and effect relationship, a decrease in the number of registered weapons will not necessarily result in a reduced murder rate.

5. College graduates tend to earn more money than non-graduates, and people who have more money are able to purchase better health care. Having more money (whether or not it resulted from having a college degree) and not studying more is the primary contributing factor toward longer life.

7. If the population of Orange County includes significantly more minority drivers than white drivers, one would expect more speeding tickets to be issued to minorities than to whites – even if the percentage of white drivers who violated the speed limit was greater than the percentage of minority drivers who did so. It is also possible that police tend to target minority drivers – so that the numbers of tickets issued to the various racial/ethnic groups does not correspond to the actual amount of speed limit violations occurring. The fact that more speeding tickets are given to minorities does not warrant the conclusion that minority persons are more likely to speed.

9. Self-reporting is not always a reliable method of gathering information. People tend to answer questions in a manner that will reflect well on them; they tend to give the answer they wish were true, or that they wish others to believe about them, rather than the true answer. The 82% rate is more likely to be accurate because it was based on objective observation rather than self-reporting.

11. The fact that the study was financed by groups with vested interests in the outcome might have influenced the conducting and/or reporting to concentrate on aspects of the study favorable to chocolate. The fact that a product contains one ingredient associated with positive health benefits may not outweigh other negative aspects of the product that were not reported.

13. There are at least two reasons why the results cannot be used to reach any conclusion about how the general population feels about keeping the United Nations in the United States. First, only viewers of the ABC "Nightline" program, which are not necessarily representative of the general population, were aware of the poll. Secondly, the 186,000 persons who responded (and there is always the possibility that some people called more than once in order to give their opinion more weight) are a voluntary response sample – which means that the responses will include an over representation of those with strong feelings and/or a personal interest in the survey.

15. People who were killed in motorcycle crashes, when a helmet may have saved their lives, were not present to testify.

17. No, this is not likely to be a good estimate of the average per capita income for all individuals in the United States. States with small populations but high wages (due to the cost of living) like Alaska and Hawaii, for example, contribute 2/50 = 4% of the average for the 50 states – but since they do not contain 4% of the US wage earners, they inflate the estimate to make it higher than the true average of all US wage earners. In general, the average of averages does not give the overall average.
NOTE: Imagine a hypothetical country of two states: one where all 4 persons earned a total of \$400,000 (for a per capita income of \$400,000/4 = \$100,000), and one where all 96 persons earned a total of \$4,800,000 (for a per capita income of \$4,800,000/96 = \$50,000). The true per capita income for the country is (400,000 + 4,800,000)/100 = 5,200,000/100 = \$52,000, but the average of the state averages is (100,000 + 50,000)/2 = 150,000/2 = \$75,000.

19. Nothing. Because the information comes from a voluntary response sample, the opinions expressed are more likely to be representative of those with strong feelings on the topic than of the general population. Even if the information had not come from a voluntary response sample, the conclusion would apply only to women who read *Good Housekeeping* magazine and not to "all women."

21. a. 5/8 = 0.625 = (0.625)(100%) = 62.5%
b. 23.4% = 23.4/100 = 0.234
c. (37%)(500) = (37/100)(500) = 185
d. 0.127 = (0.127)(100%) = 12.7%

23. a. (49%)(734) = (49/100)(734) = 360
b. 323/734 = 0.440 = (0.440)(100%) = 44.0%

25. a. (5%)(38410) = (5/100)(38410) = 1920.5
While either 1920 or 1921 seem like reasonable answers, it is not possible to give an exact answer; any number between 1729 and 2112 rounds to 5%.
b. 18053/38410 = 0.470 = (0.470)(100%) = 47.0%
c. No. Because the information comes from a voluntary response sample, the opinions expressed are more likely to be representative of those with strong feelings on the topic than of the general population. Even if the information had not come from a voluntary response sample, the conclusion would apply only to those with AOL Internet access and not to the "general population."

27. If something falls 100%, there is none remaining. For foreign investment to fall 500% it would have to decline by an amount equal to 5 times as much as it started with – which is not possible.

29. If the researcher started with 20 mice in each of the six groups, and none of the mice dropped out of the experiment for any reason, then the proportions of success could only be fractions like 0/20, 1/20, 2/20, etc. – and the success rates in percents could only be multiples of 5 like 0%, 5%, 10%, etc. A success rate of 53%, for example, would not be possible.
NOTE: This appears to be a case of miscommunication rather than falsifying data. If the 120 mice had not been divided evenly among the 6 groups, the following fractions would account for all the 19+19+19+24+21+18 = 120 mice and give the 6 reported success rates:
10/19 = 53%, 11/19 = 58%, 12/19 = 63%, 11/24 = 46%, 10/21 = 48%, 12/18 = 67%.

1-5 Collecting Sample Data

1. In a random sample, every individual member has an equal chance of being selected; while in a simple random sample of size n, every possible sample of size n has an equal chance of being selected.

3. No. A random sample is likely to be representative of the population from which it is selected. The stated sample is likely to be representative of the student's friends, but not necessarily of the general population.

5. Observational study, since the researcher (Emily) merely measured whether the therapist could identify the chosen hand and did nothing to modify the therapist.

7. Experiment, since the effect of an applied treatment (in this case a zero dose of the appropriate medicine) was measured.
NOTE: There is room for disagreement – some might argue that this is an observational study, since the intent was to learn about the effects of the disease on untreated men and specific characteristics were measured on unmodified subjects. But since the patients were given a placebo and/or were part of a larger formal study in which other persons received the proper medicine, then the given statement describes an experiment. Even if there were no such larger formal study, the spirit of the statement is that the subjects were selected and monitored for the purposes of seeing what particular effect their "treatment" had and/or comparing this effect to known results when the proper treatment was given.

9. Convenience, since the sample is those who happened to be in the student's family.

11. Cluster, since all the voters at randomly selected polling stations were surveyed.

13. Stratified, since the population of wines was subdivided into 5 different subgroups (wineries), and then samples were drawn from each subgroup.

15. Systematic, since every 100th spark plug is tested.

17. Random, since every person who filed a return had an equal chance of being selected..

19. Cluster, since all the members of six randomly selected health plans were interviewed.

21. Yes, it is a random sample because each pill has an equal chance of being selected.
Yes, it is a simple random sample because each sample of size 30 has an equal chance of being selected.

23. Yes, it is a random sample because each voter has an equal chance of being selected – and that probability is the number of precincts selected divided by the total number of precincts.
No it is not a simple random sample because some samples of a given size are not possible – e.g., a sample with at least voter from every precinct is not possible.

25. No, this is not a random sample of all New Yorkers because persons who did not visit the location had not chance of being selected.
No it is not a simple random sample because it is not a random sample.

27. Retrospective. The researchers identified groups with a common characteristic (those with and without present respiratory problems) and gathered data by going backward in time (to see how they were involved in the events of 9/11).

29. Cross-sectional. The data examined refer to a single point in time.

31. Blinding occurs when either the subject or the evaluator doesn't know if the subject received a valid treatment or a placebo. Double-blinding occurs when neither the subject nor the evaluator knows whether a treatment or a treatment was received. Blinding is important in experiments in which there is subjectivity involved in measuring the response. In this experiment, knowledge of whether a treatment or a placebo was received might influence the subject's or evaluator's assessment of whether the "treatment" was effective in dealing with the cold.

33. To see which of two statistics textbooks results in better student understanding, have Professor A use one text in all his classes and Professor B use the other text in all her classes. At the end of the semester give all the students a common exam. Confounding will occur because it will not be possible to tell if any difference between the two groups is due to the textbook or to the professor.

Statistical Literacy and Critical Thinking

1. No. Since the responders constituted a voluntary response sample, they likely represented only those with strong feelings for or against one candidate or the other and were not necessarily representative of the entire population.

2. a. Quantitative, because the data consist of numbers representing measurements
   b. Continuous, since length can be any value on a continuum – even though the values are recorded in whole minutes.
   c. Observational study, since the data were collected without involving any treatments or modifications.
   d. Nominal, since the data give names only and do not measure or rank the films.
   e. Ratio, since differences are meaningful and there is a meaningful zero.

3. The subjects must be selected so that all samples of the same size have the same chance of being selected.

4. No, this is the average travel time to work for all individuals in the United States. States with larger populations (which would likely have th greater travel times) should be given more weight.
   NOTE: Imagine a hypothetical country of two states: one where all 4 persons had an average travel time of 10 minutes, and one where all 96 persons had an average travel time of 30 minutes. The true average travel time for the country is [(4)(5) + (96)(30)]/100 = [2900]/100 = 29 minutes, but the average of the state averages is [5 + 30]/2 = [35]/2 = 17.5 minutes.

Chapter Quick Quiz

1. True. That is a possible complete set from which one might wish to take a sample.

2. Continuous. They may take on any value within a continuum.

3. False. Every combination of n names is not possible – e.g., no combination of names containing two adjacent names is possible.

4. False. A numerical value describing a characteristic of a sample is a statistic, not a parameter.

5. Experiment. The application of a treatment makes the study an experiment.

6. False. There is no natural ordering of the colors.
7. Population. For a sample, the numerical value that describes some characteristic is a statistic.
8. Categorical. The data consist of names or labels, and they are not numbers representing counts or measurements. Since there is a natural ordering to the ratings, they are categorical data at the ordinal level of measurement.
9. Nominal. The book categories are names only, and no natural ordering scheme applies.
10. No. Since the people were randomly selected, the responders constitute a random sample of the population. The typical voluntary response sample results from those choosing to respond to surveys that are sent out (whether by print media, the Internet, on the phone, etc.) indiscriminately to broad audiences. If some of the 500 people in this particular survey refused to reply, it would still be a random sample – but the response "no reply" should be used to bring the total number of responses up to 500 in order to account for all the subjects in the survey.

Review Exercises

1. a. Since the survey was based on a voluntary response sample, its results might not be representative of the population. It could be that people who respond to such surveys tend to have characteristics (e.g., being conscientious) that would affect how they squeeze their toothpaste tubes.
   b. As stated to apply to the population of all Americans, the 72% is a parameter. In truth, the figure is based on a sample and is properly a statistic. While it may be used to estimate the value in the population, it is accurate only for the sample. It is a statistic that is used to estimate the parameter, and the article should have stated that "it is estimated that 72% of Americans squeeze their toothpaste tube from the top."
   c. Observational study, since there was no attempt to modify the behavior being observed.
2. No, results based only those who agree to respond may not be representative of the entire population. Those who refuse to respond may be fundamentally different from their more cooperative peers and may possess opinions that would become under-represented in the sample. One possible strategy for dealing with those who refuse to respond would be to call back (perhaps using a different pollster with an especially agreeable manner) and hope to find them in a more cooperative mood.
3. a. Ratio, since differences are meaningful and there is a meaningful zero.
   b. Nominal, since gender is only a category with no natural ordering.
   c. Interval, since differences between temperatures are meaningful but there is no meaningful zero.
   d. Ordinal, since the categories have a natural ordering but the differences between the categories are not necessarily uniform.
4. a. Nominal, since color is only a category with no natural ordering.
   b. Ratio, since differences are meaningful and there is a meaningful zero.
   c. Ordinal, since the categories have a natural ordering but the differences between the categories are not uniform.
   d. Interval, since differences between temperatures are meaningful but there is no meaningful zero.

5. a. Discrete. While it may be possible to own ½ shares, the number of shares cannot be any value on a continuum – e.g., a person cannot own π shares.
   b. Ratio. Differences are meaningful and there is a meaningful zero.
   c. Stratified. The population has been divided into subgroups, and subjects are selected from each subgroup.
   d. Statistic. While it may be used to estimate the true value in the population (i.e., the parameter), the average in the sample is a statistic.
   e. That would be a voluntary response sample and would likely be representative only of those stockholders with strong opinions and not of the general population of stockholders.

6. NOTE: The second part of the problem involves deciding "whether the sampling scheme is likely to result in a sample that is representative of the population." This is subjective, since the term "representative" has not been well-defined. Mathematically, any random sample is representative in the sense that there is no bias and a series of such samples can be expected to average out to the true population values – even though any one particular sample may not be representative. And so we expect a random sample of size n=50 to be representative. But how about a random sample of size n=2? It has all the mathematical properties of a random sample of size n=50, but common sense suggests that any one sample of size n=2 is not necessarily likely to be representative of the population.
   a. Systematic, since the selections were made at regular intervals. Yes, there is no reason why every 500th stockholder should have some characteristic that would introduce a bias.
   b. Convenience, since those selected were the ones who happened to attend. No, those who attend would tend to be the more interested and/or more well-off stockholders.
   c. Cluster, since the stockholders were organized into groups (by stockbroker) and all the stockholders in the selected groups were surveyed. Yes, since every stockholder has the same chance of being included in the sample (assuming each stockholder works through a single stockbroker), the sample will be a random sample – and considering the large number of stockbrokers involved (see the NOTE at the beginning of the problem), it is reasonable to expect the sample to be representative of the population.
   d. Random, since each stockholder has the same chance of being selected. Yes, since every stockholder has the same chance of being included in the sample, the scheme is likely to result in a representative sample.
   e. Stratified, since the stockholders were divided into subpopulations (zip codes) from which the actual sampling was done. No, since all the zip codes were given equal weights (5 stockholders from each) – because "significant" zip codes (with large numbers of stock-holders) are counted equal with "insignificant" zip codes (with small numbers of small stockholders), stockholders from "insignificant" zip codes have a greater chance of being included and the types of stockholders that live in "significant" zip codes will be under-represented.

7. a. 12/35 = 0.343 = (0.343)(100%) = 34.3%
   b. (18%)(4544) = (18/100)(4544) = 818

8. a. Parameter, since it was calculated from the entire population.
   b. Discrete, since the numbers of votes must be whole numbers.
   c. (49.72%)(68,838,000) = (49.72/100)(68,838,000) = 34,226,000 (rounded to agree with the accuracy used for the size of the population)

9. a. To contain 100% less fat would be to contain no fat at all – 100% of the fat has been removed. It is not physically possible to contain 125% less fat.
   b. (58%)(1182) = (58/100)(1182) = 686
   c. 331/1182 = 0.280 = (0.280)(100%) = 28.0%

10. The Gallup poll used randomly selected respondents, while the AOL respondents were a voluntary response sample. In addition, those with Internet access to participate in the AOL poll would include more affluent and well-educated persons (who tended to support Obama) and fewer blue-collar workers (who tended to support Clinton). The Gallup poll is more likely to reflect the true opinions of American voters.

Cumulative Review Exercises

NOTE: Throughout the text intermediate mathematical steps will be shown to aid those who may be having difficulty with the calculations. In practice, most of the work can be done continuously on calculators and the intermediate values are unnecessary. Even when the calculations cannot be done continuously, DO NOT WRITE AN INTERMEDIATE VALUE ON YOUR PAPER AND THEN RE-ENTER IT IN THE CALCULATOR. That practice can introduce round-off and copying errors. Store any intermediate values in the calculator so that you can recall them with greater accuracy and without copying errors. In general, the degree of accuracy appropriate depends upon the particular problem – and guidelines for this will be given as needed in subsequent chapters. Unless there is reason to otherwise, answers in this section are given with 3 decimal accuracy.

1. $\frac{1.1+1.7+1.7+1.1+...+1.1}{25} = \frac{31.4}{25} = 1.256$ mg

2. $\frac{110+96+170+125+...+119}{35} = \frac{4196}{35} = 119.9$ min

3. $\frac{85-80}{3.3} = \frac{5}{3.3} = 1.52$

4. $\frac{12.13-12.00}{0.12/\sqrt{24}} = \frac{0.13}{0.0245} = 5.31$

5. $\left[\frac{(1.96)(0.25)}{0.01}\right]^2 = [49]^2 = 2401$

6. $\frac{(491-513.174)^2}{513.174} = \frac{(-22.174)^2}{513.174} = \frac{491.6863}{513.174} = 0.9581$

7. $\frac{(98.0-98.4)^2+(98.6-98.4)^2+(98.6-98.4)^2}{3-1}$

$= \frac{(-0.4)^2+(0.2)^2+(0.2)^2}{2} = \frac{0.16+0.04+0.04}{2} = \frac{0.24}{2} = 0.12$

8. $$\sqrt{\frac{(98.0\text{-}98.4)^2+(98.6\text{-}98.4)^2+(98.6\text{-}98.4)^2}{3\text{-}1}}$$

$$=\sqrt{\frac{(-0.4)^2+(0.2)^2+(0.2)^2}{2}}=\sqrt{\frac{0.16+0.04+0.04}{2}}=\sqrt{\frac{0.24}{2}}=\sqrt{0.12}=0.346$$

NOTE FOR EXERCISES 9-12: Many calculators have a display that is limited to 8 or 10 characters, and they will not display as many digits as are shown in the following solutions. Answers with fewer significant digits are acceptable.

9. $(0.4)^{12}$ = 1.6777216E-05 = 0.000016777216

10. 5^{15} = 3.0517578125E10 = 30,517,578,125

11. 9^{11} = 3.1381059609E10 = 31,381,059,609

12. $(0.25)^6$ = 2.44140625E-04 = 0.000244140625

Chapter 2

Summarizing and Graphing Data

2-2 Frequency Distributions

1. No. The first class frequency, for example, tells us only that there were 18 pennies with weights in the 2.40-2.49 grams class, but there is no way to tell the exact values of those 18 weights.

3. No. This is not a relative frequency distribution because the sum of the percentages is not 100%. It appears that each respondent was asked to indicate whether he downloaded the four types of material (and so the sum of the percentages could be anywhere from 0% to 400%), and not to place himself in one of the four categories (in which case the table would be a relative frequency distribution and the sum of the percentages would be 100%).

5. a. Class width: subtracting the first two lower class limits, 14−10 = 4.
   b. Class midpoints: the first class midpoint is (10+13)/2 = 11.5, and the others can be obtained by adding the class width to get 11.5, 15.5, 19.5, 23.5, 27.5.
   c. Class boundaries: the boundary between the first and second class is (13+14)/2 = 13.5, and the others can be obtained by adding or subtracting the class width to get 9.5, 13.5, 17.5, 21.5, 25.5, 29.5.

7. a. Class width: subtracting the first two lower class limits, 1.00−0.00 = 1.00.
   b. Class midpoints: the first class midpoint is (0.00+0.99)/2 = 0.495, and the others can be obtained by adding the class width to get 0.495, 1.495, 2.495, 3.495, 4.495.
   c. Class boundaries: the boundary between the first and second class is (0.99+1.00)/2 = 0.995, and the others can be obtained by adding or subtracting the class width to get -0.005, 0.995, 1.995, 2.995, 3.995, 4.995.

9. a. Strict interpretation: No; because there are more values at the upper end, there is not symmetry.
   b. Loose interpretation: Yes; there is a concentration of frequencies at the middle and a tapering off in both directions.

11. The requested figure is given below. Obtain each relative frequency by dividing the given frequency by 25, the total number of observations in each table. The "total" line is not necessary.

    The non-filtered cigarettes have much more tar. Yes, the filters appear to be effective in reducing the amount of tar.

Relative Frequency Comparison

	cigarette	type
tar (mg)	non-filtered	filtered
2–5	0%	8%
6–9	0%	8%
10–13	4%	24%
14–17	0%	60%
18–21	60%	0%
22–25	28%	0%
26–29	8%	0%
total	100%	100%

NOTE: For cumulative tables, this manual uses upper class boundaries in the "less than" column. Consider exercise #13, for example, to understand why is done. Conceptually, weights occur on a continuum and the integer values reported are assumed to be the nearest whole number representation of the precise measure. An exact weight of 17.7, for example, would be reported as 18 and fall in the third class. The values in the second class, therefore, are better described as "less than 17.5" (using the upper class boundary) than as "less than 18" (using the lower class limit of the next class). This distinction is crucial in the construction of pictorial representations in the next section. To present a visually simpler table, however, it is common practice to follow the example in the text and use the lower class limit of the next class. Regardless of the "less than" label, the final cumulative frequency must equal the total sample size – and the sum of the cumulative frequency column has no meaning and should never be included.

13. Obtain the cumulative frequency values by adding the given frequencies.

tar (mg) in non-filtered cigarettes	cumulative frequency
less than 13.5	1
less than 17.5	1
less than 21.5	16
less than 25.5	23
less than 29.5	25

15. Obtain the relative frequencies by dividing the given frequencies by the total of 2223.

category	relative frequency
male survivors	16.2%
males who dies	62.8%
female survivors	15.5%
females who died	5.5%
	100.0%

17. The requested table is given below. The frequency distribution of the last digits shows unusually high numbers of 0's and 5's. This is typical for data that have been rounded off to "convenient" values. It appears that the heights were reported and not actually measured.

digit	frequency
0	9
1	2
2	1
3	3
4	1
5	15
6	2
7	0
8	3
9	1
	37

19. The requested table is given below.

nicotine (mg)	frequency
1.0–1.1	14
1.2–1.3	4
1.4–1.5	3
1.6–1.7	3
1.8–1.9	1
	25

21. The requested table is given below. No, the voltages do not appear to follow a normal distribution – instead of being concentrated near the middle of the distribution, the values appear to be rather evenly distributed.

voltage (volts)	frequency
123.3–123.4	10
123.5–123.6	9
123.7–123.8	10
123.9–124.0	10
124.1–124.2	1
	40

23. The requested table is given below. While over half of the screws are within 0.01 inches of the claimed value (28 of 50 fall between 0.74 and 0.76), there are over twice as many screws below that range as there are above it (15 vs. 7). It appears that there might be a slight tendency to err on the side of making the screws too small.

length (in)	frequency
0.720–0.729	5
0.730–0.739	10
0.740–0.749	11
0.750–0.759	17
0.760–0.769	7
	50

25. The requested table is given below. The ratings appear to have a distribution that is not normal. While there is a maximum score with progressively smaller frequencies on either side of the maximum, the distribution is definitely not symmetric (i.e., the maximum score is not near the middle, but at the upper end of the distribution).

FICO score	frequency
400–449	1
450–499	1
500–549	5
550–599	8
600–649	12
650–699	16
700–749	19
750–799	27
800–849	10
850–899	1
	100

27. The requested table is given below.

weight (g)	frequency
6.0000–6.0499	2
6.0500–6.0999	3
6.1000–6.1499	10
6.1500–6.1999	8
6.2000–6.2499	6
6.2500–6.2999	7
6.3000–6.3499	3
6.3500–6.3999	1
	40

29. The requested table is given below.

blood group	frequency
O	22
A	20
B	5
AB	3
	50

31. The frequency distributions including and excluding the outlier are given below. In general, an outlier can add several rows to a frequency distribution. Even though most of the added rows have frequency zero, the table tends to suggest that these are possible values – thus distorting the reader's mental image of the distribution.

0.0111 CANS (with the outlier)

weight (lbs)	frequency
200 – 219	6
220 – 239	5
240 – 259	12
260 – 279	36
280 – 299	87
300 – 319	28
320 – 339	0
340 – 359	0
360 – 379	0
380 – 399	0
400 – 419	0
420 – 439	0
440 – 459	0
460 – 479	0
480 – 499	0
500 – 519	1
	175

0.0111 CANS (without the outlier)

weight (lbs)	frequency
200 – 219	6
220 – 239	5
240 – 259	12
260 – 279	36
280 – 299	87
300 – 319	28
	174

2-3 Histograms

1. The pulse rate data have been organized into 7 classes. Examining the frequency distribution requires consideration of 14 pieces of information: the 7 class labels, and the 7 class frequencies. The histogram efficiently presents the same information in one visual image and gives all the relevant CVDOT (center, variation, distribution shape, outlier, [time is not relevant for these data]) details in an intuitive format.

3. The data set is small enough that the individual numbers can be examined; they do not require summarization in a figure. The data set is not large enough for a histogram to reveal the true nature of the distribution; the histogram will essentially be a repeat of the individual numbers.

NOTE: For exercises 5-8, the following values are used to answer the questions. It appears that the midpoints of the first 3 classes are 5,000 and 10,000 and 15,000. It appears that the heights of the 5 bars are 2, 30, 8, 15, 5.

5. a. Adding the heights of all the bars, the total number is 2+30+8+15+5 = 60.
   b. Adding the heights of the two rightmost bars, the number over 20,000 miles is 15+5 = 20.

7. a. The minimum possible miles traveled is the lower class boundary associated with the leftmost bar, 2500 miles.

b. The maximum possible number of miles traveled is the upper class boundary associated with the rightmost bar, 42,500 miles.

9. The histogram is given below. The digits 0 and 5 occur disproportionately more than the others. This is typical for data that have been rounded off to "convenient" values. It appears that the heights were reported and not actually measured.

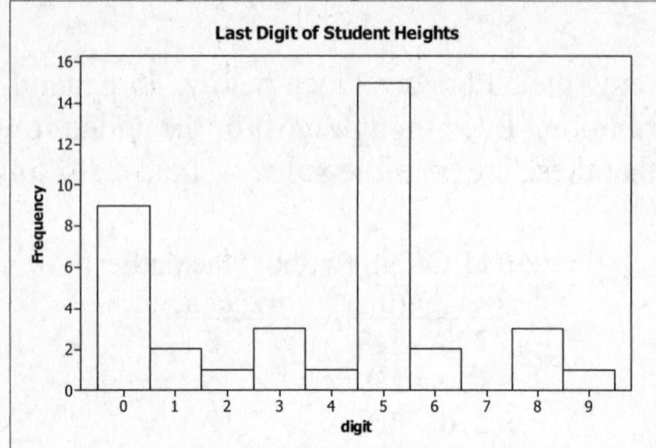

11. The histogram is given below.

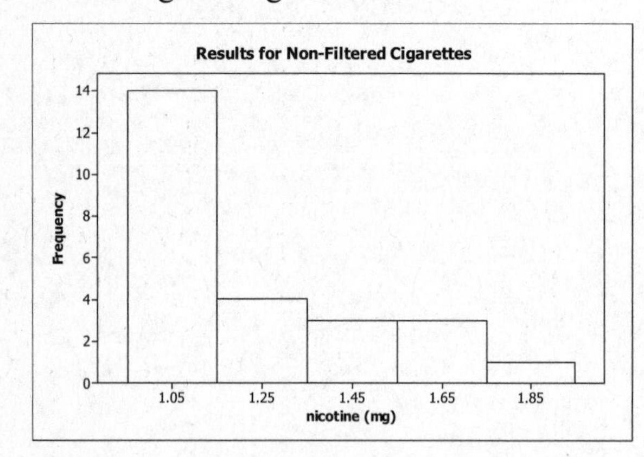

13. The histogram is given below. No, the voltages do not appear to follow a normal distribution – instead of being concentrated near the middle of the distribution, the values appear to be rather evenly distributed.

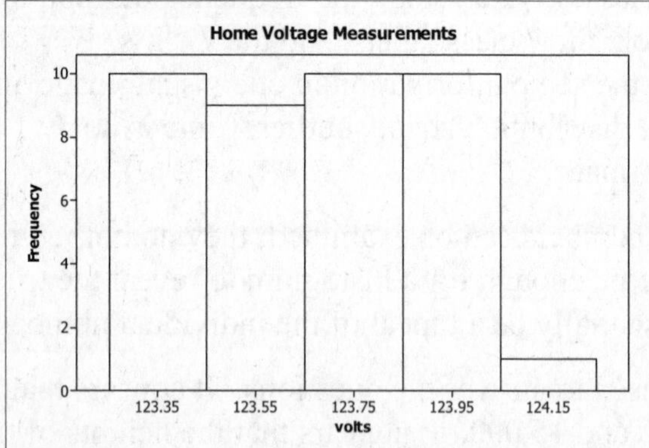

15. The histogram is given below. The true class boundaries are 0.7195, 0.7295, 0.7395, etc. The manual follows the text in presenting a histogram that communicates the information in an appropriate, though approximate, manner. While the 0.75″ label appears reasonably accurate in that all but 5 of the screws were within 0.02″ of that value, it appears that there are slightly more screws below the labeled value than above the labeled value and that the values extended farther below the labeled value than above the labeled value.

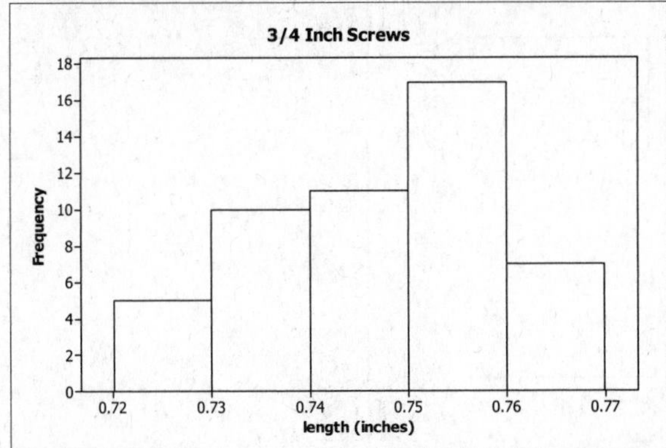

17. The histogram is given below. The true class boundaries are 399.5, 449.5, 499.5, etc. The manual follows the text in presenting a histogram that communicates the information in an appropriate, though approximate, manner. The ratings appear to have a distribution that is not normal. While there is a maximum score with progressively smaller frequencies on either side of the maximum, the distribution is definitely not symmetric (i.e., the maximum score is not near the middle, but at the upper end of the distribution).

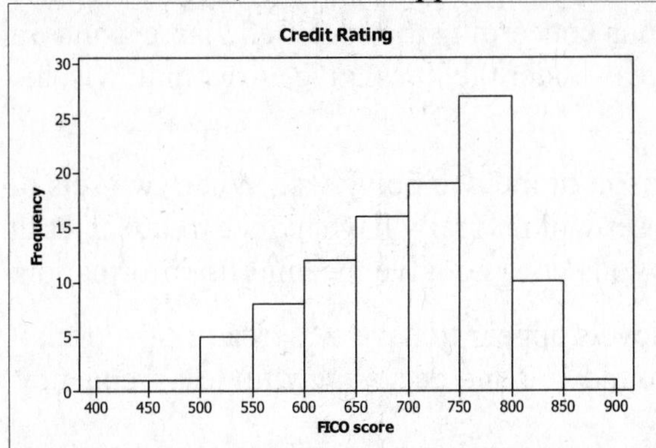

19. The histogram is given below at the left. The true class boundaries are 5.99995, 6.04995, 6.09995, etc. The manual follows the text in presenting a histogram that communicates the information in an appropriate, though approximate, manner.

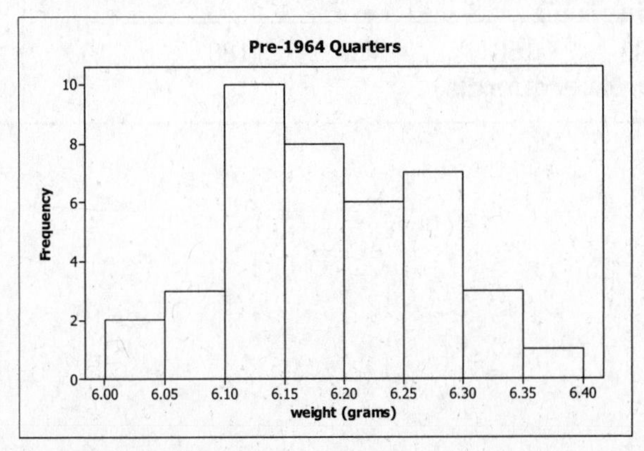

21. The back-to-back relative frequency histograms are given below. The pulse rates of the males tend to be lower than those of the females.

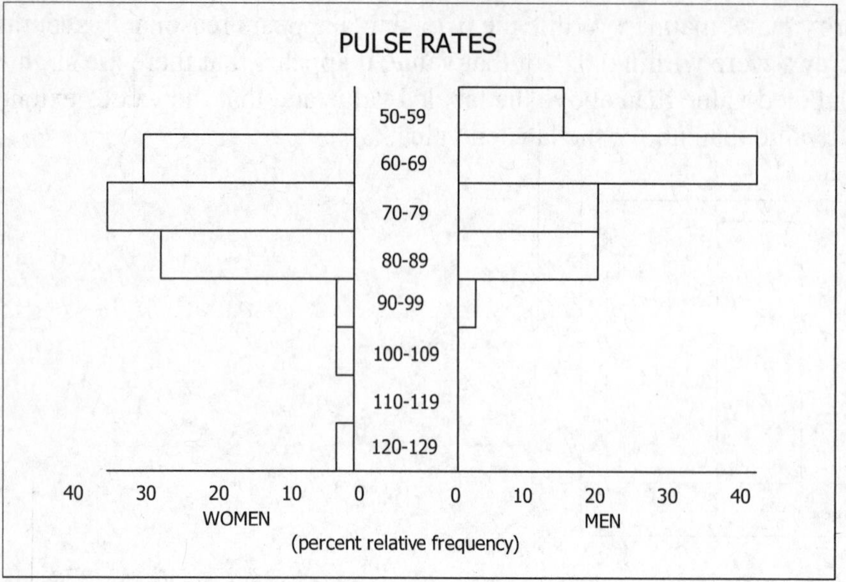

2-4 Statistical Graphics

1. The dotplot permits identification of each original value and is easier to construct. The dotplot gives an accurate visual impression of the proportion of the data within <u>any</u> selected range of values; while the polygon is limited to impressions concerning the specified classes (and only the heights at the class midpoints, and not the areas under the lines, give an accurate visual impression of those proportions).

3. Using relative frequencies allows direct comparison of the two polygons. When two sets of data have different sample sizes, the larger data set will naturally have higher frequencies and direct comparison of the heights of the two polygons does not give meaningful information.

5. The dotplot is given below. The Strontium-90 levels appear to have a "spread-out" normal distribution, a wide range of values clustered around 150 and occurring with less frequency at the extremes.

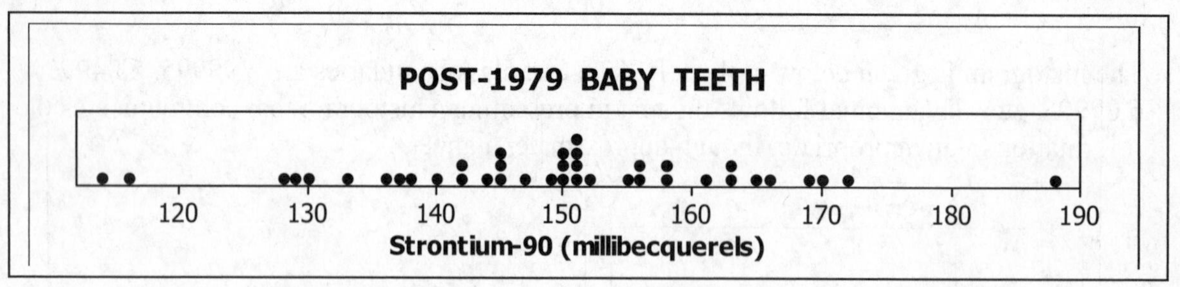

7. The frequency polygon is given below at the left.
NOTE: The frequencies are plotted at the class midpoints, which are not integer values. The polygon must begin and end at zero at the midpoints of the adjoining classes that contain no data values.

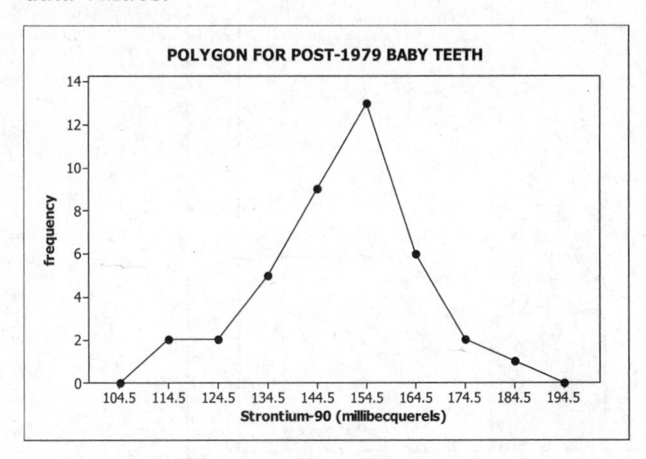

9. The stemplot is given below. The weights appear to be approximately normally distributed, except perhaps for the necessary lower truncation at zero.

<u>weight (pounds)</u>
0. | 12356677888999
1. | 1123444444445556678
2. | 0011111133346688889999
3. | 9345
4. | 36
5. | 2

11. The ogive is given below at the left. Using the figure: move up from 4 on the horizontal scale to intersect the graph, then move left to intersect the vertical scale at 59. This indicates there were approximately 59 data values which would have been recorded as being below 4, which agrees with the actual data values.
NOTE: Ogives always begin on the vertical axis at zero and end at n, the total number of data values. All cumulative values are plotted at the upper class boundaries.

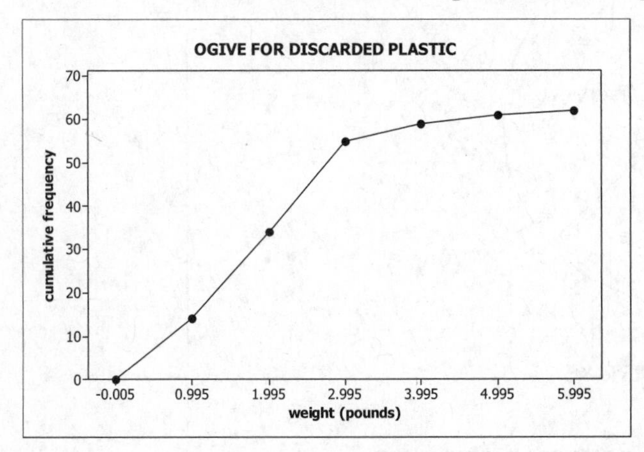

13. The Pareto chart is given below.

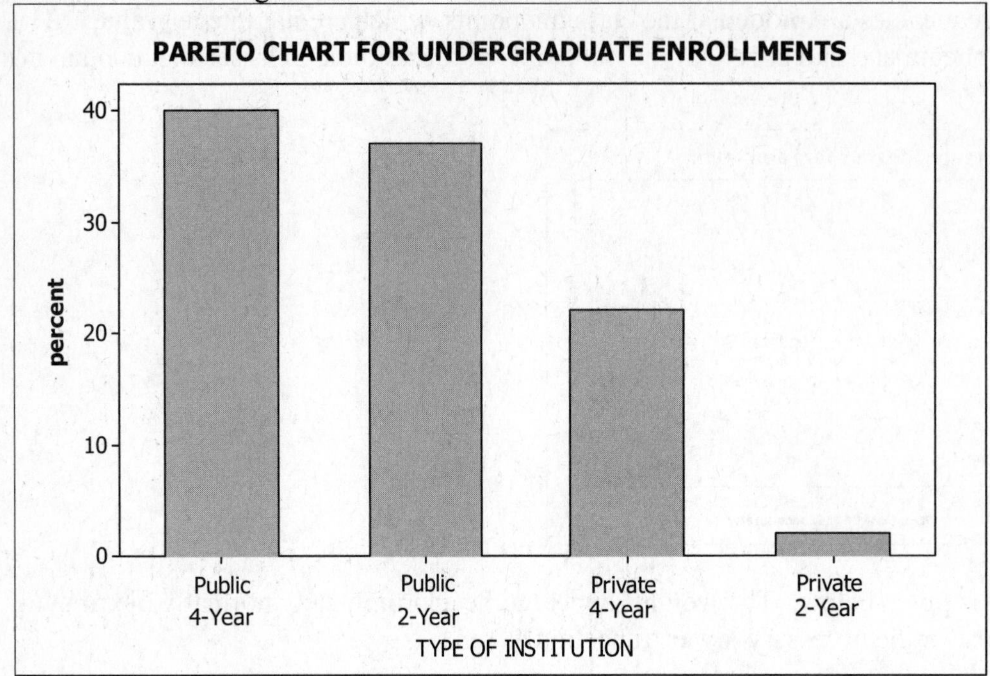

15. The pie chart is given below. The "slices" of the pie may appear in any order and in any position, but their relative sizes must be as shown. There were 1231 total responses, and the central angle of the pie chart for each category was determined as follows.

Interview: 452/1231 = 36.7%, and 36.7% of 360° is 132°
Resume: 297/1231 = 24.1%, and 24.1% of 360° is 87°
Reference Checks: 143/1231 = 11.6%, and 11.6% of 360° is 42°
Cover Letter: 141/1231 = 11/5%, and 11.5% of 360° is 41°
Interview Follow-Up: 113/1231 = 9.2%, and 9.2% of 360° is 33°
Screening Call: 85/1231 = 6.9%, and 6.9% of 360° is 35°

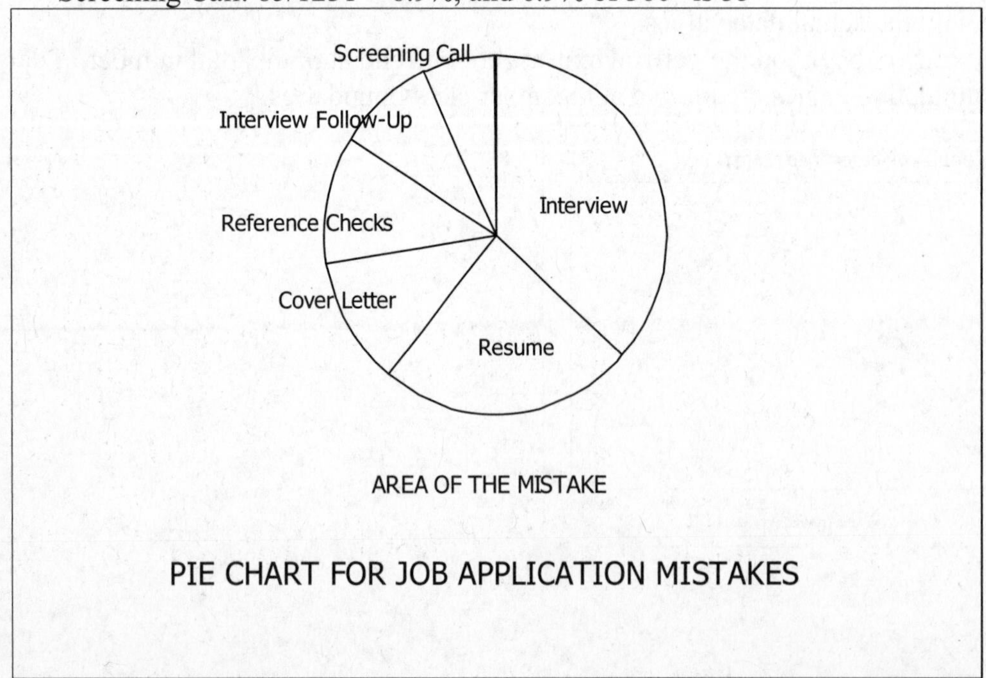

17. The pie chart is given below at the left. The "slices" of the pie may appear in any order and in any position, but their relative sizes must be as shown.

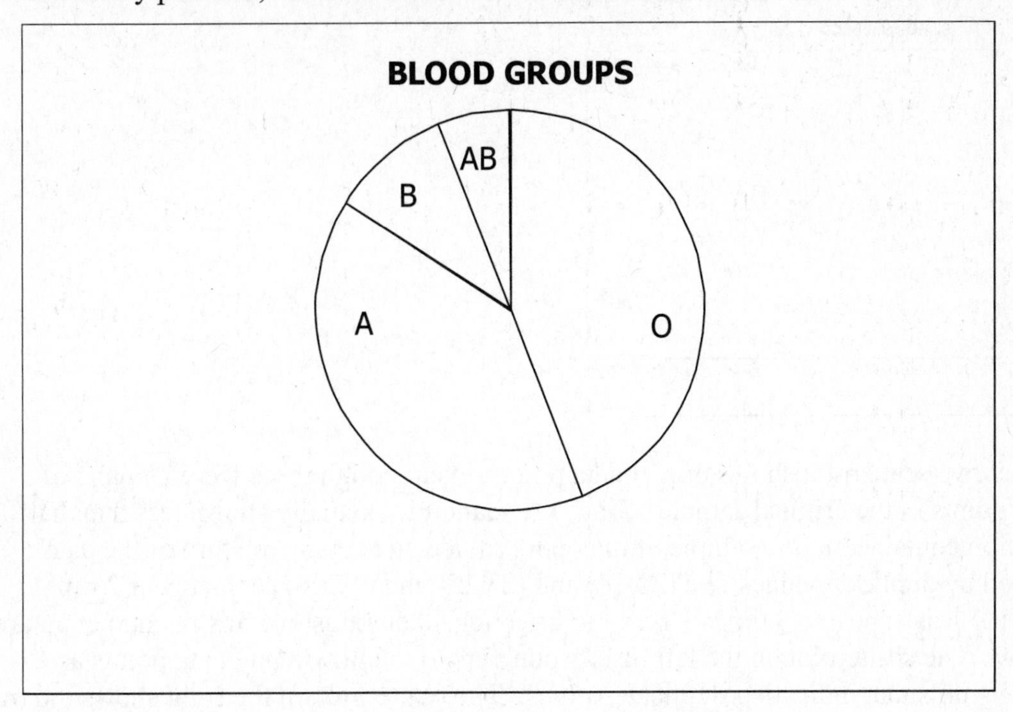

19. The Pareto chart is given below at the left.

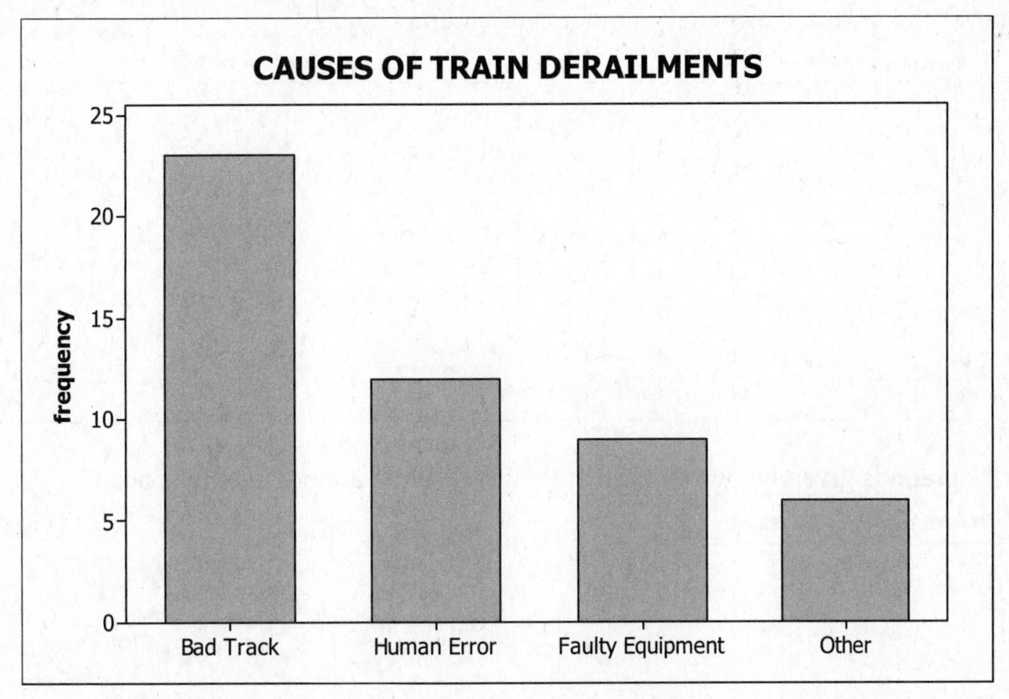

21. The simple, unmodified scatterplot is given below. There appears to be a slight tendency for cigarettes with more tar to also have more CO.

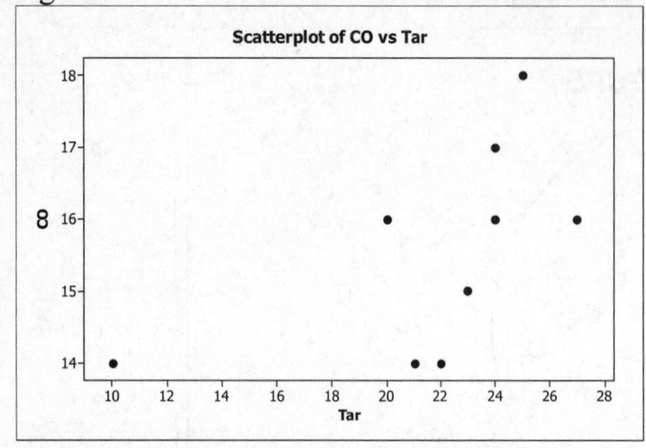

NOTE: The above scatterplot shows only 9 data points, even though there were 25 pairs of tar/CO data points in the original sample. Since the scatterplot actually shows less than half the information contained in the sample, it may not provide an accurate picture of the data. This is caused by duplicate values: the (22,14) and (23,15) and (27,16) each appear 2 times, and the (20,16) pair appears 14 times! Two modifications that adjust for this phenomenon are shown below. The scatterplot on the left inserts numbers to tell how many data points are represented by dots that indicating duplicate values. The scatterplot on the right shows the true number of dots. The same effect can also be obtained by using dots whose size is proportional to the number of duplicate values it represents. The modified scatterplots indicate that there appears to be no relationship between the amounts of tar and CO.

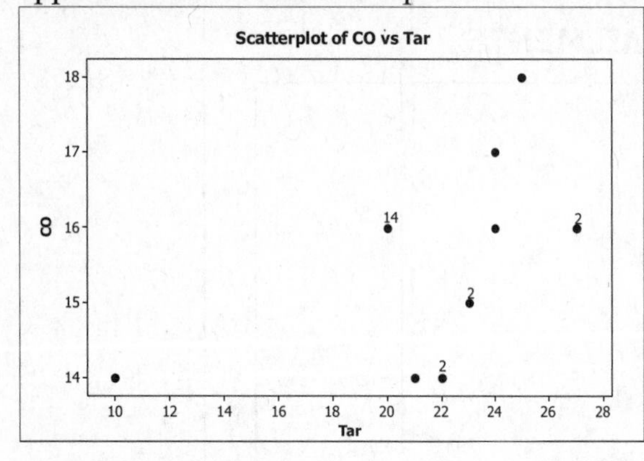

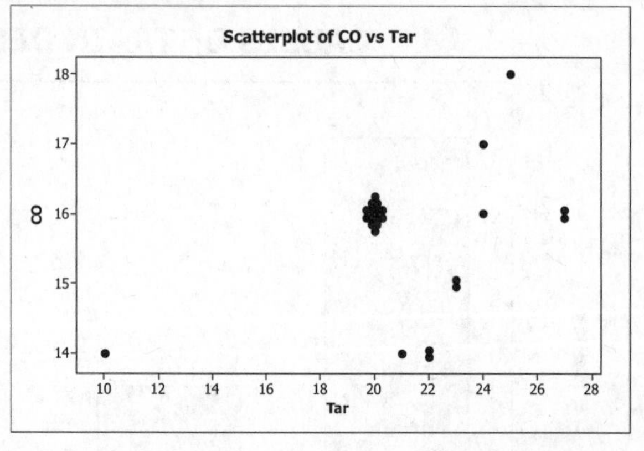

23. The time series graph is given below. Note that the given years are not evenly spaced.

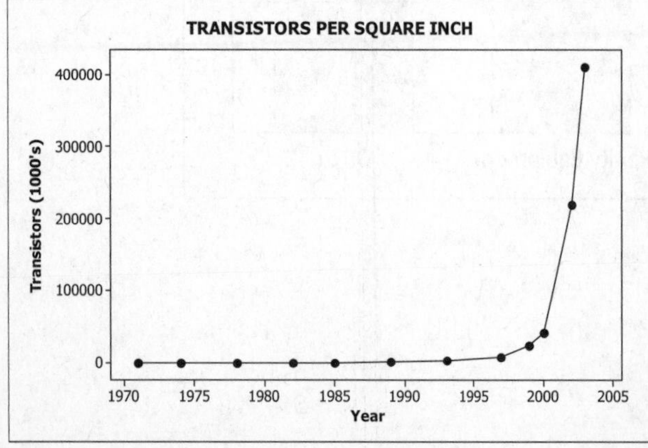

25. The multiple bar graph is given below. As the population increases, the numbers of marriages and divorces will automatically increase. To identify any change in marriage and divorce patterns, one needs to examine the rates. This is analogous to using percents (or relative frequencies) instead of frequencies to compare categories for two samples of different sizes. The marriage rate appears to have remained fairly constant, with a possible slight decrease in recent years. The divorce rate appears to have steadily grown, with a possible slight decrease in recent years.

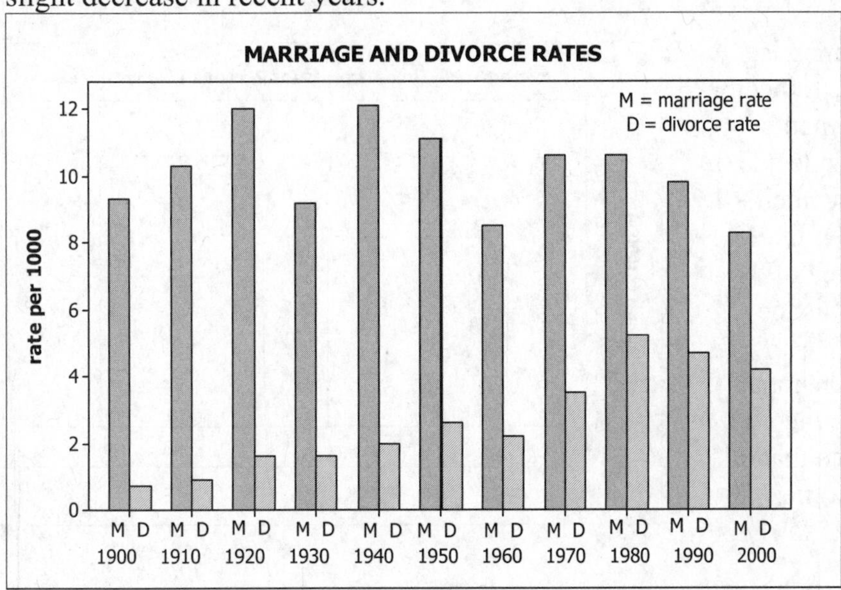

27. The back-to-back stemplot is given below. The pulse rates for men appear to be lower than the pulse rates of women.

PULSE RATES

Women		Men
	5	666666
888884444000	6	00000004444444888
66666622222222	7	22222266
88888000000	8	44448888
6	9	6
4	10	
	11	
4	12	

2-5 Critical Thinking: Bad Graphs

1. The illustration uses two-dimensional objects (dollar bills) to represent a one-dimensional variable (purchasing power). If the illustration uses a dollar bill with ½ the original length and ½ the original width to represent ½ the original purchasing power, then the illustration is misleading (because ½ the length and ½ the width translates into ¼ the area and gives the visual impression of 25% instead of 50%). But if the illustration uses a dollar bill with ½ the area (i.e, with .707 of the original length and .707 of the original width) to represent ½ the original purchasing power, then the illustration conveys the proper visual impression.

3. No. Results should be presented in a way that is fair and objective so that the reader has the reliable information necessary to reach his own conclusion.

5. No. The illustration uses two-dimensional objects to represent a one-dimensional variable (weight). The average male weight is 172/137 = 1.255 times the average female weight. Making a two-dimensional figure 1.255 times taller and 1.255 times wider increases the area by $(1.255)^2 = 1.58$ and gives a misleading visual impression.

7. The average income for men is about 1.4 times the average income for women. Making the men's pictograph 1.4 times as wide and 1.4 times as high as the women's produces a men's image with $(1.4)^2 = 1.96$ times the area of the women's image. Since it is the area that gives the visual impression in a two-dimensional figure, the men's average income appears to be almost twice that of the women's average income. A graph that depicts the data fairly is given at the right.

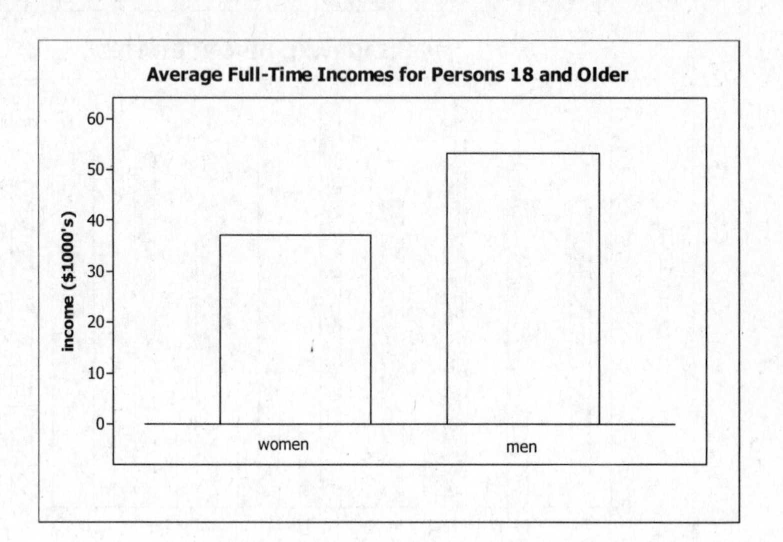

9. The graph in the text makes it appear that the braking distance for the Acura RL is more than twice that of the Volvo S80. The actual difference is about 60 feet, and the Acura RL distances is about 192/133 = 1.44 times that of the Volvo S80. The exaggeration of differences is caused by the fact that the distance scale dies not start at zero. A graph that depicts the data fairly is given at the right.

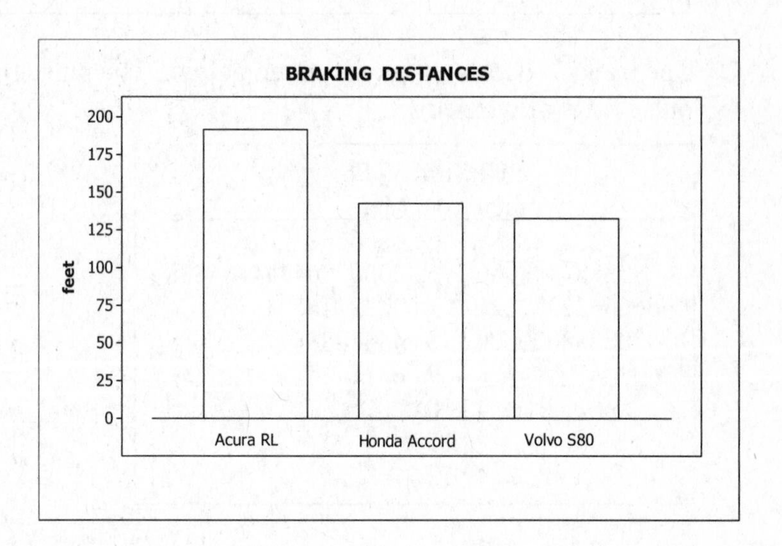

11. The given figure is misleading because the backside of the head is not visible. Categories extending to the backside of the head will not have as much area showing as comparable categories shown at the front of the head. A regular pie chart would give the relative sizes of the categories in as undistorted manner. A better graph would be a bar chart – with the vertical axis starting at 0, and the categories given in order by age. When there is a natural ordering of the categories that can be preserved with a bar chart – but it is hidden in a pie chart, which ends up placing the "first" and "last" categories side by side.

Statistical Literacy and Critical Thinking

1. When investigating the distribution of a data set, a histogram is more effective than a frequency distribution. Both figures contain the same information, but the visual impact of the histogram presents that information in a more efficient and more understandable manner.
2. When investigating changes over a period of years, a time series graph would be more effective than a histogram. A histogram would indicate the frequency with which different amounts occurred, but by ignoring the years in which those amounts occurred it would give no information about changes over time.
3. Using two-dimensional figures to compare one-dimensional variables exaggerates differences whenever the areas of the two dimensional figures are not proportional to the amounts being portrayed. Making the height and width proportional to the amounts being portrayed creates a

   distorted picture because it is area that makes the visual impression on the reader – and a two-fold increase in height and width produces a four-fold increase in area.
4. The highest histogram bars should be near the center, with the heights of the bars diminishing toward each end. The figure should be approximately symmetric.

Chapter Quick Quiz

1. 10 – 0 = 10. The class width may be found by subtracting consecutive lower class limits.
2. Assuming the data represent values reported to the nearest integer, the class boundaries for the first class are -0.5 and 9.5.
3. No. All that can be said is that there are 27 data values somewhere within that class.
4. False. A normal distribution is bell-shaped, with the middle classes having higher frequencies than the classes at the extremes. The distribution for a balanced die will be flat, with each class having about the same frequency.
5. Variation.
6. 52, 52, 59. The 5 to the left of the stem represent the tens digit associated with the ones digits to the right of the stem.
7. Scatterplot. The data is two-dimensional, requiring separate axes for each variable (shoe size and height).
8. True. The vertical scale for the relative frequency histogram will be the values of the frequency histogram divided by the sample size n.
9. A histogram reveals the shape of the distribution of the data.
10. Pareto chart. When there is no natural order for the categories, placing them in the order of their frequencies shows the relative importance without losing the nature of any significant relationships between the categories.

Review Exercises

1. The frequency distribution is given at the right. The pulse rates for the males appear to be lower than those for the females.

MALE PULSE RATES

beats per minute	frequency
50–59	6
60–69	17
70–79	8
80–89	8
90–99	1
	40

2. The histogram is given below. The basic shape is similar to the histogram for the females, but the male pulse rates appear to be lower.

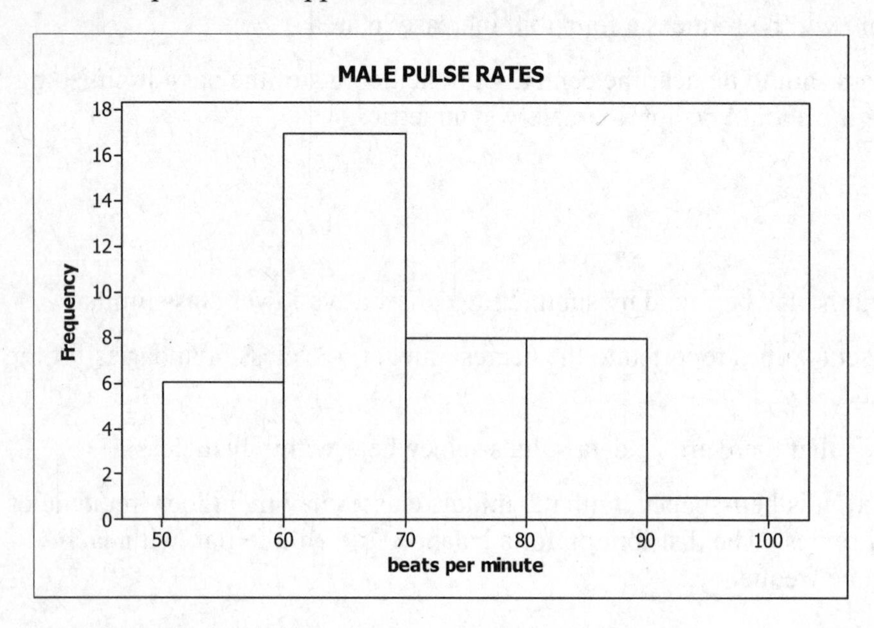

3. The dotplot is given below at the left. It shows that the male pulse rates appear to be lower than those for the females.

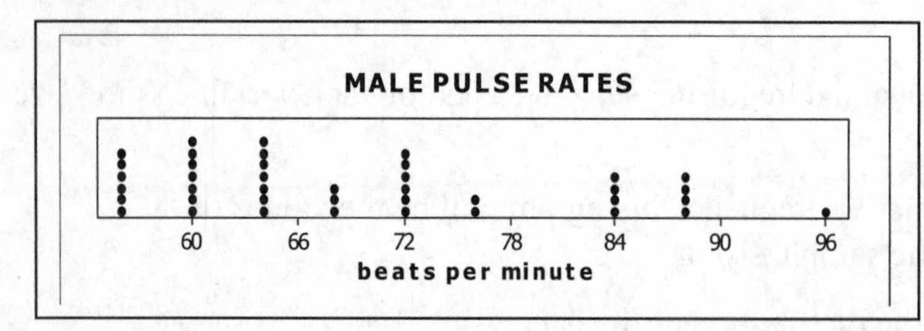

beats per minute
5 | 666666
6 | 00000004444444888
7 | 22222266
8 | 44448888
9 | 6

4. The stemplot is given above at the right. It shows that the male pulse rates appear to be lower than those of the females.

5. The scatterplot is given below.

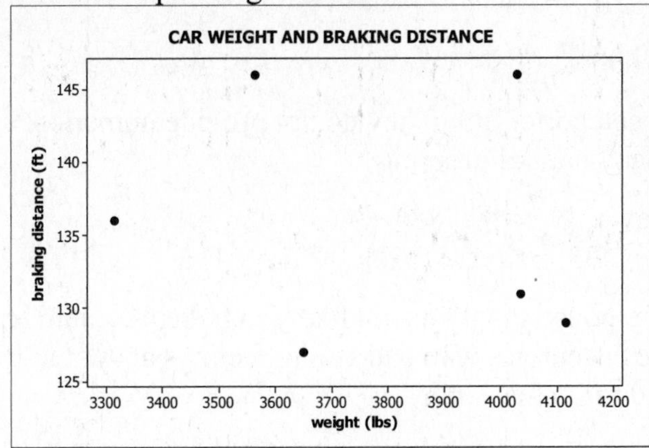

6. The time-series graph is given below.

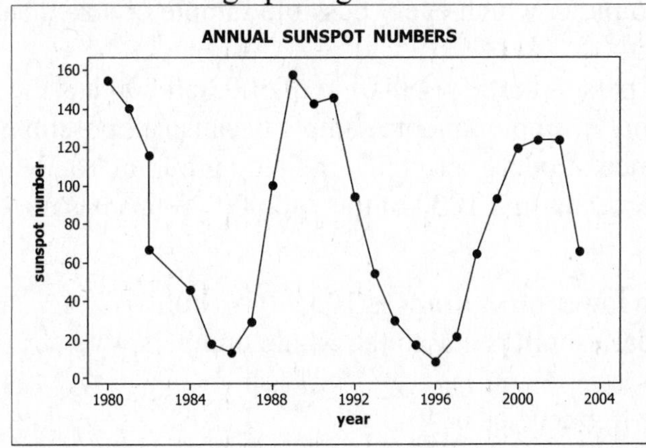

7. The graph is misleading because the vertical axis does not start at zero, causing it to exaggerate the differences between the categories. A graph that correctly illustrates the acceleration times is given below.

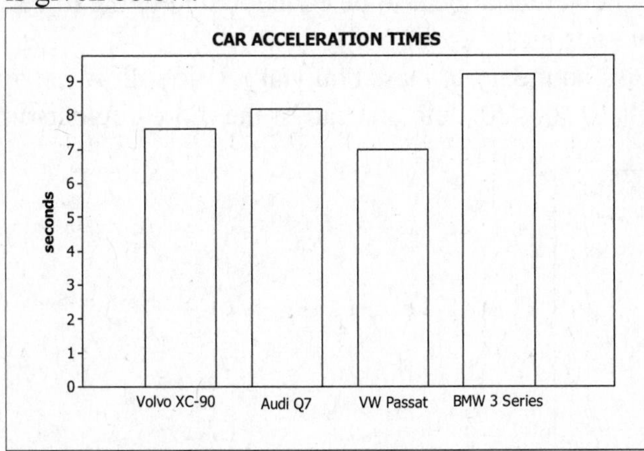

8. a. 25. The difference between the first two lower class limits is 125 – 100 = 25.
   b. 100 and 124. These are the values given in the first row of the table.
   c. 99.5 and 124.5. Values within these boundaries will round to the whole numbers given by the class limits.
   d. No. The distribution is not symmetric; the class with the largest frequency is near the right end of the distribution.

Cumulative Review Exercises

1. Yes. The sum of the relative frequencies is 100%.

2. Nominal. The yes-no-maybe responses are categories only; they do not provide numerical measures of any quantity, nor do they have any natural ordering.

3. The actual numbers of responses are as follows. Note that 884 + 433 + 416 = 1733.
   Yes: (0.51)(1733) = 884. No: (0.25)(1733) = 433. Maybe: (0.24)(1733) = 416

4. Voluntary Response Sample. A voluntary response sample is not likely to be representative of the population of all executives, but of those executives who had strong feelings about the topic and/or had enough free time to respond to such a survey.

5. a. A random sample is a sample in which every member of the population has an equal chance of being selected.
   b. A simple random sample of size n is a sample in which every possible sample of size n has an equal chance of being selected.
   c. Yes, it is a random sample because every person in the population of 300,000,000 has the same chance of being selected. No, it is not a simple random sample because all possible groups of 1000 do not have the same chance of being selected – in fact a group of 1000 composed of the oldest person in the each of the first 1000 of the 300,000 groups has no chance of being selected.

6. a. 100. The difference between the first two lower class limits is 100 – 0 = 100.
   b. -0.5 and 99.5. Values within these boundaries will round to the whole numbers given by the class limits.
   c. 11/40 = 0.275, or 27.5%. The total of the frequencies is 40.
   d. Ratio. Differences between the data values are meaningful and there is a meaningful zero.
   e. Quantitative. The data values are measurements of the cotinine levels.

7. The histogram is given below. Using a strict interpretation of the criteria, the cotinine levels do not appear to be normally distributed – the values appear to be concentrated in the lower portion of the distribution, with very few values in the upper portion.
   NOTE: The histogram bars extend from class boundary to class boundary. We follow the text and for clarity of presentation use the labels 0, 100, 200, etc. instead of the more cumbersome -0.5, 99.5, 199.5 etc.

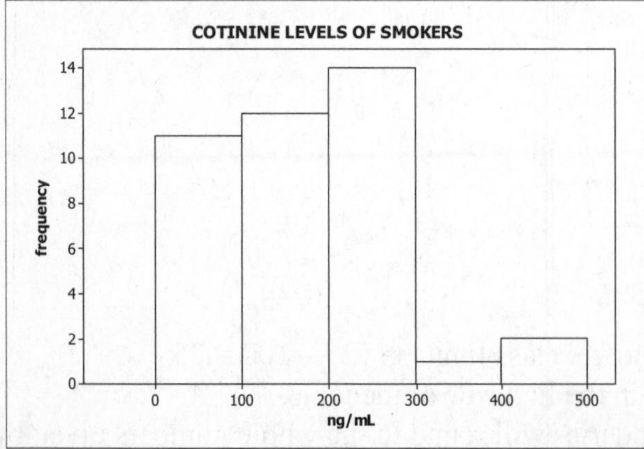

8. Statistic. A statistic is a measurement of some characteristic of a sample, while a parameter is a measurement of some characteristic of the entire population.

Chapter 3

Statistics for Describing, Exploring, and Comparing Data

3-2 Measures of Center

1. The mean, median, mode, and midrange are measures of "center" in the sense that they each attempt to determine (by various criteria – i.e., by using different approaches) what might be designated as a typical or representative value.

3. No. The price "exactly in between the highest and the lowest" would be the mean of the highest and lowest values – which is the midrange, and not the median.

NOTE: As it is common in mathematics and statistics to use symbols instead of words to represent quantities that are used often and/or that may appear in equations, this manual employs the following symbols for the various measures of center.

mean = $\bar{x}$ mode = M
median = $\tilde{x}$ midrange = m.r.

This manual will generally report means, medians and midranges accurate to one more decimal place than found in the original data. In addition, these two conventions will be employed.

(1) When there is an odd number of data, the median will be one of the original values. The manual follows example #2 in this section and reports the median as given in the original data. And so the median of 1,2,3,4,5 is reported as $\tilde{x} = 3$.

(2) When the mean falls exactly between two values accurate to one more decimal place than the original data, the round-off rule in this section gives no specific direction. This manual follows the commonly accepted convention of always rounding up. And so the mean of 1,2,3,3 is reported as $\bar{x} = 9/4 = 2.3$ [i.e., 9/4 = 2.25, rounded up to 2.3].

In addition, the median is the middle score when the scores are arranged in order, and the midrange is halfway between the first and last score when the scores are arranged in order. It is usually helpful to arrange the scores in order. This will not affect the mean, and it may also help in identifying the mode. Finally, no measure of center can have a value lower than the smallest score or higher than the largest score. Remembering this helps to protect against gross errors, which most commonly occur when calculating the mean.

5. Arranged in order, the n=10 scores are: 34 36 39 43 51 53 62 63 73 79
   a. $\bar{x} = (\Sigma x)/n = (533)/10 = 53.3$ words c. M = (none)
   b. $\tilde{x} = (51 + 53)/2 = 52.0$ words d. m.r. = (34 + 79)/2 = 56.5 words

   Using the mean of 53.3 words per page, a reasonable estimate for the total number of words in the dictionary is (53.3)(1459) = 77,765. Since the sample is a simple random sample, it should be representative of the population and the estimate of 77,675 is a valid estimate for the number of words in the dictionary – but not for the total number of words in the English language, since the dictionary does not claim to contain every word.

   NOTE: The exercise also asks whether the estimate for the total is "accurate." This is a relative term. A change of 1.0 in the estimate for the mean produces a change of 1459 in the estimate for the total. While a change of 1459 is only a change of 1459/77765 = 1.9%, this illustrates that accuracy in terms of absolute numbers is sometimes hard to attain when estimating totals.

7. Arranged in order, the n=5 scores are: 4277 4911 6374 7448 9051
   a. $\bar{x} = (\Sigma x)/n = (32061)/5 = \$6412.2$
   b. $\tilde{x} = \$6374$
   c. M = (none)
   d. m.r. = (4277 + 9051)/2 = $6664.0
   No. Even though the sample values cover a fairly wide range, the measures of center do not differ very much.

9. Arranged in order, the n=10 scores are: 8.4 9.6 10 12 13 15 27 35 36 38
   a. $\bar{x} = (\Sigma x)/n = (204)/10 = \$20.4$ million
   b. $\tilde{x} = (13 + 15)/2 = \$14.0$million
   c. M = (none)
   d. m.r. = (8.4 + 38)/2 = $23.2 million
   No, the top ten salaries provide no information (other than an upper cap) about the salaries of TV personalities in general. No, top 10 lists are generally not valuable for gaining insight into the larger population.
   NOTE: The "one more decimal place than is present in the original set of values" round-off rule given in the text sometimes requires thought before being applied. Since the data in this exercise is presented with two digit accuracy, the rule suggests reporting the appropriate measures of center with three digit accuracy.

11. Arranged in order, the n=15 scores are:
    0 73 95 165 191 192 221 235 235 244 259 262 331 376 381
    a. $\bar{x} = (\Sigma x)/n = (3260)/15 = 217.3$ hrs
    b. $\tilde{x} = 235$ hrs
    c. M = 235 hrs
    d. m.r. = (0 + 381)/2 = 190.5 hrs
    The duration time of 0 appears to be very unusual. It likely represents the Challenger disaster of January 1986, when the mission ended in an explosion shortly after takeoff.

13. Arranged in order, the n=20 scores are: 1 2 2 2 2 2 2 2 2 2 2 2 2 2 3 3 3 3 4 4
    a. $\bar{x} = (\Sigma x)/n = (47)/20 = 2.4$ mpg
    b. $\tilde{x} = (2 + 2)/2 = 2.0$ mpg
    c. M = 2 mpg
    d. m.r. = (1 + 4)/2 = 2.5 mpg
    No, there would not be much of an error. Since the amounts do not vary much, the mean appears to be a reasonable value to use for each of the older cars.

15. Arranged in order, the n=16 scores are:
    213 213 227 231 239 242 244 246 246 255 257 258 262 280 293 448
    a. $\bar{x} = (\Sigma x)/n = (4154)/16 = 259.6$ sec
    b. $\tilde{x} = (246 + 246)/2 = 246.0$ sec
    c. M = 213 sec, 246 sec (bimodal)
    d. m.r. = (213 + 448)/2 = 330.5 sec
    Yes, the time of 448 seconds appears to be very different from the others.

17. Arranged in order, the n=20 scores are:
    4 4 4 4 4 4 4.5 4.5 4.5 4.5 4.5 4.5 6 6 8 9 9 13 13 15
    a. $\bar{x} = (\Sigma x)/n = (130)/20 = 6.5$ yrs
    b. $\tilde{x} = (4.5 + 4.5)/2 = 4.5$ yrs
    c. M = 4 yrs, 4.5 yrs (bi-modal)
    d. m.r. = (4 + 14)/2 = 9.5 yrs
    Yes, it is common to earn a bachelor's degree in 4 years in that we estimate that 6/20 = 30% of the people do so – but the typical time to earn a bachelor's degree appears to be more than that.
    NOTE: The "one more decimal place than is present in the original set of values" round-off rule given in the text sometimes requires thought before being applied. In this exercise it appears that the basic measurement is being done in whole number of years, and so the appropriate measures of center should be reported with one decimal accuracy. When there are different number of decimal places in the original data, it is generally appropriate to use the least accurate data as the basis for applying the round-off rule.

19. Arranged in order, the n=12 scores are: 14 15 59 85 95 97 98 106 117 120 143 371

 a. $\bar{x} = (\Sigma x)/n = (1320)/12 = 110.0$
 b. $\tilde{x} = (97 + 98)/2 = 97.5$
 c. M = (none)
 d. m.r. = (14 + 371)/2 = 192.5

 Yes, there does appear to be a trend in the data. The numbers of bankruptcies seems to steadily increase up to a certain point and then drop off dramatically. It could be that new bankruptcy laws (that made it more difficult to declare bankruptcy) went into effect.

21. Arranged in order, the scores are as follows.

 30 days: 244 260 264 264 278 280 318
 n = 7
 $\bar{x} = (\Sigma x)/n = 1908/7 = \$272.6$
 $\tilde{x} = \$264$

 one day: 456 536 567 614 628 943 1088
 n = 7
 $\bar{x} = (\Sigma x)/n = 4832/7 = \$690.3$
 $\tilde{x} = \$614$

 The tickets purchased 30 days in advance are considerably less expensive – they appear to cost less than half of the tickets purchased one day in advance.

23. Arranged in order, the scores are as follows.

 non-filtered: 1.0 1.1 1.1 1.1 1.1 1.1 1.1 1.1 1.1 1.1 1.1 1.1 1.1
 1.1 1.2 1.2 1.3 1.3 1.4 1.4 1.5 1.6 1.7 1.7 1.8
 filtered: 0.2 0.4 0.6 0.8 0.8 0.8 0.8 0.8 0.8 0.8 0.9 1.0 1.0
 1.0 1.0 1.0 1.1 1.1 1.1 1.1 1.1 1.1 1.1 1.2 1.3

 non-filtered
 n = 25
 $\bar{x} = (\Sigma x)/n = 31.4/25 = 1.26$ mg
 $\tilde{x} = 1.1$mg

 filtered
 n = 25
 $\bar{x} = (\Sigma x)/n = 22.9/25 = 0.92$ mg
 $\tilde{x} = 1.0$ mg

 The filtered cigarettes seem to have less nicotine. Filters appear to be effective in reducing the amounts of nicotine.

25. The Minitab DESCRIBE results are as follows.

```
Variable     N    Mean  SE Mean  StDev  Minimum      Q1  Median      Q3  Maximum
BodyTemp   106  98.200   0.0605  0.623   96.500  97.800  98.400  98.600   99.600
```

 $\bar{x} = 98.20$ °F and $\tilde{x} = 98.40$ °F. These results suggest that the mean body temperature is less than 98.6 °F.

27. The Minitab DESCRIBE results are as follows.

```
Variable    N    Mean  SE Mean  StDev  Minimum      Q1  Median      Q3  Maximum
home       40  123.66   0.0380  0.240   123.30  123.43  123.70  123.90   124.20
generator  40  124.66   0.0457  0.289   124.00  124.50  124.70  124.80   125.20
UPS        40  123.59   0.0487  0.308   122.90  123.53  123.70  123.80   123.90
```

 home: $\bar{x} = 123.66$ volts and $\tilde{x} = 123.70$ volts
 generator: $\bar{x} = 124.66$ volts and $\tilde{x} = 124.70$ volts
 UPS: $\bar{x} = 123.59$ volts and $\tilde{x} = 123.70$ volts

 The three different sources appear to have about the same voltage measurements, and within each source the mean and the median are very close to each other.

29. The x values below are the class midpoints from the given frequency table, the class limits of which indicate the original data were integers.

x	f	f·x
11.5	1	11.5
15.5	0	0
19.5	15	292.5
23.5	7	164.5
27.5	2	55.0
	25	523.5

 $\bar{x} = \Sigma(f \cdot x)/\Sigma f = 523.5/25 = 20.9$ mg

 This is close to the true value of 21.1 mg.

31. The x values below are the class midpoints from the given frequency table, the class limits of which indicate the original data were integers.

x	f	f·x
43.5	25	1087.5
47.5	14	665.0
51.5	7	360.5
55.5	3	166.5
59.5	1	59.5
	50	2339.0

$\bar{x} = \Sigma(f \cdot x)/\Sigma f = 2339.0/50 = 46.8$ mph

This is close to the true value of 46.7 mph. This is considerably higher than the posted speed limit of 30 mph – but remember that this does not estimate the mean speed of all drivers, but the mean speed of drivers who receive speeding tickets.

33. The x values below are the numerical values of the letter grades, and the corresponding weights are the numbers of credit hours.

x	w	w·x
3	3	9
2	3	6
3	4	12
4	4	16
1	1	1
	15	44

$\Sigma x\ \bar{x} = \Sigma(w \cdot x)/\Sigma w = 44/15 = 2.93$

No; since 2.93 is below 3.00, the student did not make the Dean's list.

35. a. The formula $\bar{x} = (\Sigma x)/n$ implies that $n \cdot \bar{x} = \Sigma x$, the sum of the sample values.
Since $n \cdot \bar{x} = 5 \cdot (657.054) = 3285.27$, that is the sum of the sample values.
Since the 4 given sample values sum to 2642.76, the missing value is found by subtracting $3285.27 - 2642.76 = 642.51$ micrograms.

b. As above, the formula $\bar{x} = (\Sigma x)/n$ implies that $n \cdot \bar{x} = \Sigma x$.
Since n and $\bar{x}$ are given, the total Σx is determined.
For any freely assigned n−1 values, the final value is determined to be the value (whether + or −) necessary to obtain the desired total.

37. a. Arranged in order, the original 100 scores are: 444 497 502 503 517 519 532
559 568 579 579 591 591 594 598 603 604 611 611 617 618
628 630 635 636 637 638 651 654 657 660 661 664 681 681
682 689 692 693 694 696 697 698 701 706 708 709 709 711
713 714 714 714 722 729 731 732 739 741 743 744 745 751
752 753 753 755 756 756 760 768 768 777 779 781 782 782
783 784 787 789 792 793 795 795 796 797 797 798 802 809
818 824 829 830 834 835 836 849 850

$\bar{x} = (\Sigma x)/n = (70311)/100 = 703.1$

b. Trimming the highest and lowest 10% (or ten scores), the remaining 80 scores are:
579 591 591 594 598 603 604 611 611 617 618 628 630 635
636 637 638 651 654 657 660 661 664 681 681 682 689 692
693 694 696 697 698 701 706 708 709 709 711 713 714 714
714 722 729 731 732 739 741 743 744 745 751 752 753 753
755 756 756 760 768 768 777 779 781 782 782 783 784 787
789 792 793 795 795 796 797 797 798 802

$\bar{x} = (\Sigma x)/n = (56777)/80 = 709.7$

c. Trimming the highest and lowest 20% (or 20 scores), the remaining 60 scores are:

```
618   628   630   635   636   637   638   651   654   657   660   661   664   681
681   682   689   692   693   694   696   697   698   701   706   708   709   709
711   713   714   714   714   722   729   731   732   739   741   743   744   745
751   752   753   753   755   756   756   760   768   768   777   779   781   782
782   783   784   787
```

$\bar{x} = (\Sigma x)/n = (42824)/60 = 713.7$

The results are not dramatically different. Since the trimmed mean appears to increase slightly as more values are trimmed, the data may be negatively skewed – i.e., have smaller values that are not balanced by corresponding larger ones.

39. The geometric mean of the three values is the third (i.e., cube) root of their product:
$\sqrt[3]{(1.10)(1.05)(1.02)} = 1.056$, which corresponds to a 5.6% growth rate.
This is not the same as the mean of the growth rates:
$(10\% + 5\% + 2\%)/3 = (17\%)/3 = 5.7\%$.

41. There are 25 scores in the table in Exercise 29. The median, x_{13} for the ordered scores, is in the 18-21 class. The relevant values are as follows:
median lower class limit = 18 n = 25 total scores
class width = 4 m = 1 total scores preceding the median class
median class frequency = 15
$\tilde{x}$ = (median class lower limit) + (class width)·[(n+1)/2 – (m+1)]/(median class frequency)
$= 18 + 4\cdot[26/2 - 2]/15$
$= 18 + 4\cdot[11]/15$
$= 18 + 2.9$
$= 20.9$ mg
This compares favorably with the true median value of 20.0 mg calculated directly from the original list. Since frequency tables are summaries which do not contain all the information, values obtained from the original data should always be preferred to ones obtained from frequency tables.

3-3 Measures of Variation

1. Variation is a general descriptive term that refers to the fact that all the items being measured are not identical. In statistics, variance is a specific and well-defined measure of variation that has a particular mathematical formula.

3. The incomes of the adults selected from the general population should have more variation. Statistics teachers are a more homogeneous group – they would tend to have similar salaries that would not be likely to include the extremely high or extremely low salaries possible in the general population.

NOTE: Although not given in the text, the symbol R will be used for the range throughout this manual. Remember that the range is the difference between the highest and the lowest scores, and not necessarily the difference between the last and first scores as they are listed. Since calculating the range involves only the subtraction of 2 original pieces of data, the manual follows the advice in this section that "In general, the range should not be rounded" and reports the range with the same accuracy found in the original data. In general, the standard deviation and the variance will be reported with one more decimal place than the original data.

When finding the square root of the variance to obtain the standard deviation, use all the decimal places of the variance – and not the rounded value reported as the answer. To do this, keep the value on the calculator display or place it in the memory. Do not copy down all the decimal places and then re-enter them to find the square root – as that not only is a waste of time, but also could introduce copying and/or round-off errors.

5. preliminary values: $n = 10$ $\Sigma x = 533$ $\Sigma x^2 = 30615$
   $R = 79 - 34 = 45$ words
   $s^2 = [n(\Sigma x^2) - (\Sigma x)^2]/[n(n-1)]$
   $= [10(30615) - (533)^2]/[10(9)] = 22061/90 = 245.1$ words^2
   $s = \sqrt{245.1} = 15.7$ words
   There seems to be considerable variation from page to page. For small samples with n = 10, there could be considerable variation in the sample mean and, therefore, considerable variation in the projected totals for the entire dictionary. It appears that there would be a question about the accuracy of an estimate based on this sample for the total number of words.

NOTE: The above format will be used in this manual for calculation of the variance and standard deviation. To reinforce some of the key symbols and concepts, the following detailed mathematical analysis is given for the data of the above problem.

x	$x-\bar{x}$	$(x-\bar{x})^2$	x^2
34	-19.3	372.49	1156
36	-17.3	299.29	1296
39	-14.3	204.49	1521
43	-10.3	106.09	1849
51	-2.3	5.29	2601
53	-0.3	0.09	2809
62	8.7	75.69	3844
63	9.7	94.09	3969
73	19.7	388.09	5329
79	25.7	660.49	6241
533	0.0	2206.10	30615

$\bar{x} = (\Sigma x)/n = 533/10 = 53.3$ words

by formula 3-4, $s^2 = \Sigma(x-\bar{x})^2/(n-1)$
$= 2206.10/9$
$= 245.1222$ words^2

by formula 3-5, $s^2 = [n(\Sigma x^2) - (\Sigma x)^2]/[n(n-1)]$
$= [10(30615) - (533)^2]/[10(9)]$
$= 22061/90$
$= 245.1222$ words^2

NOTE: As shown in the preceding detailed mathematical analysis, formulas 3-4 and 3-5 give the same answer. The following three observations give hints for using these two formulas.
• When using formula 3-4, constructing a table having the first 3 columns shown above helps to organize the calculations and makes errors less likely. In addition, verify that $\Sigma(x-\bar{x}) = 0$ [except for a possible discrepancy at the last decimal due to rounding] before proceeding – if such is not the case, an error has been made and further calculation is fruitless.
• When using formula 3-5, the quantity $[n(\Sigma x^2) - (\Sigma x)^2]$ must be non-negative (i.e., greater than or equal to zero) – if such is not the case, an error has been made and further calculation is fruitless.
• In general, formula 3-5 is to be preferred over formula 3-4 because (1) it does not involve round-off error or messy calculations when the mean does not "come out even" and (2) the quantities Σx and Σx^2 can be found directly (on most calculators, with one pass through the data) without having to construct a table or make a second pass through the data after finding the mean.

7. preliminary values: $n = 5$ $\Sigma x = 32061$ $\Sigma x^2 = 220431831$
   $R = 9052 - 4227 = \$4774$
   $s^2 = [n(\Sigma x^2) - (\Sigma x)^2]/[n(n-1)]$
   $= [5(220431831) - (32061)^2]/[5(4)] = 74251434/20 = 3712571.7$ dollars^2
   $s = \sqrt{3712571.7} = \$1926.8$
   A value is considered unusual if it differs from the mean by more than two standard deviations. Since \$10,000 differs from the mean by (10,000 – 6,412.2)/1926.8 = 1.86 standard deviations, in this context it would not be considered an unusual value.

9. preliminary values: $n = 10$ $\Sigma x = 204$ $\Sigma x^2 = 5494.72$
   $R = 38 - 8.4 = \$29.6$ million
   $s^2 = [n(\Sigma x^2) - (\Sigma x)^2]/[n(n-1)]$
   $= [10(5494.72) - (204)^2]/[10(9)] = 13331.2/90 = 148.1$ (million dollars)^2
   $s = \sqrt{148.1} = \$12.2$ million
   No. The variation among the top ten salaries reveals essentially nothing about the variation of the salaries in general. All that can be said is that the range of salaries in the general population of TV personalities must be at least \$29.6 million.

11. preliminary values: $n = 15$ $\Sigma x = 3260$ $\Sigma x^2 = 865574$
   $R = 381 - 0 = 381$ hrs
   $s^2 = [n(\Sigma x^2) - (\Sigma x)^2]/[n(n-1)]$
   $= [15(865574) - (3260)^2]/[15(14)] = 23356010/210 = 11219.1$ hrs^2
   $s = \sqrt{11219.1} = 105.9$ hrs
   A value is considered unusual if it differs from the mean by more than two standard deviations. Since 0 hours differs from the mean by $(0 - 217.3)/105.9 = -2.05$ standard deviations, in this context it would be considered an unusual value.

13. preliminary values: $n = 20$ $\Sigma x = 47$ $\Sigma x^2 = 121$
   $R = 4 - 1 = 3$ mpg
   $s^2 = [n(\Sigma x^2) - (\Sigma x)^2]/[n(n-1)]$
   $= [20(121) - (47)^2]/[20(19)] = 211/380 = 0.6$ (mpg)^2
   $s = \sqrt{0.6} = 0.7$ mpg
   A value is considered unusual if it differs from the mean by more than two standard deviations. Since 4 mpg differs from the mean by $(4 - 2.35)/0.75 = 2.21$ standard deviations, in this context it would be considered an unusual value.

15. preliminary values: $n = 16$ $\Sigma x = 4154$ $\Sigma x^2 = 1123116$
   $R = 448 - 213 = 235$ sec
   $s^2 = [n(\Sigma x^2) - (\Sigma x)^2]/[n(n-1)]$
   $= [16(1123116) - (4154)^2]/[16(15)] = 714140/240 = 2975.6$ sec^2
   $s = \sqrt{2975.6} = 54.5$ sec
   If the highest playing time is omitted, the standard deviation drops to 22.0 sec. The change is substantial. NOTE: Use the formula again. The new Σx and Σx^2 may be found by subtracting from the above totals, and the n is reduced by 1.

17. preliminary values: $n = 20$ $\Sigma x = 130$ $\Sigma x^2 = 1078.50$
   $R = 15 - 4 = 11$ yrs
   $s^2 = [n(\Sigma x^2) - (\Sigma x)^2]/[n(n-1)]$
   $= [20(1078.50) - (130)^2]/[20(19)] = 4670/380 = 12.3$ yrs^2
   $s = 3.5$ yrs
   A value is considered unusual if it differs from the mean by more than two standard deviations. Since 12 years differs from the mean by $(12 - 6.5)/3.5 = 1.57$ standard deviations, in this context it would not be considered an unusual value.

19. preliminary values: $n = 12$ $\Sigma x = 1320$ $\Sigma x^2 = 236580$
 $R = 371 - 14 = 357$ bankruptcies
 $s^2 = [n(\Sigma x^2) - (\Sigma x)^2]/[n(n-1)]$
 $= [12(236580) - (1320)^2]/[12(11)] = 1096560/132 = 8307.3$ bankruptcies^2
 $s = 91.1$ bankruptcies
 A value is considered unusual if it differs from the mean by more than two standard deviations. Since 371 bankruptcies differs from the mean by $(371 - 110.0)/91.1 = 2.86$ standard deviations, in this context it would be considered an unusual value. None of the other given values meets that criterion.

NOTE: In exercises 21-24, $CV = s/\bar{x}$ is found by storing the value of $\bar{x}$ and dividing the unrounded s by the unrounded stored value of $\bar{x}$. Using the rounded values of $\bar{x}$ and s may produce slight different values as indicated. The CV is expressed as a percent.

21. 30 Days
 $n = 7$, $\Sigma x = 1908$, $\Sigma x^2 = 523336$
 $\bar{x} = (\Sigma x)/n = 1908/7 = \$272.6$
 $s^2 = [n(\Sigma x^2) - (\Sigma x)^2]/[n(n-1)]$
 $= [7(523336) - (1908)^2]/[7(6)]$
 $= 22888/42 = 545.0$ dollars^2
 $s = \$23.3$
 $CV = s/\bar{x} = (\$23.3)/(\$272.6) = 8.6\%$
 [8.5% using rounded values for $\bar{x}$ and s]

 One Day
 $n = 7$, $\Sigma x = 4832$, $\Sigma x^2 = 3661094$
 $\bar{x} = (\Sigma x)/n = 4832/7 = \$690.3$
 $s^2 = [n(\Sigma x^2) - (\Sigma x)^2]/[n(n-1)]$
 $= [7(3661094) - (4832)^2]/[7(6)]$
 $= 2279434/42 = 54272.2$ dollars^2
 $s = \$233.0$
 $CV = s/\bar{x} = (\$233.0)/(\$690.3) = 33.7\%$
 [33.8% using rounded values for $\bar{x}$ and s]

 There is considerably more variation among the costs for tickets purchased one day in advance.

23. Non-filtered
 $n = 25$, $\Sigma x = 31.4$, $\Sigma x^2 = 40.74$
 $\bar{x} = (\Sigma x)/n = 31.4/25 = 1.26$ mg
 $s^2 = [n(\Sigma x^2) - (\Sigma x)^2]/[n(n-1)]$
 $= [25(40.74) - (31.4)^2]/[25(24)]$
 $= 32.54/600 = 0.05$ mg^2
 $s = 0.23$ mg
 $CV = s/\bar{x} = (0.23 \text{ mg})/(1.26 \text{ mg}) = 18.5\%$
 [18.3% using rounded values for $\bar{x}$ and s]

 Filtered
 $n = 25$, $\Sigma x = 22.9$, $\Sigma x^2 = 22.45$
 $\bar{x} = (\Sigma x)/n = 22.9/25 = 0.92$ mg
 $s^2 = [n(\Sigma x^2) - (\Sigma x)^2]/[n(n-1)]$
 $= [25(22.45) - (22.9)^2]/[25(24)]$
 $= 36.84/600 = 0.06$ mg^2
 $s = 0.25$ mg
 $CV = s/\bar{x} = (0.25 \text{ mg})/(0.92 \text{ mg}) = 27.1\%$
 [27.2% using rounded values for $\bar{x}$ and s]

 The nicotine values for the filtered cigarettes appear to exhibit slightly more variation. This is reasonable, since the filter is an added element: the variability in filters adds more variation to the natural variation already present in the tobaccos.

25. The Minitab DESCRIBE results are as follows.

Variable	N	Mean	SE Mean	StDev	Minimum	Q1	Median	Q3	Maximum
BodyTemp	106	98.200	0.0605	0.623	96.500	97.800	98.400	98.600	99.600

 $R = 99.6 - 96.5 = 3.1$ °F; $s^2 = 0.623^2 = 0.39$ (°F)^2 ; $s = 0.62$ °F

27. The Minitab DESCRIBE results are as follows.

Variable	N	Mean	SE Mean	StDev	Minimum	Q1	Median	Q3	Maximum
home	40	123.66	0.0380	0.240	123.30	123.43	123.70	123.90	124.20
generator	40	124.66	0.0457	0.289	124.00	124.50	124.70	124.80	125.20
UPS	40	123.59	0.0487	0.308	122.90	123.53	123.70	123.80	123.90

 home: $R = 124.2 - 123.3 = 0.9$ volts; $s^2 = 0.240^2 = 0.06$ volts^2; $s = 0.24$ volts
 generator: $R = 125.2 - 124.0 = 1.2$ volts; $s^2 = 0.289^2 = 0.08$ volts^2; $s = 0.29$ volts
 UPS: $R = 123.9 - 122.9 = 1.0$ volts; $s^2 = 0.308^2 = 0.09$ volts^2; $s = 0.31$ volts
 The three different sources appear to have about the same amount of variation.

29. The x values below are the class midpoints from the given frequency table, the class limits of which indicate the original data were integers.

x	f	f·x	$f \cdot x^2$
11.5	1	11.5	132.25
15.5	0	0	0
19.5	15	292.5	5703.75
23.5	7	164.5	3865.75
27.5	2	55.0	1512.50
	25	523.5	11214.25

$s^2 = \{n[\Sigma(f \cdot x^2)] - [\Sigma(f \cdot x)]^2\}/[n(n-1)]$
$= \{25[11214.25] - [523.5]^2\}/[25(24)]$
$= 6304/600$
$= 10.507$
$s = 3.2$ mg

This is the same as the true value of 3.2 mg.

31. The range rule of thumb suggests $s \approx$ range/4. In this case, the range is $24 - 18 = 6$, and so we estimate $s \approx 6/4 = 1.5$ years.

33. a. The range from 154 to 168 is $\bar{x} \pm 1s$. The empirical rule suggests that about 68% of the data values should fall within those limits.
b. The range from 147 to 175 is $\bar{x} \pm 2s$. The empirical rule suggests that about 95% of the data values should fall within those limits.

35. Chebyshev's Theorem states that for any set of data there must be at least $1 - 1/2^2 = 1 - 1/4 = 3/4$ (i.e., 75%) of the data values within 2 standard deviations of the mean. In this context, the limits are $161 \pm 2(7)$: there must be at least 75% of the heights between 147 cm and 175 cm.

37. For greater accuracy and understanding, we use 3 decimal places and formula 3-4.
a. the original population

x	x-μ	$(x-\mu)^2$
1	-5	25
3	-3	9
14	8	64
18	0	98

$\mu = (\Sigma x)/N = 18/3 = 6$

$\sigma^2 = \Sigma(x-\mu)^2/N = 98/3 = 32.667$
$[\sigma = 5.715]$

b. the nine samples: using $s^2 = \Sigma(x-\bar{x})^2/(n-1)$ [for each sample, n=2]

sample	$\bar{X}$	s^2	s
1,1	1.0	0	0
1,3	2.0	2.0	1.414
1,14	7.5	84.5	9.192
3,1	2.0	2.0	1.414
3,3	3.0	0	0
3,14	8.5	60.5	7.778
14,1	7.5	84.5	9.192
14,3	8.5	60.5	7.778
14,14	14.0	0	0
	54.0	294.0	36.768

mean of the 9 s^2 values

$(\Sigma s^2)/9 = 294.0/9 = 32.667$

c. the nine samples: using $\sigma^2 = \Sigma(x-\mu)^2/N$ [for each sample, N=2]

sample	μ	σ^2
1,1	1.0	0
1,3	2.0	1.00
1,14	7.5	42.25
3,1	2.0	1.00
3,3	3.0	0
3,14	8.5	30.25
14,1	7.5	42.25
14,3	8.5	30.25
14,14	14.0	0
	54.0	147.00

mean of the 9 σ^2 values

$(\Sigma\sigma^2)/9 = 147.00/9 = 16.333$

d. The approach in (b) of dividing by n-1 when calculating the sample variance gives a better estimate of the population variance. On the average, the approach in (b) gave the correct population variance of 32.667. The approach in (c) of dividing by n underestimated the correct population variance. When computing sample variances, divide by n-1 and not by n.

e. No. An unbiased estimator is one that gives the correct answer on the average. Since the average value of s^2 in part (b) was 6, which was the correct value calculated for σ^2 in part (a), s^2 is an unbiased estimator of σ^2. Since the average value of s in part (b) is $(\Sigma s)/9 = 36.768/9 = 4.085$, which is not the correct value of 5.715 calculated for σ in part (a), s is not an unbiased estimator of σ.

NOTE: Since the average value of $\bar{x}$ in part (b) is $(\Sigma\bar{x})/9 = 54.0/9 = 6.0$, which is the correct value calculated for μ in part (a), $\bar{x}$ is an unbiased estimator of μ.

3-4 Measures of Relative Standing and Boxplots

1. For a z score of -0.61, the negative sign indicates that her age is below the mean and the numerical portion indicates that her age is 0.61 standard deviations away from the mean.

3. The values shown in the boxplot constitute the 5-number summary as follows.
   0 hours = the minimum value, the length of the shortest flight
   166 hours = the first quartile Q_1, the length below which the shortest 25% of the flights occur
   215 hours = the second quartile Q_2, the median length of the flights
   269 hours = the third quartile Q_3, the length above which the longest 25% of the flights occur
   423 hours = the maximum value, the length of the longest flight

5. a. $|x - \bar{x}| = |61 - 35.8| = |25.2| = 25.2$ years
   b. $25.2/11.3 = 2.23$
   c. $z = (x-\bar{x})/s = (61-35.8)/11.3 = 2.23$
   d. Since $2.23 > 2.00$, Helen Mirren's age is considered unusual in this context.

7. a. $|x - \bar{x}| = |110 - 245.0| = |-135.0| = 135.0$ seconds
   b. $135.0/36.4 = 3.71$
   c. $z = (x - \bar{x})/s = (110 - 245.0)/36.4 = -3.71$
   d. Since $-3.71 < -2.00$, a duration time of 110 seconds is considered unusual in this context.

9. a. $z = (x - \bar{x})/s = (101.00 - 98.20)/0.62 = 4.52$; unusual, since $4.52 > 2.00$
   b. $z = (x - \bar{x})/s = (96.90 - 98.20)/0.62 = -2.10$; unusual, since $-2.10 < -2.00$
   c. $z = (x - \bar{x})/s = (96.98 - 98.20)/0.62 = -1.97$; usual, since $-2.00 < -1.97 < 2.00$

11. $z = (x - \bar{x})/s = (308 - 268)/15 = 40/15 = 2.67$
   Yes, since $2.67 > 2.00$, a pregnancy of 308 days is considered unusual. "Unusual" is just what the word implies – out of the ordinary. Other exercises in this section note that there are unusual ages and heights, and so there can be unusual pregnancies.

13. SAT: $z = (x - \bar{x})/s = (1840 - 1518)/325 = 322/325 = 0.99$
   ACT: $z = (x - \bar{x})/s = (26.0 - 21.1)/4.8 = 4.9/4.8 = 1.02$
   Since $1.02 > 0.99$, the ACT score of 26.0 is the relatively better score.

For exercises 15-27, refer to the list of ordered scores at the right. The units are total points.

#	points
1	36
2	37
3	37
4	39
5	39
6	41
7	43
8	44
9	44
10	47
11	50
12	53
13	54
14	55
15	56
16	56
17	57
18	59
19	61
20	61
21	65
22	69
23	69
24	75

15. Let b = # of scores below x; n = total number of scores
    In general, the percentile of score x is $(b/n)\cdot100$.
    The percentile score of 47 is $(9/24)\cdot100 = 38$.

17. Let b = # of scores below x; n = total number of scores
    In general, the percentile of score x is $(b/n)\cdot100$.
    The percentile score of 54 is $(12/24)\cdot100 = 50$.

19. To find P_{20}, $L = (20/100)\cdot24 = 4.8$, rounded up to 5.
    Since the 5th score is 39, $P_{20} = 39$.

21. To find $Q_3 = P_{75}$, $L = (75/100)\cdot24 = 18$, a whole number.
    The mean of the 18th and 19th scores, $Q_3 = (59+61)/2 = 60$.

23. To find P_{50}, $L = (50/100)\cdot24 = 12$, a whole number.
    The mean of the 12th and 13th scores, $P_{50} = (53+54)/2 = 53.5$.

25. To find P_{25}, $L = (25/100)\cdot24 = 6$, a whole number.
    The mean of the 6th and 7th scores, $P_{25} = (41+43)/2 = 42$.

27. Arrange the 24 values in order. The 5-number summary and boxplot are given below. All measurements are in total points.

$$\min = x_1 = 36$$
$$Q_1 = (x_6+x_7)/2 = (41+43)/2 = 42$$
$$\text{median} = (x_{12}+x_{13})/2 = (53+54)/2 = 53.5$$
$$Q_3 = (x_{18}+x_{19})/2 = (59+61)/2 = 60$$
$$\max = x_{24} = 75$$

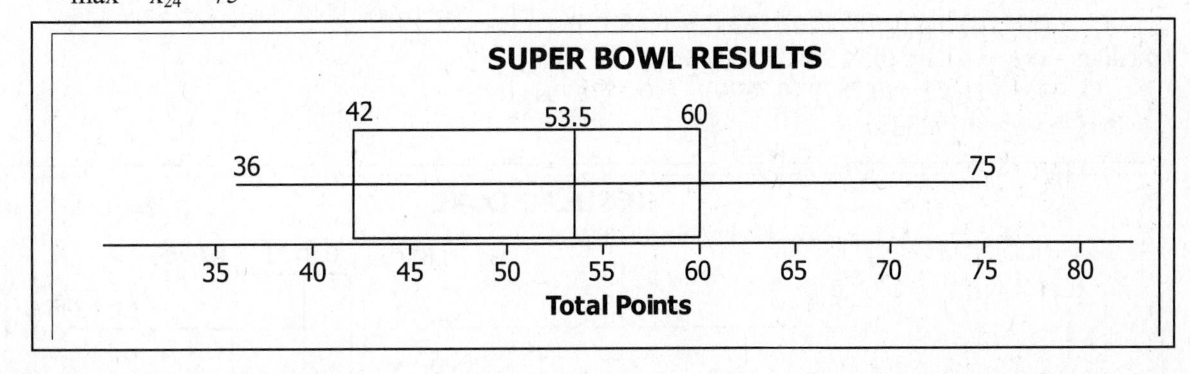

GENERAL NOTE ABOUT BOXPLOTS: There is no universally accepted format for boxplots, and the final appearance is often determined by the software used to generate the figure. The examples in the text generated by STATDISK, for example, include a vertical "barrier" at the minimum and maximum values even though such a feature is not included in the text's definition of a boxplot. If the boxplot includes an adequate horizontal scale, the 5-number summary values can be inferred from the scale and do not need to clutter up the figure – in fact, the purpose of boxplots and other figures is to give a visual image of the data without such clutter. While this manual includes the 5-number summary values in the figure, it is more for student understanding than for good display. In addition, some formats extend the line connecting the minimum and maximum values through the box and others do not. In general, follow the format preferred by the instructor.

29. Arrange the 12 values in order. The 5-number summary and boxplot are given below. All measurements are in FICO rating units.

min = x_1 = 664
$Q_1 = (x_3+x_4)/2 = (698+714)/2 = 706$
median = $(x_6+x_7)/2 = (753+779)/2 = 766$
$Q_3 = (x_9+x_{10})/2 = (802+818)/2 = 810$
max = x_{12} = 836

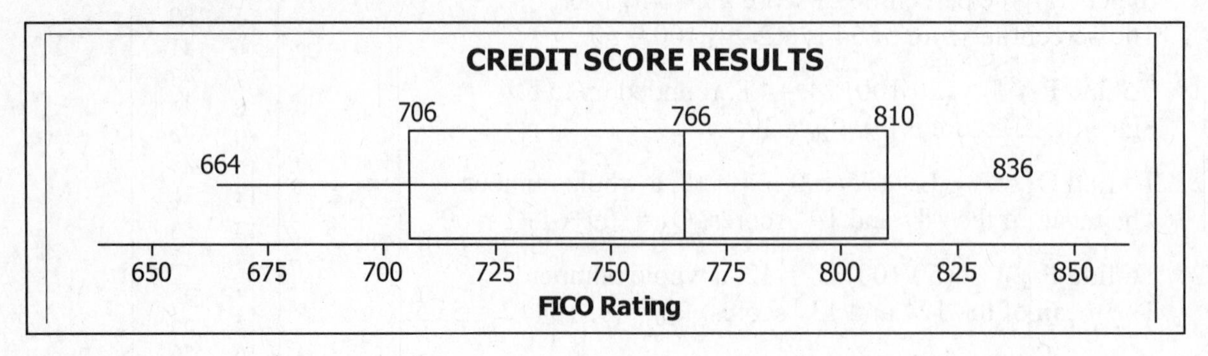

31. Arrange the 36 values in order. The 5-number summaries and boxplots are given below.

Regular Coke

min = x_1 = 0.7901 lbs
$Q_1 = (x_9+x_{10})/2 = (0.8143+0.8150)/2 = 0.81465$ lbs
median = $(x_{18}+x_{19})/2 = (0.8170+0.8172)/2 = 0.8171$ lbs
$Q_3 = (x_{27}+x_{28})/2 = (0.8207+0.8211)/2 = 0.8209$ lbs
max = x_{36} = 0.8295 lbs

Diet Coke

min = x_1 = 0.7758 lbs
$Q_1 = (x_9+x_{10})/2 = (0.7822+0.7822)/2 = 0.7822$ lbs
median = $(x_{18}+x_{19})/2 = (0.7852+0.7852)/2 = 0.7852$ lbs
$Q_3 = (x_{27}+x_{28})/2 = (0.7879+0.7879)/2 = 0.7879$ lbs
max = x_{36} = 0.7923 lbs

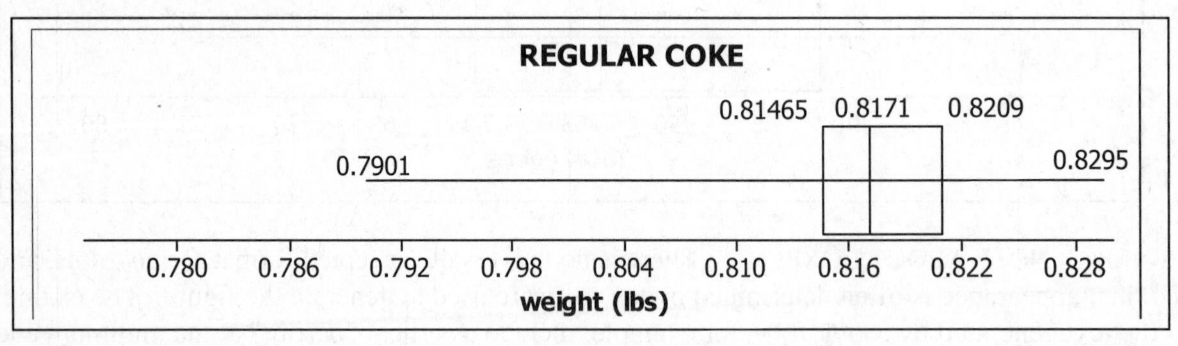

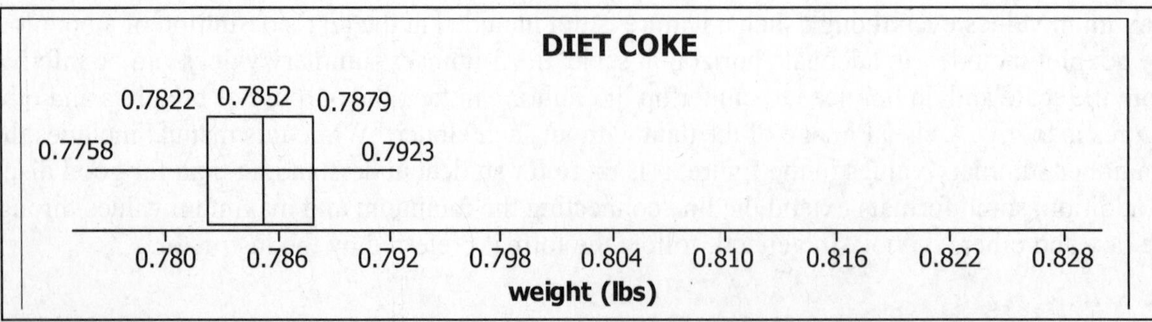

The weights of the diet Coke appear to be substantially less than those of the regular Coke.

33. Arrange the 40 values in order. The 5-number summaries and boxplots are given below.

Pre-1964 Quarters

$\min = x_1 = 6.0002$ grams

$Q_1 = (x_{10}+x_{11})/2 = (6.1309+6.1352)/2 = 6.13305$ grams

$\text{median} = (x_{20}+x_{21})/2 = (6.1940+6.1947)/2 = 6.19435$ grams

$Q_3 = (x_{30}+x_{31})/2 = (6.2647+6.2674)/2 = 6.26605$ grams

$\max = x_{40} = 6.3669$ grams

Post-1964 Quarters

$\min = x_1 = 5.5361$ grams

$Q_1 = (x_{10}+x_{11})/2 = (5.5928+5.5941)/2 = 5.59345$ grams

$\text{median} = (x_{20}+x_{21})/2 = (5.6274+5.6449)/2 = 5.63615$ grams

$Q_3 = (x_{30}+x_{31})/2 = (5.6841+5.6848)/2 = 5.68445$ grams

$\max = x_{40} = 5.7790$ grams

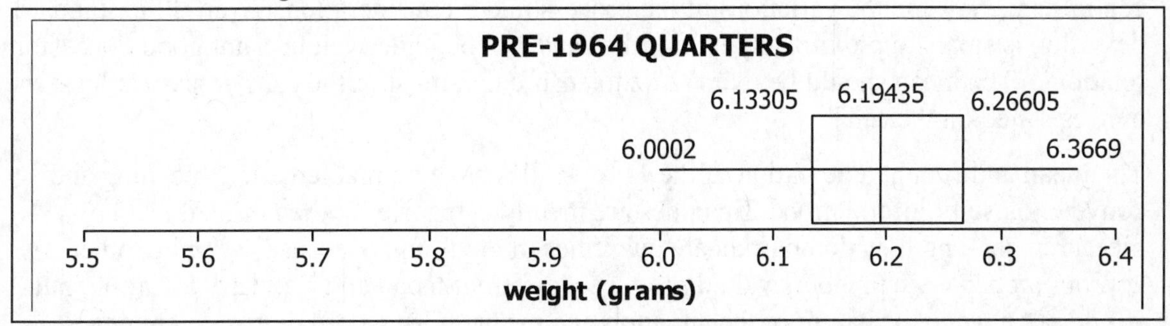

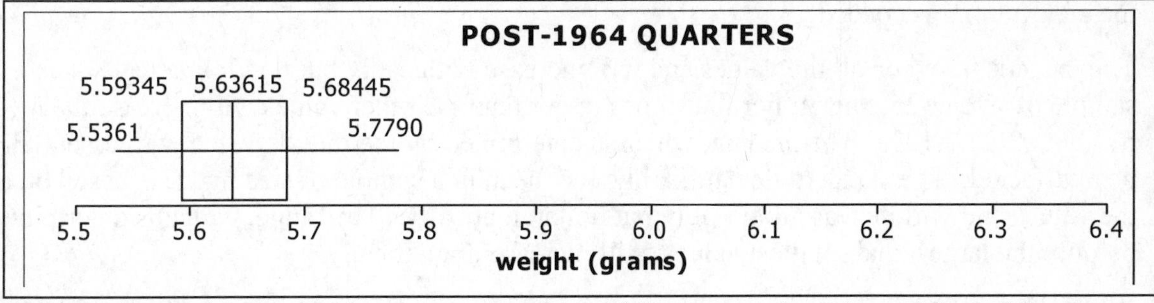

The weights of the pre-1964 quarters are substantially greater than the weights of the post-1964 quarters.

35. Arrange the 40 values in order. The 5-number summary and boxplot are given below. All measurements are in centimeters

$\min = x_1 = 27.0$

$Q_1 = (x_{10}+x_{11})/2$
$= (37.5+38.0)/2 = 37.75$

$\text{median} = (x_{20}+x_{21})/2$
$= (39.0+39.1)/2 = 39.05$

$Q_3 = (x_{30}+x_{31})/2$
$= (41.0+41.0)/2 = 41.0$

$\max = x_{40} = 48.6$

FEMALE LEG RESULTS

25 30 35 40 45 50

length (cm)

Outliers are identified as follows.

$IQR = Q_3 - Q_1$
$= 41.0 - 37.75 = 3.25$

$1.5(IQR) = 1.5(3.25) = 4.875$

lower limit $= Q_1 - 1.5(IQR)$
$= 37.75 - 4.875 = 32.875$

upper limit $= Q_3 + 1.5(IQR)$
$= 41.0 + 4.875 = 45.875$

4 outliers: 27.0, 31.1, 32.1, 48.6

37. There are 35 scores. To find $Q_1 = P_{25}$, $L = (25/100)\cdot 35 = 8.75$. Since $x_8 = 30$ and $x_9 = 35$, we estimate $x_{8.75}$ to be 30 + (0.75)(35-30) = 30 + 3.75 = $33.75 million. This close to the result x_9 = $35 million found using Figure 3-5 without interpolation.

Statistical Literacy and Critical Thinking

1. No, the correct mean amount of Coke in each can is not sufficient to guarantee that the production process is proceeding as it should. If the standard deviation of the amounts is large, then unacceptable numbers of under-filled and over-filled cans are being produced. Quality control engineers need to monitor both the mean and the standard deviation (i.e., both the central tendency and the variation) of the process under consideration. Even if the standard deviation is small, approximately half of the cans will be underweight – not good for customer relations. The mean should be over 12 oz to ensure that most of the cans contain at least the 12 ounces stated on the can.
2. The mean and standard deviation of the 11 zip codes have no mathematical meaning and convey no useful information. Zip codes are merely categories at the nominal level of measurement, and they do not indicate quantities of anything. Because assigning other appropriate zip code numbers would produce a different mean and standard deviation, but would not change the nature of the situation, the mean and standard deviation are not meaningful in this context.
3. The mean is based on all the values and will increase noticeably but not dramatically – in a sample of size n=25, one outlier four times larger than the other values will increase the mean by about 3/25 = 12%. The median, which is determined by the middle value, will essentially be unaffected. The standard deviation, like the mean in a sample of size n=25, is based on all the values and will increase noticeably but not dramatically. The range, which is determined by only the largest and smallest values, will increase dramatically.
4. When the data are presented sorted by magnitude, any trends or patterns over time are lost.

Chapter Quick Quiz

1. $\bar{x} = (\Sigma x)/n = 20/5 = 4.0$ cm
2. Arranged in order, the values are 2,2,3,5,8. The median is the middle value, 3 cm.
3. The mode is the most frequently occurring value, 2 cm.
4. The variance is the square of the standard deviation. $s^2 = 5.0^2 = 25.0$ ft^2
5. $z = (x - \bar{x})/s = (4.0 - 10.0)/2.0 = -3.00$
6. The range, standard deviation, and variance are all measures of variation.
7. In general, sample statistics are indicated by English letters and population parameters are indicated by the corresponding Greek letter. The symbol for the standard deviation of a sample is s, and the symbol for the standard deviation of a population is σ.
8. The symbol for the mean of a sample is $\bar{x}$, and the symbol for the mean of a population is μ.

9. The 25th percentile separates the lowest 25% from the highest 75%. Approximately 75 percent of the values in a sample are greater than or equal to the 25th percentile.

10. True. Both the median and the 50th percentile are defined as the middle value – the value with at least 50% of the values less than or equal to it, and with at least 50% of the values greater than or equal to it.

Review Exercises

1. Arranged in order, the values are: 17 18 18 19 19 20 20 20 21 21
   summary statistics: $n = 10$ $\Sigma x = 193$ $\Sigma x^2 = 3741$
   a. $\bar{x} = (\Sigma x)/n = 193/10 = 19.3$ oz
   b. $\tilde{x} = (x_5+x_6)/2 = (19+20)/2 = 19.5$ oz
   c. M = 20 oz
   d. m.r. $= (x_1+x_n)/2 = (17+21)/2 = 19.0$ oz
   e. $R = x_n - x_1 = 21 - 17 = 4$ oz
   f. $s = 1.3$ oz [the square root of the answer given in part g]
   g. $s^2 = [n(\Sigma x^2) - (\Sigma x)^2]/[n(n-1)] = [10(3741) - (193)^2]/[10(9)] = 161/90 = 1.8 \text{ oz}^2$
   h. $Q_1 = x_3 = 18$ oz [L = (25/100)(10) = 2.5, rounded up to 3]
   i. $Q_3 = x_8 = 20$ oz [L = (75/100)(10) = 7.5, rounded up to 8]

2. Arrange the 10 values in order. The 5-number summary and boxplot are given below.
   min $= x_1 = 17$ oz
   $Q_1 = x_3 = 18$ oz
   median $= (x_5+x_6)/2 = (19+20)/2 = 19.5$ oz
   $Q_3 = x_8 = 20$ oz
   max $= x_{10} = 21$ oz

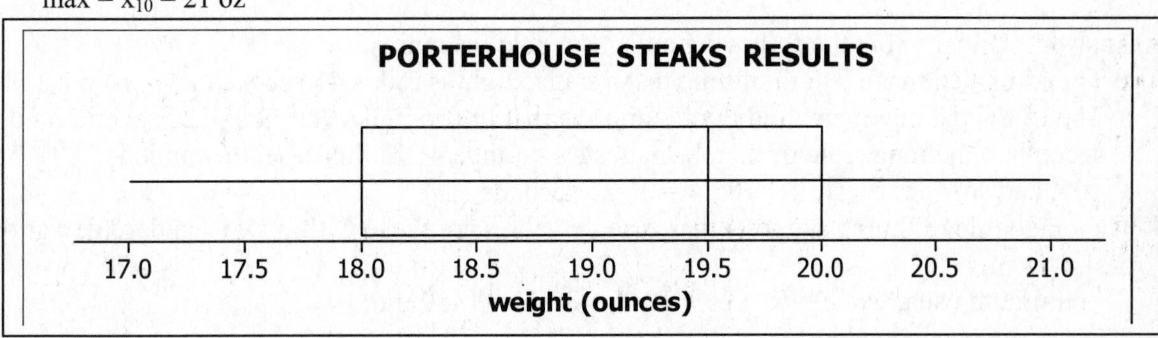

3. Arranged in order, the values are: 878 880 923 924 928 930 934 936 936
   summary statistics: $n = 9$ $\Sigma x = 8269$ $\Sigma x^2 = 7601621$
   a. $\bar{x} = (\Sigma x)/n = 8269/9 = 918.8$ mm
   b. $\tilde{x} = x_5 = 928$ mm
   c. M = 936 mm
   d. m.r. $= (x_1+x_n)/2 = (878+936)/2 = 907$ mm
   e. $R = x_n - x_1 = 936 - 878 = 58$ mm
   f. $s = 23.0$ mm [the square root of the answer given in part g]
   g. $s^2 = [n(\Sigma x^2) - (\Sigma x)^2]/[n(n-1)] = [9(7601621) - (8269)^2]/[9(8)] = 38228/72 = 530.9 \text{ mm}^2$
   h. $Q_1 = x_3 = 923$ mm [L = (25/100)(9) = 2.25, rounded up to 3]
   i. $Q_3 = x_7 = 934$ mm [L = (75/100)(9) = 6.75, rounded up to 7]

4. $z = (x - \bar{x})/s = (878 - 918.8)/23.0 = -1.77$
   No. Since $-2 < -1.77 < 2$, the sitting height of 878 mm is not unusual.

5. Arrange the 10 values in order. The 5-number summary and boxplot are given below.
   min = x_1 = 878 mm
   $Q_1 = x_3$ = 923 mm
   median = x_5 = 928 mm
   $Q_3 = x_7$ = 934 mm
   max = x_9 = 936 mm

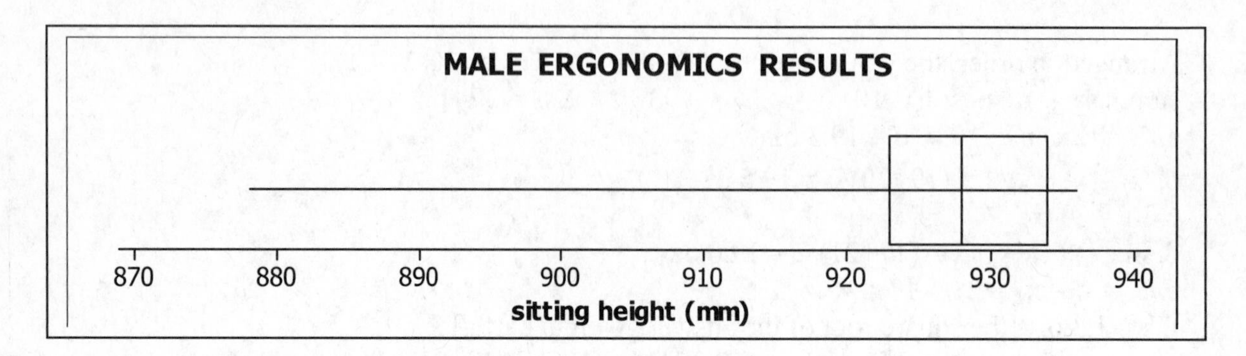

6. SAT: $z = (x - \bar{x})/s = (1030 - 1518)/325 = -488/325 = -1.50$
   ACT: $z = (x - \bar{x})/s = (14.0 - 21.1)/4.8 = -7.1/4.8 = -1.48$
   Since $-1.48 > -1.50$, the ACT score of 14.0 is the relatively better score.

7. Answers will vary, but the following are reasonable responses.
   a. The estimated mean age of cars driven by students is 6.0 years.
   b. The estimated maximum and minimum age for cars driven by students is 12 years and 0 years. The range rule of thumb suggests $s \approx$ range/4. In this case, the range is $12 - 0 = 12$, and so we estimate $s \approx 12/4 = 3.0$ years.

8. Answers will vary, but the following are reasonable responses.
   a. The estimated mean length of time that a traffic light is red is 45 seconds.
   b. The estimated maximum and minimum time that traffic lights are red is 120 seconds and 15 seconds. The range rule of thumb suggests $s \approx$ range/4. In this case, the range is $120 - 15 = 105$, and so we estimate $s \approx 105/4 = 26.25$ seconds.

9. The range rule of thumb suggests that "usual" values are those within two standard deviations of the mean.
   minimum usual value = $\bar{x} - 2s = 1212 - 2(51) = 1110$ mm
   maximum usual value = $\bar{x} + 2s = 1212 + 2(51) = 1314$ mm
   Since accessibility to controls is the issue, the minimum usual value of 1110 mm is the more relevant value. No controls should be placed beyond the reach of the person with the smallest likely grip reach.

10. The range rule of thumb suggests that "usual" values are those within two standard deviations of the mean.
    minimum usual value = $\bar{x} - 2s = 97.5 - 2(6.9) = 83.7$ cm
    maximum usual value = $\bar{x} + 2s = 97.5 + 2(6.9) = 111.3$ cm
    No; since $83.7 < 87.8 < 111.3$, a height of 87.8 cm is not considered unusual. On this basis, the physician should not be concerned.

Cumulative Review Exercises

1. a. Since the values can be any number on a continuum, the sitting heights are continuous data.
   b. Since differences between the values are meaningful and there is a meaningful zero, the sitting heights are data at the ratio level of measurement.

2. The frequency distribution is given below.

sitting height (mm)	frequency
870 – 879	1
880 – 889	1
890 – 899	0
900 – 909	0
910 – 919	0
920 – 929	3
930 – 939	4
	9

3. The histogram is given at the right. The distribution appears to be skewed.
   NOTE: The true boundaries for the bars are as given in the figure, but for visual simplicity it is permissible to use the labels 870, 880, etc. In general, visual images are intended to convey a concise summary of the data and not technical detail.

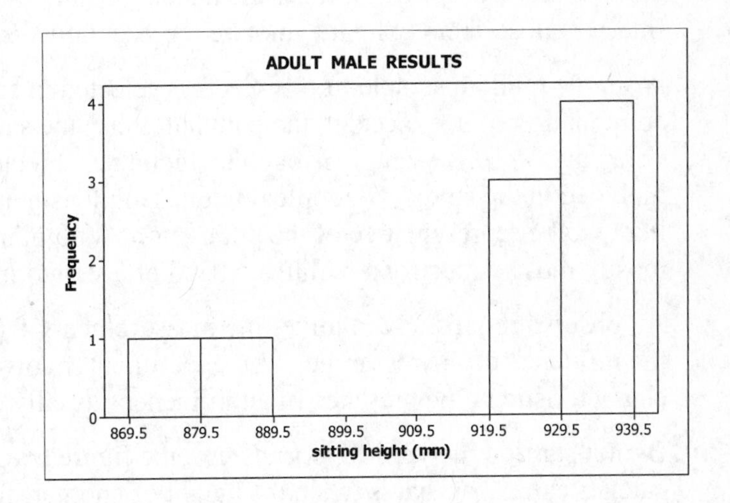

4. The dotplot is given below.

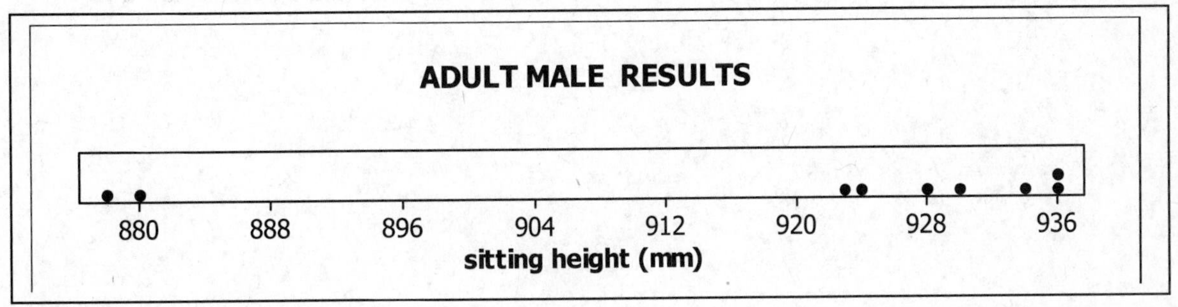

5. The stemplot is given below.
   sitting height (mm)
   87 | 8
   88 | 0
   89 |
   90 |
   91 |
   92 | 348
   93 | 0466

6. a. The mode is the only appropriate measure of central tendency for data at the nominal level of measurement. All the other measures require that the data have some underlying order, which does not exist for data at the nominal level.
   b. This is a convenience sample, because the sample was obtained without detailed planning from those people conveniently at hand. Had the botanist divided the phone book into groups of 250, selected one of those groups at random, and happened to get the first such group, then he would have a cluster sample.
   c. This is a cluster sample, since the population was divided into groups, groups were selected at random, and everyone in each selected group was sampled.
   d. The problem is that there appears to be too much variation. Of the measures listed, only the standard deviation is a measure of variation. The is the statistic that should be monitored, with the hope lowering its value.

7. No, the responses cannot be considered representative of the population of the United States for at least two reasons: the responders were a voluntary response sample (i.e., persons who tend to feel strongly about the issue under consideration), and the sample was limited to AOL Internet subscribers (who may not be representative in education, incomes, etc.).

8. A simple random sample of n subjects is selected in such a way that every possible combination of n subjects in the population has the same chance of being selected. A voluntary response sample is one in which the subjects themselves decide whether to be included in the study. A simple random sample is generally a better sample because it is more likely to be representative of the population. Voluntary response samples tend to include mostly those with strong feelings on and/or a vested interest in the issue under consideration.

9. An observational study involves the measurement of some response or characteristic not under the influence of the researcher. An experiment involves the measurement of some response or characteristic in the presence of an influence (usually called a "treatment") from the researcher.

10. By not starting the vertical axis at zero, the figure exaggerates the differences between the categories. A bar that is twice the height of another, for example, gives the faulty visual impression of twice the frequency count – when the actual difference is less than that.

Chapter 4

Probability

4-2 Basic Concepts of Probability

1. To say that the probability of being injured while using recreation equipment in 1/500 means that approximately one injury occurs for every 500 times that recreation equipment is used. Note that this is an aggregate result, and not necessarily true for any one particular piece of equipment – since some pieces of equipment are more dangerous than others. Because the probability 1/500 = 0.002 is small, such an injury is considered unusual.

3. If A denotes the fact that some event occurs, then $\bar{A}$ denotes the fact that the event does not occur. If P(A) = 0.995, then P($\bar{A}$) = 1 – 0.995 = 0.005. If P(A) = 0.995, then A is very likely to occur and it would be unusual for $\bar{A}$ to occur – i.e., it would be unusual for A not to occur.

5. "4 in 21" = 4/21 = 0.190

7. "50–50 chance" = 50% = 50/100 = 0.50

9. 6/36 = 0.167

11. "impossible" = 0% = 0/100 = 0

13. All probability must be between 0 and 1 inclusive.
    The value 3:1 = 3/1 = 3 cannot be a probability because 3 > 1.
    The value 5/2 = 2.5 cannot be a probability because 2.5 > 1.
    The value -0.5 cannot be a probability because -0.5 < 0.
    The value 321/123 = 2.610 cannot be a probability, because 2.610 > 1.

15. a. Let having exactly one girl be the set with 3 simple events A = {bbg,bgb,gbb}.
       P(A) = 3/8 = 0.375
    b. Let having exactly two girls be the set with 3 simple events B = {bgg,gbg,ggb}.
       P(B) = 3/8 = 0.375
    c. Let having all three girls be the set with 1 simple event C = {ggg}.
       P(C) = 1/8 = 0.125

17. a. The total number of responses summarized in the table is 15 + 42 + 32 + 9 = 98.
    b. The number of negative responses (the test says the person did not lie) is 32 + 9 = 41.
    c. Let N = selecting one of the negative test responses. P(N) = 41/98.
    d. 41/98 = 0.418

19. Let F = selecting a false positive response. P(F) = 15/98 = 0.153.
    The probability of this type of error is high enough that such an occurrence would not be considered unusual; the test is not highly accurate.

21. There are 84 + 16 = 100 total senators.
    Let W = selecting a woman. P(W) = 16/100 = 0.16.
    No; this probability is too far below 0.50 to agree with the claim that men and women have equal opportunities to become a senator.

23. Let L = getting struck by lightning. By the relative frequency approximation of probability, the estimate is P(L) = 281/290,789,000 = 0.000000966. No; engaging in at risk behavior increases the probability of experiencing negative consequences.

25. a. Let B = the birthdate is correctly identified. P(B) = 1/365 = 0.003.
    b. Yes. Since 0.003 ≤ 0.05, it would be unusual to guess correctly on the first try.
    c. Most people would probably believe that Mike's correct answer was the result of having inside information – and not making a lucky guess.
    d. Fifteen years is a big error. If the guess were serious, Kelly would likely think that Mike was not knowledgeable and/or be personally insulted – and a second date is unlikely. If the guess were perceived as being made in jest, Kelly might well appreciate his sense of humor and/or handling of a potentially awkward situation – and a second date is likely.

27. Let F = a pacemaker malfunction is caused by firmware.
    Based on the given results, P(F) = 504/8834 = 0.0571.
    No. Since 0.0571 > 0.05, a firmware malfunction is not unusual among pacemaker malfunctions.

29. There were 795 + 10 = 805 total persons executed.
    Let W = a randomly selected execution was that of a woman. P(W) = 10/805 = 0.0124.
    Yes. Since 0.0124 ≤ 0.05, it is unusual for an executed person to be a woman. This is due to the fact that more crimes worthy of the death penalty are committed by men than by women.

31. There were 211+ 288 + 366 + 144 + 89 = 1098 total persons in the survey.
    Let F = a household has 4 or more cellphones in use.
    Based on these results, P(F) = 89/1098 = 0.0811.
    No. Since 0.0811 > 0.05, it is not unusual for a household to have 4 or more cellphones in use.
    NOTE: Technically, the given survey cannot be used to answer the question as posed. A survey made of persons selected at random cannot be used to answer questions about households. If the goal is to make statements about households, then households (and not persons) should be selected at random. Selecting persons at random means that households with more people have a higher probability of being in the survey than households with fewer people. Since households with more people are more likely to have more cellphones, the survey will tend to overestimate the number of cellphones per household.

33. a. Listed by birth order, the sample space contains 4 simple events: bb bg gb gg.
    b. P(gg) = 1/4 = 0.25
    c. P(bg or gb) = 2/4 = 0.50

35. a. Listed with the father's contribution first, the sample space has 4 simple events:
       brown/brown brown/blue blue/brown blue/blue.
    b. P(blue/blue) = 1/4 = 0.25
    c. P(brown eyes) = P(brown/brown or brown/blue or blue/brown) = 3/4 = 0.75

37. The odd against winning are P(not winning)/P(winning) = (423/500)/(77/500) = 423/77, or 423:77 – which is approximately 5.5:1, or 11:2.

39. a. If a $2 bet results in a net return of $14.20, the net profit is 14.20 – 2.00 = $12.20.
    b. If you get back your bet plus $12.20 for every $2 you bet, the payoff odds are 12.2:2 [typically expressed as either 6.1:1 or 61:10], or approximately 6:1.
    c. odds against W = P(not win)/P(W) = (443/500)/(57/500) = 443/57 or 443:57 [typically expressed as 7.77:1], or approximately 8:1.

d. If the payoff odds are 7.77:1, a win gets back your bet plus $7.77 for every $1 bet. If you bet $2 and win, you get back 2 + 15.54 = $17.54. The ticket would be worth $17.54 and your profit would be 17.54 – 2.00 = $15.54.

41. Making a table like the one on the right organizes the information and helps to understand the concepts involved.

		Headache? no	yes	
Treatment	Nasonex	2077	26	2103
	control	1649	22	1671
		3276	48	3774

$p_t = 26/2103 = 0.0124$
$p_c = 22/1671 = 0.0132$

a. The relative risk is the P(risk) for the treatment as a percent of the P(risk) for the control.
relative risk = $p_t/p_c = 0.0124/0.0132 = 0.939 = 93.9\%$

b. The odds in favor of the risk are P(risk)/P(not risk).
for the treatment: odds in favor = $p_t/(1\text{-}p_t) = (26/2103)/(2077/2103) = 26/2077 = 0.012518/1$
for the control: odds in favor = $p_c/(1\text{-}p_c) = (22/1671)/(1649/1671) = 22/1649 = 0.013341/1$
The odds ratio is the odds in favor of the risk for the treatment as a percent of the odds in favor of the risk for the control.
odds ratio = $[p_t/(1\text{-}p_t)/[p_c/(1\text{-}p_c)] = 0.012518/0.013314 = 0.938 = 93.8\%$
Both the relative risk and the odds ratio are less than 1.00, suggesting that the risk of headache is actually smaller for Nasonex than it is for the control. Nasonex does not appear to pose a risk of headaches.

43. This difficult problem will be broken into two events and a conclusion. Let L denote the length of the stick. Label the midpoint of the stick M, and label the first and second breaking points X and Y.

- Event A: X and Y must be on opposite sides of M. If X and Y are on the same side of M, the side without X and Y would be longer than 0.5L and no triangle will be possible. Once X is set, P(Y on the same side as X) = P(Y on opposite side from X) = ½.
- Event B: The distance |XY| must be less than 0.5L. Assuming X and Y are on opposite sides of M, let Q denote the end closest to X and R denote the end closest to Y – so that the following sketch applies.

```
|--------|------------|----|----------------|
Q        X            M    Y                R
```

In order to form a triangle, it must be true that |XY| = |XM| + |MY| < 0.5L. This happens only when |MY| < |QX|. With all choices at random, there is no reason for |QX| to be larger than |MY|, or vice-versa. This means that P(|MY|<|QX|) = ½.

- Conclusion: For a triangle, we need both events A and B.
Since P(A) = ½ and P(B occurs assuming A has occurred) = ½, events A and B together will happen only ½ of ½ of the time: the final answer would be ½ of ½, or ¼. In the notation of section 4-4, P(A and B) = P(A)·P(B|A) = (0.5)(0.5) = 0.25.

NOTE: In situations where the laws of science and mathematics are difficult to apply (like finding the likelihood that a particular thumbtack will land point up when it is dropped), probabilities are frequently estimated using Rule #1 of this section. Suppose an instructor gives 100 sticks to a class of 100 students and says, "Break your stick at random two times so that you have 3 pieces." Does this mean that only about 25% of the students should be able to form a triangle with their 3 pieces? In theory, yes. In practice, no. It is likely the students would break their sticks "conveniently" and not "at random" – e.g., there would likely be fewer (if any) pieces less than one inch than you would expect by chance, because such pieces would be difficult to break off by hand.

4-3 Addition Rule

1. Two events in a single trial are disjoint if they are outcomes that cannot happen at the same time.

3. In the context of the addition rule defined in this section, P(A and B) is the probability that both event A and event B occur at the same time in a single trial.

5. Not disjoint, since a physician can be a female surgeon.

7. Disjoint, since a car that is free of defects cannot have a dead battery – which is a defect.

9. Not disjoint, since a subject who believes the evidence for global warming can also be opposed to stem cell research.

11. Not disjoint, since an R-rated movie can receive a four-star review.

13. $P(\bar{M})$ is the probability that a STATDISK user is not a Macintosh computer user

$$\begin{aligned} P(\bar{M}) &= 1 - P(M) \\ &= 1 - 0.05 \\ &= 0.95 \end{aligned}$$

15. Let M = a selected American believes it morally wrong not to report all income. P(M) = 0.79.

$$\begin{aligned} P(\bar{M}) &= 1 - P(M) \\ &= 1 - 0.79 \\ &= 0.21 \end{aligned}$$

NOTE: For exercises 17-20, refer to the table at the right and use the following notation.
Let P = the polygraph test indicates a positive
Let L = the person actually lied.

		Subject Lie? no	yes	
Test Result	positive	15	42	57
	negative	32	9	41
		47	51	98

17. $$\begin{aligned} P(P \text{ or } \bar{L}) &= P(P) + P(\bar{L}) - P(P \text{ and } \bar{L}) \\ &= 57/98 + 47/97 - 15/98 \\ &= 89/98 = 0.908 \end{aligned}$$

19. $P(\bar{L} \text{ and } \bar{P}) = 32/98 = 0.327$

NOTE: For exercises 21-26, refer to the table at the right and use the following notation.
Let M = the challenge was made by a male.
Let S = the challenge was successful.

		Successful? yes	no	
Gender	male	201	288	489
	female	126	224	350
		327	512	839

21. $P(\bar{S}) = 512/839 = 0.610$

23. $$\begin{aligned} P(M \text{ or } S) &= P(M) + P(S) - P(M \text{ and } S) \\ &= 489/839 + 327/839 - 201/839 \\ &= 615/839 = 0.733 \end{aligned}$$

25. $$\begin{aligned} P(M \text{ or } \bar{S}) &= P(M) + P(\bar{S}) - P(M \text{ and } \bar{S}) \\ &= 489/839 + 512/839 - 288/839 \\ &= 713/839 = 0.850 \end{aligned}$$

NOTE: For exercises 27-32, refer to the following table and notation.
For a person selected at random, let the following designations apply.
Y = there is a response
N = there is a refusal
A = the age is 18-21
B = the age is 22-29
C = the age is 30-39
D = the age is 40-49
E = the age is 50-59
F = the age is 60+

		Age						
		18-21	22-29	30-39	40-49	50-59	60+	
Response?	yes	73	255	245	136	138	202	1049
	no	11	20	33	16	27	49	156
		84	275	278	152	165	251	1205

27. P(N) = 156/1205 = 0.129
Yes. Persons who refuse to answer form a subpopulation whose opinions will not be counted, thus preventing the survey from being representative of the entire population.

29. P(Y or A) = P(Y) + P(A) – P(Y and A)
= 1049/1205 + 84/1205 – 73/1205
= 1060/1205 = 0.880

31. Use the intuitive approach rather than the formal addition rule.
P(Y or B or C) = P(Y) + P(B and N) + P(C and N)
= 1049/1205 + 20/1205 + 33/1205
= 1102/1205 = 0.915

NOTE: For exercises 33-36 refer to the table at the right and use the following notation.
Let P = the drug test indicates a positive
Let M = the person actually uses marijuana

		Use Marijuana?		
		yes	no	
Test Result	positive	119	24	143
	negative	3	154	157
		122	178	300

33. a. 300
b. 178
c. $P(\overline{M})$ = 178/300 = 0.593

35. $P(\overline{P} \text{ or } \overline{M}) = P(\overline{P}) + P(\overline{M}) - P(\overline{P} \text{ and } \overline{M})$
= 157/300 + 178/300 – 154/300
= 181/300 = 0.603

37. P[(false positive) or (false negative)]
$= P[(P \text{ and } \overline{M}) \text{ or } (\overline{P} \text{ and } M)]$
$= P(P \text{ and } \overline{M}) + (\overline{P} \text{ and } M) - P[(P \text{ and } \overline{M}) \text{ and } (\overline{P} \text{ and } M)]$
= 24/300 + 3/300 – 0/300
= 27/300 = 0.090
Using the 0.05 guideline, the occurrence of a false positive or a false negative is not unusual. It appears that the test is not highly accurate.

39. Assuming girls and boys are equally likely, the birth-toss sample space consists of 4 equally likely simple events: GH GT BH BT.
Using the concepts of the formal addition rule,
P(G or H) = P(G) + P(H) – P(G and H)
= 2/4 + 2/4 – 1/4
= 3/4 = 0.75
Using the concepts of the intuitive addition rule,
P(G or H) = 3/4 = 0.75

41. If the *exclusive or* is used instead of the *inclusive or*, then the double-counted probability must be completely removed (i.e., it must be subtracted twice) and the formula becomes as follows.
$P(A \text{ or } B) = P(A) + P(B) - 2 \cdot P(A \text{ and } B)$

43. The following two English/logic facts are used in this exercise.
• not (A or B) = not A and not B Is your sister either artistic or bright?
No? Then she is not artistic and she is not bright.
• not (A and B) = not A or not B Is your brother artistic and bright?
No? Then either he is not artistic or he is not bright.

a. $P(\overline{A \text{ or } B}) = P(\bar{A} \text{ and } \bar{B})$ from the first fact above
or (from the rule for complementary events)
$P(\overline{A \text{ or } B}) = 1 - P(A \text{ or } B) = 1 - [P(A) + P(B) - P(A \text{ and } B)] = 1 - P(A) - P(B) + P(A \text{ and } B$
b. $P(\bar{A} \text{ or } \bar{B}) = 1 - P(A \text{ and } B)$ from the second fact above
c. They are different: part (a) gives the complement of "A or B"
part (b) gives the complement of "A and B"
NOTE: The following example illustrates this concept. Consider the following 5 students and their majors: Rob (art), Steve (art), Tom (art and biology), Uriah (biology), Vern (math). Select one student at random and consider the major.
$P(A) = 3/5$ $P(B) = 2/5$ $P(A \text{ or } B) = 4/5$ $P(A \text{ and } B) = 1/5$.
a. $P(\overline{A \text{ or } B}) = 1 - P(A \text{ or } B)$
$= 1 - 4/5$
$= 1/5$ [only Vern is "neither an art major nor a biology major"]
b. $P(\bar{A} \text{ or } \bar{B}) = P(\bar{A}) + P(\bar{B}) - P(\bar{A} \text{ and } \bar{B})$
$= 2/5 + 3/5 - 1/5$
$= 4/5$ [everyone but Tom is either "not an art major" or "not a biology major"]
c. No, $1/5 \neq 4/5$.

4-4 Multiplication Rule: Basics

1. Answers will vary, but the following are possibilities.
Independent events: Getting an even number when a die is rolled and getting a head when a coin is tossed.
Dependent events: Selecting aces on two consecutive draws without replacement from an ordinary deck of cards.

3. The events of selecting adults are dependent, because the adults are selected without replacement. On any given pick, the probability of a particular person being selected depends upon who and how many persons have already been selected.

5. Independent, since the population is so large that the 5% guideline applies.

7. Dependent, if there is visual contact during the invitation, since visual impressions affect responses.
Independent, if there is no visual contact during the invitation, since the invitee would have no knowledge of the inviter's clothing.

9. Dependent, since the two items operate from the same power supply.

11. Independent, since the population is so large that the 5% guideline applies.

13. Let F = a selected person had false positive results.
$P(F_1 \text{ and } F_2) = P(F_1)\cdot P(F_2|F_1)$
$= (15/98)(14/97) = 0.0221$
Yes; since $0.0221 \leq 0.05$, getting two subjects who had false positives would be unusual.
NOTE: Since n=2 is $2/98 = 0.0204 \leq 0.05$ of the population, one could use the 5% guideline and treat the selections as independent to get $(15/98)^2 = 0.0234$. Typically the 5% guideline is invoked only when the calculations for dependent events are extremely tedious.

15. Let P = a selected person tested positive
$P(P_1 \text{ and } P_2 \text{ and } P_3 \text{ and } P_4) = P(P_1)\cdot P(P_2|P_1)\cdot P(P_3|P_1 \text{ and } P_2)\cdot P(P_4|P_1 \text{ and } P_2 \text{ and } P_3)$
$= (57/98)(56/97)(55/96)(54/95) = 0.109$
No; since $0.109 > 0.05$, getting four subjects who had positive results would not be unusual.

NOTE: For exercises 17-20 refer to the table at the right.

		Blood Group				
		O	A	B	AB	
Type	Rh+	39	35	8	4	86
	Rh-	6	5	2	1	14
		45	40	10	5	100

17. P(O and Rh+) = 39/100
a. (39/100)(39/100) = 0.152
b. (39/100)(38/99) = 0.150

19. P(O and Rh-) = 6/100
a. (6/100)(6/100)(6/100)(6/100) = 0.0000130
b. (6/100)(5/99)(4/98)(3/97) = 0.00000383

21. a. Yes; since the correctness of the first response has no effect on the correctness of the second, the events are independent.
b. $P(C_1 \text{ and } C_2) = P(C_1)\cdot P(C_2|C_1)$
$= (1/2)(1/4) = 1/8 = 0.125$
c. No; since $0.125 > 0.05$, getting both answers correct would not be unusual – but it certainly isn't likely enough to recommend such a practice.

23. Let A = a public opinion poll is accurate within its margin of error.
P(A) = 0.95 for each polling organization
$P(\text{all 9 are accurate}) = P(A_1 \text{ and } A_2 \text{ and} \ldots \text{and } A_9)$
$= P(A_1)\cdot P(A_2)\cdot \ldots \cdot P(A_9)$
$= (0.95)^9 = 0.630$
No. With 9 independent polls the probability that at least one of them is not accurate within its margin of error is 1 – P(all accurate) = 1 – 0.630 = 0.370.

25. Let G = getting a girl.
P(G) = 1/2 for each birth
$P(G_1 \text{ and } G_2 \text{ and } G_3) = P(G_1)\cdot P(G_2)\cdot P(G_3)$
$= (1/2)(1/2)(1/2) = 1/8 = 0.125$
No; since $0.125 > 0.05$, getting 3 girls by chance alone is not an unusual event. The results are not necessarily evidence that the gender-selection method is effective.

27. Let A = the alarm works.
P(A) = 0.9 for each alarm
a. $P(\bar{A}) = 1 - P(A) = 1 - 0.9 = 0.1$
b. $P(\bar{A}_1 \text{ and } \bar{A}_2) = P(\bar{A}_1)\cdot P(\bar{A}_2)$
$= (0.1)(0.1) = 0.01$

c. $\text{P(being awakened)} = \text{P}(\overline{\overline{A}_1 \text{ and } \overline{A}_2})$
$= 1 - \text{P}(\overline{A}_1 \text{ and } \overline{A}_2)$
$= 1 - 0.01 = 0.99$

NOTE: The above parts (b) and (c) assume that the alarm clocks work independently of each other. This would not be true if they are both electric alarm clocks.

29. Let G = getting a tire that is good
P(G) = 4800/5000 = 0.96
a. $P(G_1 \text{ and } G_2 \text{ and } G_3 \text{ and } G_4) = P(G_1)\cdot P(G_2|G_1)\cdot P(G_3|G_1 \text{ and } G_2)\cdot P(G_4|G_1 \text{ and } G_2 \text{ and } G_3)$
$= (4800/5000)(4799/4999)(4798/4998)(4797/4997) = 0.849$
b. Since n = 100 represents 100/5000 = 0.02 ≤ 0.05 of the population, use the 5% guideline and treat the repeated selections as being independent.
$P(G_1 \text{ and } G_2 \text{ and} \ldots \text{and } G_{100}) = P(G_1)\cdot P(G_2)\cdot \ldots \cdot P(G_{100})$
$= (0.96)(0.96)\ldots(0.96) = (0.96)^{100} = 0.0169$
Yes; since 0.0169 ≤ 0.05, getting 100 good tires would be an unusual event and the outlet should plan on dealing with returns of defective tires.

31. Let W = the surge protector works correctly
P(W) = 0.99 for each of p and q
a. The TV is protected if at least one of the surge protectors works correctly.
$P(W_p \text{ or } W_q) = P(W_p) + P(W_q) - P(W_p \text{ and } W_q)$
$= 0.99 + 0.99 - 0.9801$ [.9801 comes from part (b)]
$= 0.9999$
b. The TV is protected only if both surge protectors work correctly.
$P(W_p \text{ and } W_q) = P(W_p)\cdot P(W_q)$ [since they operate independently]
$= (.99)(.99) = 0.9801$
c. The series arrangement in part (a) provided the better coverage. The parallel arrangement in part (b) actually provides poorer protection than a single protector alone.

33. This problem can be done by two different methods. In either case, let
A = getting an ace
S = getting a spade.
- Consider the sample space
The first card could be any of 52 cards, and for each first card there are 51 possible second cards. This makes a total of 52·51 = 2652 equally likely outcomes in the sample space. How many of them are A_1 and S_2?
The aces of hearts, diamond and clubs can be paired with any of the 13 spades for a total of 3·13 = 39 favorable possibilities. The ace of spades can only be paired with any of the remaining 12 members of that suit for a total of 12 favorable possibilities. Since there are 39 + 12 = 51 total favorable possibilities among the equally likely outcomes, $P(A_1 \text{ and } S_2) = 51/2652 = 0.0192$
- Use the formulas (and express the event as two mutually exclusive possibilities)
$P(A_1 \text{ and } S_2) = P([\text{spade}A_1 \text{ and } S_2] \text{or } [\text{other}A_1 \text{ and } S_2])$
$= P(\text{spade}A_1 \text{ and } S_2) + P(\text{other}A_1 \text{ and } S_2)$
$= P(\text{spade}A_1)\cdot P(S_2|\text{spade}A_1) + P(\text{other}A_1)\cdot P(S_2|\text{other}A_1)$
$= (1/52)(12/51) + (3/52)(13/51)$
$= 12/2652 + 39/2652 = 51/2652 = 0.0192$

4-5 Multiplication Rule: Complements and Conditional Probability

1. If "at least one" of the 10 pacemakers is defective, the exact number of defective pacemakers must be one of the following: 1,2,3,4,5,6,7,8,9,10.

3. The probability of a particular outcome is equal to one divided by the total number of outcomes only when all the outcomes are equally likely and mutually exclusive. In this case the two outcomes are not equally likely, and so that reasoning cannot be used. The actual survival rate was not considered, but it should have been.

5. If it is not true that at least one of the 15 tests positive, then all 15 of them test negative

7. If it is not true that none of the 4 has the recessive gene, then at least one of the 4 has the recessive gene.

9. P(at least one girl) = 1 – P(all boys)
$= 1 - P(B_1 \text{ and } B_2 \text{ and } B_3 \text{ and } B_4 \text{ and } B_5 \text{ and } B_6)$
$= 1 - P(B_1)\cdot P(B_2)\cdot P(B_3)\cdot P(B_4)\cdot P(B_5)\cdot P(B_6)$
$= 1 - (½)(½)(½)(½)(½)(½)$
$= 1 - 1/64$
$= 63/64 = 0.984$

Yes, that probability is high enough for the couple to be very confident that they will get at least one girl in six children.

11. P(at least one correct) = 1 – (all wrong)
$= 1 - P(W_1 \text{ and } W_2 \text{ and } W_3 \text{ and } W_4)$
$= 1 - P(W_1)\cdot P(W_2)\cdot P(W_3)\cdot P(W_4)$
$= 1 - (4/5)(4/5)(4/5)(4/5)$
$= 1 - 256/625$
$= 269/625 = 0.590$

Since there is a greater chance of passing than of failing, the expectation is that such a strategy would lead to passing. In that sense, one can reasonably expect to pass by guess. But while the expectation for a single test may be to pass, such a strategy can be expected to lead to failing about 4 times every 10 times it is applied.

13. Since each birth is an independent event, $P(G_4|G_1 \text{ and } G_2 \text{ and } G_3) = P(G_4) = P(G) = ½$.
One can also consider the 16-outcome sample space S for 4 births and use the formula for conditional probability as follows.

S = {BBBG, BBGG, BGBG, GBBG, BGGG, GBGG, GGBG, GGGG,
BBBB, BBGB, BGBB, GBBB, BGGB, GBGB, GGBB, GGGB}

Let F = the first 3 children are girls
Let G_4 = the fourth child is a girl
P(F) = 2/16
$P(F \text{ and } G_4) = 1/16$

$$P(G_4|F) = \frac{P(G_4 \text{ and } F)}{P(F)} = \frac{1/16}{2/16} = 1/2$$

Note: Using the sample space S also verifies directly that $P(G_4) = 8/16 = 1/2 = P(G)$, independent of what occurred during the first three births.
No. This probability is not the same as P(GGGG) = 1/16.

15. Let P(C) = P(crash) = 0.0480, and P(N) = P(no crash) = 0.9520.
P(at least one crash) = 1 – P(no crashes)
$= 1 - P(N_1 \text{ and } N_2 \text{ and } N_3 \text{ and } N_4)$
$= 1 - P(N_1)\cdot P(N_2)\cdot P(N_3)\cdot P(N_4)$
$= 1 - (0.9520)(0.9520)(0.9520)(9.8520)$
$= 1 - 0.821 = 0.179$
Yes, the above probability does not apply to the given story problem. The above solution assumes that the cars are selected at random – so that, for example, $P(C_2|C_1) = P(C) = 0.0480$. These cars were not selected at random, but they were all taken from the same family. Since the cars were driven by the same people, or at least by people likely to have similar driving habits, knowing that one of the cars has a crash would reasonably make a person think that it was more likely that another of the cars would crash – i.e., the events are not independent and $P(C_2|C_1) > P(C) = 0.0480$.

17. P(at least one with vestigial wings) = 1 – P(all have normal wings)
$= 1 - P(N_1 \text{ and } N_2 \text{ and } N_3 \text{ and } \ldots \text{ and } N_{10})$
$= 1 - P(N_1)\cdot P(N_2)\cdot P(N_3)\cdot \ldots \cdot P(N_{10})$
$= 1 - (3/4)(3/4)(3/4)\ldots(3/4)$
$= 1 - (3/4)^{10} = 1 - 0.0563 = 0.9437$, rounded to 0.944
Yes, the researchers can be 94.4% certain of getting at least one such offspring.

NOTE: For exercises 19-22, refer to the table at the right and use the following notation.
Let P = the lie detector test indicates a positive
Let Y = the person actually lied

Test Result	Actually Lie? no	Actually Lie? yes	
positive	15	42	57
negative	32	9	41
	47	51	98

19. $P(P|\overline{Y}) = 15/47 = 0.319$
This case is problematic because it represents the test indicating incorrectly that a person lied when he was telling the truth.

21. $P(Y|\overline{P}) = 9/41 = 0.220$
This result (the probability that a person is really lying when the test fails to accuse him) is different from the one in Exercise 20 (the probability that the test fails to accuse a person when

NOTE: For exercises 23-26, refer to the table at the right.

type	Twin Genders BB	BG	GB	GG	
identical	5	0	0	5	10
fraternal	5	5	5	5	20
	10	5	5	10	30

23. a. P(identical) = 10/30 = 1/3 = 0.333
b. P(identical|BB) = 5/10 = 1/2 = 0.5

25. P("BG or GB"|fraternal) = 10/20 = 1/2 = 0.5

27. P(at least one works) = 1 – P(all fail)
$= 1 - P(F_1 \text{ and } F_2 \text{ and } F_3)$
$= 1 - P(F_1)\cdot P(F_2)\cdot P(F_3)$
$= 1 - (0.10)(0.10)(0.10)$
$= 1 - 0.001 = 0.999$
Yes, the probability of a working clock rises from 90% with just one clock to 99.9% with 3 clocks. If the alarm clocks run on electricity instead of batteries, then the clocks do not operate independently and the failure of one could be the result of a power failure or interruption and may be related to the failure of another – i.e., $P(F_2|F_1)$ is no longer P(F) = 0.90.

29. P(HIV positive result) = P(at least one person is HIV positive)
$= 1 - $ P(all persons are HIV negative)
$= 1 - P(N_1 \text{ and } N_2 \text{ and } N_3)$
$= 1 - P(N_1) \cdot P(N_2) \cdot P(N_3)$
$= 1 - (0.9)(0.9)(0.9)$
$= 1 - 0.729 = 0.271$

NOTE: This plan is very efficient. Suppose, for example, there were 3000 people to be tested. Only in 0.271 = 27.1% of the groups would a retest need to be done for each of the three individuals. Those (0.271)(1000) = 271 groups would generate (271)(3) = 813 retests. The total number of tests required would be 1813 (the 1000 original + the 813 retests), only 60% of the 3,000 tests that would have been required to test everyone individually.

4-6 Counting

1. The permutations rule applies when different orderings of the same collection of objects represent a different solution. The combinations rule applies when all orderings of the same collection of objects represent the same solution.

3. Permutations. Because order makes a difference, and a different ordering of the top 3 finishers represents a different race outcome, the trifecta involves permutations and not combinations.

5. $5! = 5 \cdot 4 \cdot 3 \cdot 2 \cdot 1 = 120$

7. $_{52}C_2 = \dfrac{52!}{50!2!} = \dfrac{52 \cdot 51}{2!} = 1326$

NOTE: This technique of reducing the problem by "canceling out" the 50! from both the numerator and denominator can be used in most combinations and permutations problems. In general, a smaller factorial in the denominator can be completely divided into a larger factorial in the numerator to leave only the "excess" factors not appearing in the denominator. Furthermore, $_nC_r$ and $_nP_r$ will always be integers. More generally, a non-integer answer to <u>any counting problem</u> (but not a probability problem) indicates that an error has been made.

9. $_9P_5 = 9!/4! = 9 \cdot 8 \cdot 7 \cdot 6 \cdot 5 = 15{,}120$

11. $_{42}C_6 = \dfrac{42!}{36!6!} = \dfrac{42 \cdot 41 \cdot 40 \cdot 39 \cdot 38 \cdot 37}{6!} = 5{,}245{,}786$

13. Let W = winning the described lottery with a single selection.
The total number of possible combinations is

$$_{54}C_6 = \frac{54!}{48!6!} = \frac{54 \cdot 53 \cdot 52 \cdot 51 \cdot 50 \cdot 49}{6!} = 25{,}827{,}165.$$

Since only one combination wins, P(W) = 1/25,827,165.

15. Let W = winning the described lottery with a single selection.
The total number of possible combinations is

$$_{36}C_5 = \frac{36!}{31!5!} = \frac{36 \cdot 35 \cdot 34 \cdot 33 \cdot 32}{5!} = 376{,}992$$

Since only one combination wins, P(W) = 1/376,992.

17. a. Let G = generating a given social security number in a single trial.
    The total number of possible sequences is
    $10 \cdot 10 \cdot 10 \cdot 10 \cdot 10 \cdot 10 \cdot 10 \cdot 10 \cdot 10 = 10^9 = 1{,}000{,}000{,}000$
    Since only one sequence is correct, P(G) = 1/1,000,000,000.
    b. Let F = generating the first 5 digits of a given social security number in a single trial.
    The total number of possible sequences is
    $10 \cdot 10 \cdot 10 \cdot 10 \cdot 10 = 10^5 = 100{,}000$
    Since only one sequence is correct, P(F) = 1/100,000. Since this probability is so small, one need not worry about the given scenario.

19. Since order makes a difference, use permutations.
    ${}_{20}P_6 = 20!/14! = 20 \cdot 19 \cdot 18 \cdot 17 \cdot 16 \cdot 15 = 27{,}907{,}200$

21. Since the order of the two wires being tested is irrelevant, use combinations.
    $${}_5C_2 = \frac{5!}{3!2!} = \frac{5 \cdot 4}{2!} = 10$$

23. There are 8 tasks to perform, and each task can be performed in either of 2 ways.
    The total number of possible sequences is $2 \cdot 2 \cdot 2 \cdot 2 \cdot 2 \cdot 2 \cdot 2 \cdot 2 = 2^8 = 256$.
    Yes. The typical keyboard has approximately 50 keys, each with a lower an upper case possibility, representing about 100 commonly used characters.

25. The number of possible sequences of n different objects is n!, and $5! = 5 \cdot 4 \cdot 3 \cdot 2 \cdot 1 = 120$
    The unscrambled word is JUMBO. Since there is only one correct sequence, the probability of finding it with one random arrangement is 1/120.

27. a. Since order makes a difference, as there are 4 different offices, use permutations.
    ${}_{11}P_4 = 11!/7! = 11 \cdot 10 \cdot 9 \cdot 8 = 7920$
    b. Since the order in which the 4 are picked makes no difference, use combinations.
    $${}_{11}C_4 = \frac{11!}{7!4!} = \frac{11 \cdot 10 \cdot 9 \cdot 8}{4!} = 330$$

29. a. There are 14 tasks to perform, and each task can be performed in either of 2 ways.
    The total number of possible sequences is $2 \cdot 2 \cdot 2 \cdot 2 \cdot 2 \cdot 2 \cdot 2 \cdot 2 \cdot 2 \cdot 2 \cdot 2 \cdot 2 \cdot 2 \cdot 2 = 2^{14} = 16{,}384$.
    b. The number of possible sequences of n objects is when some are alike is
    $$\frac{n!}{n_1!n_2!\cdots n_k!}, \text{ and } \frac{14!}{13!1!} = 14$$
    c. P(13G,1B) = (# of ways to get 13G,1B)/(total number of ways to get 14 babies)
    = 14/16,384 = 0.000854
    d. Yes. Since P(13G,1B) is so small, and since 13G,1B so far (only the 14G,0B result is more extreme) from the expected 7G,7B result, the gender-selection method appears to yield results significantly different from those of chance alone.

31. Since order makes no difference when one is choosing groups, use combinations.
    The number of possible treatment groups is $${}_{10}C_5 = \frac{10!}{5!5!} = \frac{10 \cdot 9 \cdot 8 \cdot 7 \cdot 6}{5!} = 252.$$
    The number of ways to select a group of 5 men from the 5 males is ${}_5C_5 = 1$.
    The number of ways to select a group of 5 women from the 5 females is ${}_5C_5 = 1$.
    Since the number of possible same sex groups is 1 + 1 = 2, the probability that random selection determines a treatment group of all the same sex is 2/252 = 0.00794.

Yes. Having a treatment group all of the same sex would create confounding – the researcher could not say whether any difference in the treatment group was a treatment effect or a gender effect.

33. Let A be selecting the correct 5 numbers from 1 to 55. The number of possible selections is

$$_{55}C_5 = \frac{55!}{50!5!} = \frac{55 \cdot 54 \cdot 53 \cdot 52 \cdot 51}{5!} = 3{,}478{,}761.$$

Since there is only one winning selection, P(A) = 1/3,478,761.
Let B be selecting the correct winning number from 1 to 42. There are 42 possible selections.
Since there is only one winning selection, P(B) = 1/42.
P(winning Powerball)
  = P(A and B) = P(A)·P(B) = (1/3,478,761)(1/42) = 1/146,107,962 = 0.00000000684

35. The first digit could be any of the 8 numbers 2,3,4,5,6,7,8,9.
The second digit could be either of the 2 numbers 0,1.
The third digit could be any of 9 numbers as follows.
  1,2,3,4,5,6,7,8,9 if the second digit is 0.
  0,2,3,4,5,6,7,8,9 if the second digit is 1.
By the fundamental counting rule, the number of possible 3-digit area codes is 8·2·9 = 144.

37. There are 26 possible first characters, and 36 possible characters for the other positions.
Find the number of possible names using 1,2,3,…,8 characters and then add to get the total.

characters	possible names		
1	26	=	26
2	26·36	=	936
3	26·36·36	=	33,696
4	26·36·36·36	=	1,213,056
5	26·36·36·36·36	=	43,670,016
6	26·36·36·36·36·36	=	1,572,120,576
7	26·36·36·36·36·36·36	=	56,596,340,736
8	26·36·36·36·36·36·36·36	=	2,037,468,266,496
		total =	2,095,681,645,538

39. a. The calculator factorial key gives 50! = $3.04140932 \times 10^{64}$
Using the approximation, K = (50.5)·log(50) + 0.39908993 – 0.43429448(50)
  = 85.79798522 + 0.39908993 – 21.71472400
  = 64.48235225
and then 50! = 10^K
  = $10^{64.48235225}$
  = $3.036345215 \times 10^{64}$

NOTE: The two answers differ by 0.0051×10^{64} – i.e., 5.1×10^{61}, or 51 followed by 60 zeros. While such an error of "zillions and zillions" may seem quite large, it is only an error of $(5.1 \times 10^{61})/(3.04 \times 10^{64})$ = 1.7%.

b. The number of possible routes is 300!
Using the approximation, K = (300.5)·log(300) + 0.39908993 – 0.43429448(300)
  = 744.374937 + 0.39908993 – 130.288344
  = 614.485683
and then 300! = 10^K
  = $10^{614.485683}$
Since the number of digits in 10^x is the next whole number above x, 300! has 615 digits.

41. This problem cannot be solved directly using permutations, combinations or other techniques presented in this chapter. It is best solved by listing all the possible solutions in an orderly fashion and then counting the number of solutions. Often this is the most reasonable approach to a counting problem.

While you are encouraged to develop your own systematic approach to the problem, the following table represents one way to organize the solution. The table is organized by rows, according to the numbers of pennies in each way that change can be made. The numbers in each row give the numbers of the other coins as explained in the footnotes below the table.

The three numbers in the 80 row, for example, indicate that there are three ways to make change using 80 pennies. The 4 in the "only 5¢" column represents one way (80 pennies, 4 nickels). The 2 in the 10¢ column of the "nickels and one other coin" columns represents another way (80 pennies, 2 dimes). The 1 in the 10¢ column of the "nickels and one other coin" columns represents a third way (80 pennies, 1 dime, and [by default] 2 nickels).

The bold-face **3** in the 20 row and the 10¢/25¢ column of the "nickels and two other coins" columns represents 20 pennies, 3 dimes, one quarter, and [by default] 5 nickels.

Using the following table, or the system of your own design, you should be able to (1) take a table entry and determine what way to make change it represents and (2) take a known way to make change and find its representation in the table.

	only	nickels and one other coin[a]			nickels and two other coins[b]			all[c]
1¢	5¢	10¢	25¢	50¢	10¢ 25¢	10¢ 50¢	25¢ 50¢	10¢ 25¢ 50¢
100	0	.	.	.	.	.	.	.
95	1	.	.	.	.	.	.	.
90	2	1	.	.	.	.	.	.
85	3	1	.	.	.	.	.	.
80	4	2,1	.	.	.	.	.	.
75	5	2,1	1	.	.	.	.	.
70	6	3,2,1	1	.	.	.	.	.
65	7	3,2,1	1	.	1	.	.	.
60	8	4,3,2,1	1	.	1	.	.	.
55	9	4,3,2,1	1	.	2,1	.	.	.
50	10	5,4,3,2,1	2,1	1	2,1	.	.	.
45	11	5,4,3,2,1	2,1	1	3,2,1	.	.	.
40	12	6,5,4,3,2,1	2,1	1	3,2,1	1	.	.
35	13	6,5,4,3,2,1	2,1	1	4,3,2,1	1	.	.
30	14	7,6,5,4,3,2,1	2,1	1	4,3,2,1	2,1	.	.
25	15	7,6,5,4,3,2,1	3,2,1	1	5,4,3,2,1	2,1	1	.
20	16	8,7,6,5,4,3,2,1	3,2,1	1	5,4,**3**,2,1	3,2,1	1	.
15	17	8,7,6,5,4,3,2,1	3,2,1	1	6,5,4,3,2,1	3,2,1	1	1
10	18	9,8,7,6,5,4,3,2,1	3,2,1	1	6,5,4,3,2,1	4,3,2,1	1	1
5	19	9,8,7,6,5,4,3,2,1	3,2,1	1	7,6,5,4,3,2,1	4,3,2,1	1	2,1
0	20	10,9,8,7,6,5,4,3,2,1	4,3,2,1	2,1	7,6,5,4,3,2,1	5,4,3,2,1	2,1	2,1
ways	21	100	34	12	56	25[d]	7	6[e]

[a]The possible numbers of the non-nickel coin are given. The numbers of nickels are found by default.

[b]This is for a single occurrence of the second coin listed. The possible numbers of the leading non-nickel coin are given. The numbers of nickels are found by default.

[c]This is for a single occurrence of the second and third coin listed. The possible numbers of the leading non-nickel coin are given. The numbers of nickels are found by default.

[d]There are another 25 ways when the 50¢ is in the form of 2 quarters.

[e]There are another 6 ways when the 50¢ is in the form of 2 quarters.

The total number of ways identified is 21+100+34+12+56+25+25+7+6+6 = 292.
If a one-dollar coin is also considered as change for a dollar, there are 293 ways.
The one dollar bill itself should not be counted.

Statistical Literacy and Critical Thinking

1. An event that has probability 0.004 is unlikely to occur by chance. It would occur by chance only about 4 times in 1000 opportunities.
2. No. The detectors would be related in the sense that they would all be inoperable whenever the home loses its electricity.
3. No, her reasoning is incorrect for two reasons. (1) Since she is a new recruit she has no reason to assume that the historical rate of 12.7% will apply to her. (2) Even if the 12.7% rate applies, the $(0.127)(0.127) = 0.0161$ calculation applies only to burglaries that are selected at random and/or handled independently. Since the cases were handled by the same person, knowing that one of the cases closes with an arrest would reasonably make a person think that it was more likely the other case would close with an arrest – i.e., the events are not independent and $P(C_2|C_1) > P(C) = 0.127$.
4. No. Lottery numbers are selected so that past results have no effect on future drawings. Each drawing is statistically independent of the all the others.

Chapter Quick Quiz

1. $P(\bar{C}) = 1 - P(C)$
   $= 1 - 0.70 = 0.30$
2. $P(C_1 \text{ and } C_2) = P(C_1) \cdot P(C_2)$
   $= (0.70)(0.70) = 0.49$
3. Answers will vary, but such an occurrence is not very common. A reasonable guess for such an interruption might be $P(I) = 0.005$.
4. No. A result with probability 0.342 could easily have occurred by chance.
5. $P(\bar{A}) = 1 - P(A) = 1 - 0.4 = 0.6$

NOTE: For exercises 6-10, refer to the table at the right and use A,B,P,F to identify in the natural way the groups and qualifying exam results.

		Result: passed	Result: failed	
Group	A	10	14	24
	B	417	145	562
		427	159	586

6. $P(P) = 427/586 = 0.729$
7. $P(B \text{ or } P) = P(B) + P(P) - P(B \text{ and } P)$
   $= 562/586 + 427/586 - 417/586$
   $= 572/586 = 0.976$
8. $P(A_1 \text{ and } A_2) = P(A_1) \cdot P(A_2|A_1)$
   $= (24/586)(23/585)$
   $= 0.00161$
9. This problem may be worked either of two ways.
   directly from chart: $P(A \text{ and } P) = 10/586 = 0.0171$
   by formula: $P(A \text{ and } P) = P(A) \cdot P(P|A)$
   $= (24/586)(10/24) = 10/586 = 0.0171$
10. $P(P|A) = 10/24 = 0.417$

Review Exercises

NOTE: For exercises 1-10, refer to the table at the right and use the following notation.
Let H = the rider wore a helmet.
Let I = the rider experienced a head injury.

		Head Injury?		
		yes	no	
Wore Helmet?	yes	96	656	752
	no	480	2330	2810
		576	2986	3562

1. P(I) = 576/3562 = 0.162

2. P(H) = 752/3562 = 0.211

3. P(I or H) = P(I) + P(H) – P(I and H)
= 576/3562 + 752/3562 – 96/3562
= 1232/3562 = 0.346

4. $P(\bar{H} \text{ or } \bar{I}) = P(\bar{H}) + P(\bar{I}) - P(\bar{H} \text{ and } \bar{I})$
= 2810/3562 + 2986/3562 – 2330/3562
= 3466/3562 = 0.973

5. This problem may be worked either of two ways.
directly from chart: P(H and I) = 96/3562 = 0.0270
by formula: P(H and I) = P(H)·P(I|H)
= (752/3562)(96/752)
= 96/3562 = 0.0270

6. This problem may be worked either of two ways.
directly from chart: $P(\bar{H} \text{ and } \bar{I})$ = 2330/3562 = 0.654
by formula: $P(\bar{H} \text{ and } \bar{I}) = P(\bar{H}) \cdot P(\bar{I}|\bar{H})$
= (2810/3562)(2330/2810)
= 2330/3562 = 0.654

7. $P(H_1 \text{ and } H_2) = P(H_1) \cdot P(H_2|H_1)$
= (752/3562)(751/3561)
= 0.0445

8. $P(I_1 \text{ and } I_2) = P(I_1) \cdot P(I_2|I_1)$
= (576/3562)(575/3561)
= 0.0261

9. $P(\bar{H}|I)$ = 480/576 = 0.833

10. $P(\bar{I}|H)$ = 656/752 = 0.872

11. Answers will vary. Black cars are not popular, but they are not rare enough to be considered unusual. P(B) = 0.08 seems like a reasonable guess.

12. a. $P(\bar{B}) = 1 - P(B)$
= 1 – 0.35
= 0.65

b. $P(B_1 \text{ and } B_2 \text{ and } B_3 \text{ and } B_4) = P(B_1) \cdot P(B_2) \cdot P(B_3) \cdot P(B_4)$
= (0.35)(0.35)(0.35)(0.35)
= $(0.35)^4$ = 0.0150

c. Yes. Since 0.015 ≤ 0.05, selecting four people at random and finding they all have blue eyes would be considered an unusual event.

13. a. P(O18) = 1/365 = 0.00274
 b. P(O) = 31/365 = 0.0849
 c. This would be a very rare event, probably occurring less than one time in a million tries. A reasonable guess would be P(E) = 0.0000001
 d. Yes, based on the answer in (c). Since $0.0000001 \leq 0.05$, randomly selecting an adult American and fifing someone who knows that date would be considered an unusual event.

14. a. P(D) = 15.2/100,000 = 0.000152
 b. $P(D_1 \text{ and } D_2) = P(D_1) \cdot P(D_2)$
 $= (0.000152)(0.000152)$
 $= (0.000152)^2$
 $= 0.0000000231$
 c. $P(\bar{D}) = 1 - P(D)$
 $= 1 - 0.000152$
 $= 0.999848$
 $P(\bar{D}_1 \text{ and } \bar{D}_2) = P(\bar{D}_1) \cdot P(\bar{D}_2)$
 $= (0.999848)(0.999848)$
 $= (0.999848)^2$
 $= 0.999696$

15. a. $P(\bar{A}) = 1 - P(A)$
 $= 1 - 0.40$
 $= 0.60$
 b. No, for two reasons. First, the sample was limited to America OnLine subscribers – who are not necessarily representative of the general population. Second, the poll was based on a voluntary response sample – and not on people selected at random. People who respond in a voluntary response sample are typically those with strong feelings one way or the other (i.e., either people who really like Sudoku, or those who can't stand it) and are not necessarily representative of the general population.

16. For each person, P(negative) = 1 – P(positive)
 = 1 – 0.00320
 = 0.99680.
 P(positive group result) = P(at least one person is positive)
 = 1 – P(all persons are negative)
 $= 1 - P(N_1 \text{ and } N_2 \text{ and} \ldots \text{and } N_{10})$
 $= 1 - P(N_1) \cdot P(N_2) \cdot \ldots \cdot P(N_{10})$
 $= 1 - (0.99680)(0.99680)\ldots(0.99680)$
 $= 1 - (0.99680)^{10}$
 $= 1 - 0.9685$
 $= 0.0315$
 No. Since $0.0315 \leq 0.05$, a positive group result is not considered likely.
 NOTE: This plan is very efficient. Suppose, for example, there were 3000 people to be tested. Only in 0.0315 = 3.15% of the 300 groups would a retest need to be done for each of the ten individuals. Those $(0.0315)(300) \approx 9$ groups would generate (9)(10) = 90 retests. The total number of tests required would be 390 (the 300 original + the 90 retests), only about 13% of the 3,000 tests that would have been required to test everyone individually.

17. Let D = getting a Democrat.
 $P(D) = 0.85$ for each selection
 $P(D_1 \text{ and } D_2 \text{ and} \ldots \text{and } D_{30}) = P(D_1) \cdot P(D_2) \cdot \ldots \cdot P(D_{30})$
 $= (0.85)(0.85)\ldots(0.85)$
 $= (0.85)^{30}$
 $= 0.00763$
 Yes; since $0.00763 \le 0.05$, getting 30 Democrats by chance alone is an unusual event. Since the probability 0.00763 is so small, the results are evidence that the pollster is not telling the truth.

18. Let $\bar{M}$ = a male in that age bracket not surviving. $P(\bar{M}) = 114.4/100{,}000 = 0.001144$
 Let $\bar{F}$ = a female in that age bracket not surviving. $P(\bar{F}) = 44/100{,}000 = 0.00044$
 a. $P(M) = 1 - P(\bar{M})$
 $= 1 - 0.001144$
 $= 0.998856$
 b. $P(M_1 \text{ and } M_2) = P(M_1) \cdot P(M_2)$
 $= (.998856)(0.998856)$
 $= (0.998856)^2$
 $= 0.998$
 c. $P(F) = 1 - P(\bar{F})$
 $= 1 - 0.00044 = 0.99956$
 $P(F_1 \text{ and } F_2) = P(F_1) \cdot P(F_2)$
 $= (.99956)(0.99956)$
 $= (0.99956)^2$
 $= 0.999$
 d. Males are more likely than females to be involved in situations where death is a possible result (e.g., military combat, violent crimes, hazardous occupations, etc.).

19. The number of possible selections is $_{38}C_5 = \dfrac{38!}{33!5!} = \dfrac{38 \cdot 37 \cdot 36 \cdot 35 \cdot 34}{5!} = 501{,}942$.
 Since there is only one winning selection, $P(W) = 1/501{,}942 = 0.00000199$.
 Since $0.00000199 \le 0.05$, winning the jackpot with a single selection is an unusual event. But since hundreds of thousands of tickets are purchased, it is not unusual for there to be a winner.

20. There are 10 possibilities for each digit: 0,1,2,3,4,5,6,7,8,9.
 By the fundamental counting rule, the number of possibilities is now
 $10 \cdot 10 \cdot \ldots \cdot 10 = 10^{13} = 10{,}000{,}000{,}000{,}000 = 10$ trillion

Cumulative Review Exercises

1. values in order are: 17 18 18 18 18 19 19 19 19 19 20 20 20 20 20 20 20 20 21 21
   The summary statistics are: n = 20 $\Sigma x = 386$ $\Sigma x^2 = 7472$

   a. $\bar{x} = (\Sigma x)/n = 386/20 = 19.3$ oz

   b. $\tilde{x} = (19+20)/2 = 19.5$ oz

   c. $s^2 = [n(\Sigma x^2) - (\Sigma x)^2]/[n(n-1)]$
   $= [20(7472) - (386)^2]/[20(19)]$
   $= 444/380 = 1.168$
   s = 1.081, rounded to 1.1 oz

   d. $s^2 = 1.168$, rounded to 1.2 oz^2 [from part (c) above]

   e. No. The mean and the median are below the supposed weight of 21 oz, as are 18/20 = 90% of the individual values.

2. The total number of responses is 3042 + 2184 = 5226
   a. 3042/5226 = 58.2%
   b. P(Y) = 3042/5226 = 0.582
   c. This is a voluntary response sample. If the intended population is all people, this is also a convenience sample (of those who could be easily reached by AOL since they were AOL subscribers). Neither voluntary response samples nor convenience samples are suitable for making statements about the general population.
   d. A simple random sample is one for which every sample of size n (in this case, n = 5226) from the population has an equal chance of being selected. A simple random sample would be better than a voluntary response sample or a convenience sample because it is more likely to be representative of the general population.

3. Organize the data as follows.
   values in order
   regular: 370 370 371 372 372 374
   diet: 352 353 356 357 357 358
   summary statistics
   regular: n = 6 $\Sigma x = 2229$ $\Sigma x^2 = 828085$
   diet: n = 6 $\Sigma x = 2133$ $\Sigma x^2 = 758311$

   a. $\bar{x} = (\Sigma x)/n$
   regular: 2229/6 = 371.5 g
   diet: 2133/6 = 355.5 g
   The mean weight for the diet Coke appears to be significantly smaller.

   b. $\tilde{x} = (x_3+x_4)/2$
   regular: (371+372)/2 = 371.5 g
   diet: (356+357)/2 = 356.5 g
   The median weight for the diet Coke appears to be significantly smaller.

c. $s^2 = [n(\Sigma x^2) - (\Sigma x)^2]/[n(n-1)]$
regular: $s^2 = [6(828085) - (2229)^2]/[6(5)]$
$= 69/30 = 2.3$
$s = 1.5$ g
diet: $s^2 = [6(758311) - (2133)^2]/[6(5)]$
$= 177/30 = 5.9$
$s = 2.4$ g
The diet Coke weights appear to have significantly more variability.

d. The answers for the variance are taken from part (c) above.
regular: $s^2 = 2.3\ g^2$
diet: $s^2 = 5.9\ g^2$

e. No. The weights for the diet Coke appear to be significantly smaller and have more variability.

4. a. According to one criterion, unusual values are those that are more than two standard deviations from the mean. A score of 38 is unusual because it is $(38-67.4)/11.6 = -2.53$ standard deviations from the mean.

b. According to one criterion, unusual events are those for which $P(E) \leq 0.05$. Getting all ten T-F questions correct by guessing is unusual because $P(E) = (1/2)10 = 1/1024 = 0.000977$.

5. a. This is a convenience sample because the data was gathered from those who happened to be at hand.

b. If eye color is related to ethnicity and the college is not ethnically representative of the United States, then this particular sample might have a sample bias preventing it from being representative of the general population.

c. P(Brown or Blue) = P(Brown) + P(Blue) – P(Brown and Blue)
$= 0.40 + 0.35 - 0$
$= 0.75$

d. $P(\text{Brown}_1 \text{ and Brown}_2) = P(\text{Brown}_1)\cdot P(\text{Brown}_2|\text{Brown}_1) = (0.40)(0.40) = 0.16$
$P(\text{at least one Brown}) = P(\text{Brown}_1 \text{ or Brown}_2)$
$= P(\text{Brown}_1) + P(\text{Brown}_2) - P(\text{Brown}_1 \text{ and Brown}_2)$
$= 0.40 + 0.40 - 0.16$
$= 0.64$

6. The first note is given by *. There are 3 possibilities for each of the next 15 notes. By the fundamental counting rule, there are
$3\cdot3\cdot3\cdot3\cdot3\cdot3\cdot3\cdot3\cdot3\cdot3\cdot3\cdot3\cdot3\cdot3\cdot3 = 3^{15} = 14{,}348{,}907$ possible sequences.
NOTE: This assumes that each song has at least 16 notes, and it does not guarantee that two different melodies will not have the same representation – if one goes up two steps every time the other goes up one step, for example, they both will show a U in that position.

Chapter 5

Discrete Probability Distributions

5-2 Random Variables

1. As defined in the text, a random variable is a variable that takes on a single numerical value, determined by chance, for each outcome of a procedure. In this exercise, the random variable is the number of winning lottery tickets obtained over a 52 week period. The possible values for this random variable are 0,1,2,…,52.

3. The probability distribution gives each possible non-overlapping outcome O_i of a procedure and the probability $P(O_i)$ associated with each of those outcomes.
Since it is a certainty that one of the outcomes will occur,
$P(O_1 \text{ or } O_2 \text{ or } O_3 \text{ or} \ldots) = 1$.
Since the outcomes are mutually exclusive,
$P(O_1 \text{ or } O_2 \text{ or } O_3 \text{ or} \ldots) = P(O_1) + P(O_2) + P(O_3) + \ldots$.
Therefore,
$P(O_1) + P(O_2) + P(O_3) + \ldots = \Sigma P(O_i) = 1$.

5. a. Discrete, since such a number is limited to certain values – viz., the non-negative integers.
b. Continuous, since weight can be any value on a specified continuum.
c. Continuous, since height can be any value on a specified continuum.
d. Discrete, since such a number is limited to certain values – viz., the non-negative integers.
e. Continuous, since time can be any value on a specified continuum.

NOTE: If one of the conditions for a probability distribution does not hold, the formulas do not apply – and they produce numbers that have no meaning. When working with probability distributions and formulas in the exercises that follow, always keep the following important facts in mind.

- $\Sigma P(x)$ must always equal 1.000
- $\Sigma[x \cdot P(x)]$ gives the mean of the x values and must be a number between the highest and lowest x values.
- $\Sigma[x^2 \cdot P(x)]$ gives the mean of the x^2 values and must be a number between the highest and lowest x^2 values.
- Σx and Σx^2 have no meaning and should not be calculated
- The quantity "$\Sigma[x^2 \cdot P(x)] - \mu^2$" cannot possibly be negative – if it is, there is a mistake.
- Always be careful to use the unrounded mean in the calculation of the variance, and to take the square root of the unrounded variance to find the standard deviation.

7. The given table is a probability distribution since $0 \leq P(x) \leq 1$ for each x and $\Sigma P(x)=1$.

x	P(x)	x·P(x)	x^2	$x^2 \cdot P(x)$
0	0.125	0	0	0
1	0.375	0.375	1	0.375
2	0.375	0.750	4	1.500
3	0.125	0.375	9	1.125
	1.000	1.500		3.000

$\mu = \Sigma[x \cdot P(x)]$
$= 1.500$, rounded to 1.5 children
$\sigma^2 = \Sigma[x^2 \cdot P(x)] - \mu^2$
$= 3.000 - (1.500)^2$
$= 0.750$
$\sigma = 0.866$, rounded to 0.9 children

9. The given table is not a probability distribution since $\Sigma P(x) = 0.984 \neq 1$.

11. The given table is a probability distribution since $0 \leq P(x) \leq 1$ for each x and $\Sigma P(x)=1$.

x	P(x)	x·P(x)	x^2	$x^2 \cdot P(x)$
0	0.02	0	0	0
1	0.15	0.15	1	0.15
2	0.29	0.58	4	1.16
3	0.26	0.78	9	2.34
4	0.16	0.64	16	2.56
5	0.12	0.60	25	3.00
	1.00	2.75		9.21

$\mu = \Sigma[x \cdot P(x)]$
$= 2.75$, rounded to 2.8 TV sets
$\sigma^2 = \Sigma[x^2 \cdot P(x)] - \mu^2$
$= 9.21 - (2.75)^2$
$= 1.6475$
$\sigma = 1.284$, rounded to 1.3 TV sets

13. The given table is a probability distribution since $0 \leq P(x) \leq 1$ for each x and $\Sigma P(x)=1$.

x	P(x)	x·P(x)	x^2	$x^2 \cdot P(x)$
0	0+	0	0	0
1	0+	0	1	0
2	0.004	0.008	4	0.016
3	0.023	0.069	9	0.207
4	0.087	0.348	16	1.382
5	0.208	1.040	25	5.200
6	0.311	1.866	36	11.196
7	0.267	1.869	49	13.083
8	0.100	0.800	64	6.400
	1.000	6.000		37.484

$\mu = \Sigma[x \cdot P(x)]$
$= 6.000$, rounded to 6.0 peas
$\sigma^2 = \Sigma[x^2 \cdot P(x)] - \mu^2$
$= 37.484 - (6.000)^2$
$= 1.484$
$\sigma = 1.218$, rounded to 1.2 peas

15. a. $P(x=7) = 0.267$

b. $P(x \geq 7) = P(x=7 \text{ or } x=8)$
$= P(x=7) + P(x=8)$
$= 0.267 + 0.100$
$= 0.367$

c. The answer in part (b) is the one which determines whether 7 is an unusually high result. When there are several different possible outcomes, the probability of getting any one of them exactly (even the more likely ones near the middle of the distribution) may be small. An outcome is unusually high if it is in the upper tail of the distribution – i.e., if the probability of getting it or a higher value is small.

d. No; since $P(x \geq 7) = 0.367 > 0.05$, 7 is not an unusually high number of peas with green pods.

17. a. The given table is a probability distribution since $0 \leq P(x) \leq 1$ for each x and $\Sigma P(x)=1$.

x	P(x)	x·P(x)	x^2	$x^2 \cdot P(x)$
4	0.1919	0.7676	16	3.0704
5	0.2121	1.0605	25	5.3025
6	0.2222	1.3332	36	7.9992
7	0.3737	2.6159	49	18.3113
	0.9999	5.7772		34.6834

$\mu = \Sigma[x \cdot P(x)]$
$= 5.7772$, rounded to 5.8
$\sigma^2 = \Sigma[x^2 \cdot P(x)] - \mu^2$
$= 34.6834 - (5.7772)^2$
$= 1.3074$
$\sigma = 1.1434$, rounded to 1.1

b. $\mu = 5.8$ games and $\sigma = 1.1$ games

c. No; since $P(x=4) = 0.1919 > 0.05$, winning in 4 games is not an unusual event.

19. a. The given table is a probability distribution since $0 \le P(x) \le 1$ for each x and $\Sigma P(x)=1$.

x	P(x)	x·P(x)	x^2	$x^2 \cdot P(x)$
0	0.824	0	0	0
1	0.083	0.083	1	0.083
2	0.039	0.078	4	0.156
3	0.014	0.042	9	0.126
4	0.012	0.048	16	0.192
5	0.008	0.040	25	0.200
6	0.008	0.048	36	0.288
7	0.004	0.028	49	0.196
8	0.004	0.032	64	0.256
9	0.004	0.036	81	0.324
	1.000	0.435		1.821

$\mu = \Sigma[x \cdot P(x)]$
$= 0.435$, rounded to 0.4
$\sigma^2 = \Sigma[x^2 \cdot P(x)] - \mu^2$
$= 1.821 - (0.435)^2$
$= 1.635$
$\sigma = 1.279$, rounded to 1.3

b. $\mu = 0.4$ bumper stickers and $\sigma = 1.3$ bumper stickers

c. The range rule of thumb suggests that "usual" values are those within two standard deviations of the mean.

minimum usual value = $\mu - 2\sigma = 0.4 - 2(1.3) = -2.2$, truncated to 0
maximum usual value = $\mu + 2\sigma = 0.4 + 2(1.3) = 3.0$

The range of values for usual numbers of bumper stickers is from 0 to 3.0.

d. No; since $0 < 1 < 3.0$, it is not unusual to have more than one bumper sticker.

21. There are eight equally likely possible outcomes: GGG GGB GBG BGG GBB BGB BBG BBB. The following table describes the situation, where x is the number of girls per family of 3.

x	P(x)	x·P(x)	x^2	$x^2 \cdot P(x)$
0	0.125	0	0	0
1	0.375	0.375	1	0.375
2	0.375	0.750	4	1.500
3	0.125	0.375	9	1.125
	1.000	1.500		3.000

$\mu = \Sigma[x \cdot P(x)]$
$= 1.500$, rounded to 1.5 girls
$\sigma^2 = \Sigma[x^2 \cdot P(x)] - \mu^2$
$= 3.000 - (1.500)^2$
$= 0.75$
$\sigma = 0.8660$, rounded to 0.9 girls

No. Since $P(x=3) = 0.125 > 0.05$, it is not unusual for a family of 3 children to have all girls.

23. The following table describes the situation.

x	P(x)	x·P(x)	x^2	$x^2 \cdot P(x)$
0	0.1	0	0	0
1	0.1	0.1	1	0.1
2	0.1	0.2	4	0.4
3	0.1	0.3	9	0.9
4	0.1	0.4	16	1.6
5	0.1	0.5	25	2.5
6	0.1	0.6	36	3.6
7	0.1	0.7	49	4.9
8	0.1	0.8	64	6.4
9	0.1	0.9	81	8.1
	1.0	4.5		28.5

$\mu = \Sigma[x \cdot P(x)]$
$= 4.5$
$\sigma^2 = \Sigma[x^2 \cdot P(x)] - \mu^2$
$= 28.5 - (4.5)^2$
$= 8.25$
$\sigma = 2.8723$, rounded to 2.9

The probability histogram for this distribution would be flat, with each bar having the same height.

25. a. Since each of the 3 positions could be filled (with replacement) by any of the 10 digits 0,1,2,3,4,5,6,7,8,9, there are $10 \cdot 10 \cdot 10 = 1000$ possible different selections.

b. Let W = winning (i.e., matching the selection drawn).
Since only one of the possible 1000 selections is a inner, $P(W) = 1/1000 = 0.001$

c. The net profit is the payoff minus the original bet, in this case $250.00 – $0.50 = $249.50.

d. The following table describes the situation.

x	P(x)	x·P(x)
-0.50	0.999	-0.4995
249.50	0.001	0.2495
	1.000	-0.2500

E = Σ[x·P(x)] = -0.2500 [i.e., a loss of 25¢]

The expected value is -25¢.

e. Since both games have the same expected value, neither bet is better than the other.

27. a. The following table describes the situation.

x	P(x)	x·P(x)
-5	33/38	-165/38
30	5/38	150/38
	38/38	-15/38

E = Σ[x·P(x)] = -15/38 = -0.3947, rounded to -39.4¢

b. Since -26 > -39.4, wagering $5 on the number 13 is the better bet.

29. a. From the 30-year-old male's perspective, the two possible outcome values are
-$161, if he lives
100,000 – 161 = $99,839, if he dies

b. The following table describes the situation.

x	P(x)	x·P(x)
-161	0.9986	-160.7746
99,839	0.0014	139.7746
	1.0000	-21.0000

E = Σ[x·P(x)] = -21.0000, rounded to -$21.0

c. Yes; the insurance company can expect to make an average of $21.0 per policy.

31. For every $1000, Bond A gives a profit of (0.06)($1000) = $60 with probability 0.99.
The following table describes the situation.

x	P(x)	x·P(x)
-1000	0.01	-10.00
60	0.99	59.40
	1.00	49.40

E = Σ[x·P(x)] = 49.40, rounded to 49.4 [i.e., a profit of $49.4]

For every $1000, Bond B gives a profit of (0.08)($1000) = $80 with probability 0.95.
The following table describes the situation.

x	P(x)	x·P(x)
-1000	0.05	-50.00
80	0.95	76.40
	1.00	26.00

E = Σ[x·P(x)] = 26.00, rounded to 26.0 [i.e., a profit of $26.0]

Bond A is the better bond since it has the higher expected value – i.e., 49.4 > 26.0. Since both bonds have positive expectations, either one would be a reasonable selection. Although Bond A has a higher expectation, a person willing to assume more risk in hope of a higher payoff might opt for Bond B.

33. Since each die has 6 faces, there will be 6·6 = 36 possible outcomes. If each of the 12 sums 1,2,3,…,12 is to be equally likely, each sum must occur exactly 3 times. As a starting point, suppose one of the dice is normal. If one die contains the usual digits 1,2,3,4,5,6, the other die
(1) must have three 0's to pair with the 1 to get three sums of 1.
(2) must have three 6's to pair with the 6 to get three sums of 12.

The 36 possible outcomes generated by such dice are Given in the box at the right. Inspection indicates that each of the sums 1,2,3,…,12 appears exactly 3 times so that P(x) = 3/36 = 1/12 for x = 1,2,3,…,12. A solution, therefore, is to mark one die normally 1,2,3,4,5,6 and mark the other die 0,0,0,6,6,6.

1-0	2-0	3-0	4-0	5-0	6-0
1-0	2-0	3-0	4-0	5-0	6-0
1-0	2-0	3-0	4-0	5-0	6-0
1-6	2-6	3-6	4-6	5-6	6-6
1-6	2-6	3-6	4-6	5-6	6-6
1-6	2-6	3-6	4-6	5-6	6-6

5-3 Binomial Probability Distributions

1. Getting two people with blue eyes followed by three people with eyes that are not blue is only one way of getting exactly two people with blue eyes in a sample of five randomly selected people. All the ways of getting exactly two people as described must be considered.

3. Yes. Since $30/1236 = 0.024 \leq 0.05$, the 5% rule can be applied to determine that the repeated selections can be treated as if they were independent.

5. Yes. All four requirements are met.

7. No. Requirement (3) is not met. There are more than two possible outcomes for the ages of the parents.

9. No. Requirements (2) and (4) are not met. Since $20/100 = 0.20 > 0.05$, and the selections are done without replacement, the trials cannot be considered independent. The value p of obtaining a success changes from trial to trial as each selection without replacement changes the population from which the next selection is made.

11. Yes. All four requirements are met. Since $500/2{,}800{,}000 = 0.00018 \leq 0.05$, the selections can considered to be independent – even though the they are made without replacement.

13. a. P(WWC) = P(W)·P(W)·P(C)
 = (4/5)(4/5)(1/5) = 16/125 = 0.128
 b. There are three possible arrangements: WWC, WCW, CWW
 P(WWC) = P(W)·P(W)·P(C) = (4/5)(4/5)(1/5) = 16/125
 P(WCW) = P(W)·P(C)·P(W) = (4/5)(1/5)(4/5) = 16/125
 P(CWW) = P(C)·P(W)·P(W) = (1/5)(4/5)(4/5) = 16/125
 c. P(exactly one correct) = P(WWC or WCW or CWW)
 = P(WWC) + P(WCW) + P(CWW)
 = 16/125 + 16/125 + 16/125
 = 48/125 = 0.384

15. From Table A-1 in the .30 column and the 2-1 row, P(x=1) = 0.420.

17. From Table A-1 in the .99 column and the 15-11 row, P(x=11) = 0+.

19. From Table A-1 in the .05 column and the 10-2 row, P(x=2) = 0.075.

21. $P(x) = \dfrac{n!}{(n-x)!x!} p^x q^{n-x}$

$P(x=10) = [12!/(2!10!)](0.75)^{10}(0.25)^2$
$= [66](0.0563)(0.0625)$
$= 0.232$

NOTE: The intermediate values 66, 0.0563, and 0.0625 are given in Exercise 21 to help those with an incorrect answer to identify the portion of the problem in which the mistake was made. In the future, only the value n!/[(n-x)!x!] will be given separately. In practice, all calculations can be done in one step on a calculator. You may choose (or be asked) to write down intermediate values for your own (or the instructor's) benefit, but…

- Never round off in the middle of a problem.
- Do not write the values down on paper and then re-enter them in the calculator.
- Use the memory to let the calculator remember with complete accuracy any intermediate values that will be used in subsequent calculations.

In addition, always make certain that n!/[(n-x)!x!] is a whole number and that the final answer is between 0 and 1.

23. $P(x) = \frac{n!}{(n-x)!x!} p^x q^{n-x}$

$P(x=4) = [20!/(16!4!)](0.15)^4(0.85)^{16}$
$= [4845](0.15)^4(0.85)^{16} = 0.182$

25. $P(x \geq 1) = 1 - P(x=0)$
$= 1 - 0.050328$
$= 0.949672$, rounded to 0.0950

Yes, it is reasonable to expect that at least one group O donor will be obtained.

27. $P(x=5) = 0.018453$, rounded to 0.018

Yes. Since $0.018 \leq 0.05$, getting all 5 donors from group O would be considered unusual.

29. Let x = the number of consumers who recognize the Mrs. Fields brand name.

binomial problem: n = 10 and p = 0.90, use Table A-1

a. $P(x=9) = 0.387$

b. $P(x \neq 9) = 1 - P(x=9)$
$= 1 - 0.387 = 0.613$

31. Let x = the number of people with brown eyes.

binomial problem: n = 14 and p = 0.40, use Table A-1

$P(x \geq 12) = P(x=12) + P(x=13) + P(x=14)$
$= 0.001 + 0^+ + 0^+ = 0.001$

Yes. Since $0.001 \leq 0.05$, getting at least 12 persons with brown eyes would be unusual.

33. Let x = the number of offspring peas with green pods.

binomial problem: n = 10 and p = 0.75, use the binomial formula

$P(x) = \frac{n!}{(n-x)!x!} p^x q^{n-x}$

$P(x \geq 9) = P(x=9) + P(x=10)$
$= [10!/(1!9!)](0.75)^9(0.25)^1 + [10!/(0!10!)](0.75)^{10}(0.25)^0$
$= [10](0.75)^9(0.25)^1 + [1](0.75)^{10}(0.25)^0$
$= 0.1877 + 0.0563 = 0.2440$, rounded to 0.244

No. Since $0.244 > 0.05$, getting at least 9 peas with green pods is not unusual.

35. Let x = the number of special program students who graduated.
binomial problem: n = 10 and p = 0.94, use the binomial formula

$$P(x) = \frac{n!}{(n-x)!x!} p^x q^{n-x}$$

a. $P(x \geq 9) = P(x=9) + P(x=10)$
$= [10!/(1!9!)](0.94)^9(0.06)^1 + [10!/(0!10!)](0.94)^{10}(0.06)^0$
$= [10](0.94)^9(0.06)^1 + [1](0.94)^{10}(0.06)^0$
$= 0.3438 + 0.5386$
$= 0.8824$, rounded to 0.882

b. $P(x=8) = [10!/(2!8!)](0.94)^8(0.06)^2$
$= [45](0.94)^8(0.06)^2$
$= 0.0988$

$P(x \geq 8) = P(x=8) + P(x=9) + P(x=10)$
$= 0.0988 + 0.3438 + 0.5386$
$= 0.9812$

$P(x \leq 7) = 1 - P(x \geq 8)$
$= 1 - 0.9812$
$= 0.0188$

Yes. Since $P(x \leq 7) = 0.0188 \leq 0.05$, getting only 7 that graduated would be unusual.
NOTE: Remember that in situations involving multiple ordered outcomes, the unusualness of a particular outcome is generally determined by the probability of getting that outcome or a more extreme outcome.

37. Let x = the number of households tuned to the NBC game if the stated share is correct.
binomial problem: n = 20 and p = 0.22, use the binomial formula

$$P(x) = \frac{n!}{(n-x)!x!} p^x q^{n-x}$$

a. $P(x=0) = [20!/(20!0!)](0.22)^0(0.78)^{20}$
$= [1](0.22)^0(0.78)^{20}$
$= 0.00695$

b. $P(x \geq 1) = 1 - P(x=0)$
$= 1 - 0.00695$
$= 0.99305$, rounded to 0.993

c. $P(x=1) = [20!/(19!1!)](0.22)^1(0.78)^{19}$
$= [20](0.22)^1(0.78)^{19}$
$= 0.0392$

$P(\text{at most } 1) = P(x \leq 1)$
$= P(x=0) + P(x=1)$
$= 0.0069 + 0.0392$
$= 0.0461$

d. Yes. Since $0.0461 \leq 0.05$, getting at most one household tuned to the NBC game would be unusual. Either an unusual event has occurred or the 22% share value is incorrect.

39. Let x = the number who stayed less than one year.
binomial problem: n = 15 and p = 0.36, use the binomial formula

$$P(x) = \frac{n!}{(n-x)!x!} p^x q^{n-x}$$

a. $P(x=5) = [15!/(10!5!)](0.36)^5(0.64)^{10}$
$= [3003](0.36)^5(0.64)^{10} = 0.209$

b. The binomial distribution requires that the repeated selections be independent. Since these persons are selected from the original group of 320 without replacement, the repeated selections are not independent and the binomial distribution should not be used. In part (a), however, the sample size is 15/320 = 4.6% ≤ 5% of the population and the repeated samples may be treated as though they are independent. If the sample size is increased to 20, the sample is 20/320 = 6.25% > 5% of the population and the criteria for using independence to get an approximate probability is no longer met.

41. Let x = the number of aspirin tablets that do not meet the specifications.
binomial problem: n = 40 and p = 0.03, use the binomial formula

$$P(x) = \frac{n!}{(n-x)!x!}p^x q^{n-x}$$

$$\begin{aligned}P(\text{accept shipment}) &= P(x=0) + P(x=1)\\ &= [40!/(40!0!)](0.03)^0(0.97)^{40} + [40!/(39!1!)](0.03)^1(0.97)^{39}\\ &= [1](0.03)^0(0.97)^{40} + [40](0.03)^1(0.97)^{39}\\ &= 0.2957 + 0.3658\\ &= 0.6615\text{, rounded to } 0.662\end{aligned}$$

Approximately 2/3 of such shipments will be accepted.

43. Let x = the number of females hired if there is no discrimination.
binomial problem: n = 24 and p = 0.50, use the binomial formula

$$P(x) = \frac{n!}{(n-x)!x!}p^x q^{n-x}$$

$$\begin{aligned}P(x \le 3) &= P(x=0) + P(x=1) + P(x=2) + P(x=3)\\ &= [24!/(24!0!)](0.5)^0(0.5)^{24} + [24!/(23!1!)](0.5)^1(0.5)^{23} + [24!/(22!2!)](0.5)^2(0.05)^{22}\\ &\qquad + [24!/(21!3!)](0.5)^3(0.5)^{21}\\ &= [1](0.5)^0(0.5)^{24} + [24](0.5)^1(0.5)^{23} + [276](0.5)^2(0.5)^{22} + [2024](0.5)^3(0.05)^{21}\\ &= 0.00000006 + 0.00000143 + 0.00001645 + 0.00012063\\ &= 0.\ 00013857\text{, rounded to } 0.000139\end{aligned}$$

Yes. Since 0.000139 ≤ 0.05, such an event would be unusual. Either a very unusual event has occurred or there is gender discrimination.

45. Let x = the number of peas with green pods
binomial problem: n = 580 and p = 0.75, use the binomial formula

$$P(x) = \frac{n!}{(n-x)!x!}p^x q^{n-x}$$

$$\begin{aligned}P(x=428) &= [580!/(152!428!)](0.75)^{428}(0.25)^{152}\\ &= 0.0301231\text{, rounded to } 0.0301\end{aligned}$$

Since 0.0301 ≤ 0.05, this particular result might be considered unusual – but so would any of the 581 possible individual outcomes. In situations involving multiple ordered outcomes, the unusualness of a particular outcome is generally determined by the probability of getting that outcome or a more extreme outcome. In this instance the appropriate probability is $P(x \le 428) = 0.264956$, rounded to 0.265. Since 0.265 > 0.05, the obtained result is not considered unusual – and there is no suggestion that Mendel's probability value of 0.75 is wrong.

NOTE: The large factorials required are beyond the capabilities of most calculators. The above answers [$P(x=428) = 0.0301231$, and $P(x \le 428) = 0.264956$] were obtained using the Minitab software package.

47. Let x = the number of winners selected.
hypergeometric problem: A=6 (winners), B=53 (losers) and n=6 (selections)

$$P(x) = \frac{A!}{(A-x)!x!} \cdot \frac{B!}{(B-n+x)!(n-x)!} \div \frac{(A+B)!}{(A+B-n)!n!}$$

a. P(x=6) = [6!/(0!6!)]·[53!/(53!0!)] ÷ [59!/(53!6!)]
= [1]·[1] ÷ [45,057,474]
= 0.0000000222

b. P(x=5) = [6!/(1!5!)]·[53!/(52!1!)] ÷ [59!/(53!6!)]
= [6]·[53] ÷ [45,057,474]
= 0.00000706

c. P(x=3) = [6!/(3!3!)]·[53!/(50!3!)] ÷ [59!/(53!6!)]
= [20]·[23,426] ÷ [45,057,474]
= 0.0104

d. P(x=0) = [6!/(6!0!)]·[53!/(47!6!)] ÷ [59!/(53!6!)]
= [1]·[22,957,480] ÷ [45,057,474]
= 0.510

NOTE: This formula is actually $[{}_AC_x]\cdot[{}_BC_{n-x}] \div [{}_{A+B}C_n]$ and follows directly from the methods of section 4-7.

49. Let x = the number of major earthquakes per year.
Poisson problem: μ = 93/100 = 0.93, use $P(x) = \mu^x e^{-\mu}/x!$

a. $P(x=0) = (0.93)^0(e^{-0.93})/0! =$ (1)(0.3946)/1 = 0.3946
b. $P(x=1) = (0.93)^1(e^{-0.93})/1! =$ (0.93)(0.3946)/1 = 0.3669
c. $P(x=2) = (0.93)^2(e^{-0.93})/2! =$ (0.8649)(0.3946)/2 = 0.1706
d. $P(x=3) = (0.93)^3(e^{-0.93})/3! =$ (0.8044)(0.3946)/6 = 0.05289
e. $P(x=4) = (0.93)^4(e^{-0.93})/4! =$ (0.7481)(0.3946)/24 = 0.01230
f. $P(x=5) = (0.93)^5(e^{-0.93})/5! =$ (0.6957)(0.3946)/120 = 0.002287
g. $P(x=6) = (0.93)^6(e^{-0.93})/6! =$ (0.6470)(0.3946)/720 = 0.0003545
h. $P(x=7) = (0.93)^7(e^{-0.93})/7! =$ (0.6071)(0.3946)/5040 = 0.00004710

The following table compares the actual relative frequencies to the Poisson probabilities. The comparison can be made either by changing the actual frequencies to relative frequencies (by dividing by n = Σf) and comparing them to the predicted Poisson probabilities, or by changing the Poisson probabilities to predicted frequencies (by multiplying by n = Σf) and comparing them to the actual frequencies. We use the first approach, but the choice is arbitrary.

x	f	r.f.	P(x)
0	47	0.4700	0.3946
1	31	0.3100	0.3669
2	13	0.1300	0.1706
3	5	0.0500	0.0529
4	2	0.0200	0.0123
5	0	0.0000	0.0023
6	1	0.0100	0.0004
7	1	0.0100	0.0000^+
8 or more	0	0.0000	0.0000^+ (by subtraction)
	280	1.0000	1.0000

NOTE: r.f. = f/Σf = f/100

Yes. The agreement between the relative frequencies and the probabilities predicted by the Poisson formula is relatively good. Earthquakes may not meet the Poisson requirements – due to the dynamics of geological faults, they may tend to occur in clusters and not randomly.

5-4 Mean, Variance, and Standard Deviation for the Binomial Distribution

1. Yes. By definition $q = 1 - p$. The two expressions represent the same numerical quantity, and they may be used interchangeably

3. Using the exact formula, $\sigma^2 = npq = 1236(0.05)(0.95) = 58.71$, rounded to 58.7 people^2.
   Using the rounded value given for σ, $\sigma^2 = (\sigma)^2 = (7.7)^2 = 59.29$, rounded to 59.3 people^2.

5. $\mu = np = (50)(0.2) = 10.0$
   $\sigma^2 = npq = (50)(0.2)(0.8) = 8$
   $\sigma = 2.828$, rounded to 2.8
   minimum usual value $= \mu - 2\sigma = 10 - 2(2.828) = 4.3$
   maximum usual value $= \mu + 2\sigma = 10 + 2(2.828) = 15.7$

7. $\mu = np = (300)(0.48) = 144.0$
   $\sigma^2 = npq = (300)(0.48)(0.52) = 74.88$
   $\sigma = 8.653$, rounded to 8.7
   minimum usual value $= \mu - 2\sigma = 144.0 - 2(8.653) = 126.7$
   maximum usual value $= \mu + 2\sigma = 144.0 + 2(8.653) = 161.3$

9. Let x = the number of correct answers.
   binomial problem: $n = 75$ and $p = 0.5$
   a. $\mu = np = (75)(0.5) = 37.5$
      $\sigma^2 = npq = (75)(0.5)(0.5) = 18.75$
      $\sigma = 4.330$, rounded to 4.3
   b. Unusual values are those outside $\mu \pm 2\sigma$
      $37.5 \pm 2(4.330)$
      37.5 ± 8.7
      28.8 to 46.2 [using rounded values, 28.9 to 46.1]

   No. Since 45 is within the above limits, it would not be unusual for a student to pass by getting at least 45 correct answers.

11. Let x = the number of green M&M's.
    binomial problem: $n = 100$ and $p = 0.16$
    a. $\mu = np = (100)(0.16) = 16.0$
       $\sigma^2 = npq = (100)(0.16)(0.84) = 13.44$
       $\sigma = 3.666$, rounded to 3.7
    b. Unusual values are those outside $\mu \pm 2\sigma$
       $16.0 \pm 2(3.666)$
       16.0 ± 7.3
       8.7 to 23.3 [using rounded values, 8.6 to 23.4]

    No. Since 19 is within the above limits, it would not be unusual for 100 M&M's to include 19 green ones. This is not evidence that the claimed rate of 16% is wrong.

13. Let x = the number of baby girls.
    binomial problem: $n = 574$ and $p = 0.5$
    a. $\mu = np = (574)(0.5) = 287.0$
       $\sigma^2 = npq = (574)(0.5)(0.5) = 143.5$
       $\sigma = 11.979$, rounded to 12.0

b. Unusual values are those outside $\mu \pm 2\sigma$

$$287.0 \pm 2(11.979)$$
$$287.0 \pm 24.0$$
$$263.0 \text{ to } 311.0$$

Yes. Since 525 is not within the above limits, it would be unusual for 574 births to include 525 girls. The results suggest that the gender selection method is effective.

15. Let x = the number who stay at their first job less than two years.
binomial problem: n = 320 and p = 0.5

a. $\mu = np = (320)(0.5) = 160.0$
$\sigma^2 = npq = (320)(0.5)(0.5) = 80.0$
$\sigma = 8.944$, rounded to 8.9

b. Unusual values are those outside $\mu \pm 2\sigma$

$$160.0 \pm 2(8.944)$$
$$160.0 \pm 17.9$$
$$142.1 \text{ to } 177.9$$ [using rounded values, 142.2 to 177.8]

c. x = (0.78)(320) = 250 [actually 248,249,250,251 all round to 78%]
Since 250 is not within the above limits, it would be unusual for 320 graduates to include 250 persons who stayed at their first job less than two years if the true proportion were 50%. Since 250 is greater than the above limits, the true proportion is most likely greater than 50%. The result suggests that the headline is justified.

d. The statement suggests that the 320 participants were a voluntary response sample, and so the results might not be representative of the target population.

17. Let x = the number of persons who voted.
binomial problem: n = 1002 and p = 0.61

a. $\mu = np = (1002)(0.61) = 611.22$
$\sigma^2 = npq = (1002)(0.61)(0.39) = 238.3758$
$\sigma = 15.439$, rounded to 15.4

b. Unusual values are those outside $\mu \pm 2\sigma$

$$611.22 \pm 2(15.439)$$
$$611.22 \pm 30.88$$
$$580.3 \text{ to } 642.1$$ [using rounded values, 580.4 to 642.0]

No. Since 701 is not within the above limits, it would be unusual for 1002 potential voters to include 701 who actually voted. The results are not consistent with the actual voter turnout and suggest that those polled may not be telling the truth.

c. No. It appears that asking persons how they acted may not yield accurate results.

19. Let x = the number of persons receiving the drug who experience nausea.
binomial problem: n = 821 and p = 0.0124

a. $\mu = np = (821)(0.0124) = 10.18$
$\sigma^2 = npq = (821)(0.0124)(0.9876) = 10.0542$
$\sigma = 3.171$, rounded to 3.2

b. Unusual values are those outside $\mu \pm 2\sigma$

$$10.18 \pm 2(3.171)$$
$$10.18 \pm 6.34$$
$$3.8 \text{ to } 16.5$$ [using rounded values, 3.8 to 16.6]

Yes. Since 30 is not within the above limits, it is an unusual result.

c. Perhaps. The results suggest that Chantix does increase the likelihood of experiencing nausea, but at 30/821 = 3.65% that probability is still relatively low.

21. Refer to the Exercise 47 in Section 5-3 and the NOTE at the end of the problem.
Let x = the number of females selected.

$P(x) = [_{10}C_x][_{30}C_{12-x}]/[_{40}C_{12}]$

$P(0) = [_{10}C_0][_{30}C_{12}]/[_{40}C_{12}] = 1 \cdot 86493225/5586853480 = 86{,}493{,}225/5{,}586{,}853{,}480$

$P(1) = [_{10}C_1][_{30}C_{11}]/[_{40}C_{12}] = 10 \cdot 54627300/5586853480 = 546{,}273{,}000/5{,}586{,}853{,}480$

$P(2) = [_{10}C_2][_{30}C_{10}]/[_{40}C_{12}] = 45 \cdot 30045015/5586853480 = 1{,}352{,}025{,}675/5{,}586{,}853{,}480$

$P(3) = [_{10}C_3][_{30}C_9]/[_{40}C_{12}] = 120 \cdot 14307150/5586853480 = 1{,}716{,}858{,}000/5{,}586{,}853{,}480$

$P(4) = [_{10}C_4][_{30}C_8]/[_{40}C_{12}] = 210 \cdot 5852925/5586853480 = 1{,}229{,}114{,}250/5{,}586{,}853{,}480$

$P(5) = [_{10}C_5][_{30}C_7]/[_{40}C_{12}] = 252 \cdot 2035800/5586853480 = 513{,}021{,}600/5{,}586{,}853{,}480$

$P(6) = [_{10}C_6][_{30}C_6]/[_{40}C_{12}] = 210 \cdot 593775/5586853480 = 124{,}692{,}750/5{,}586{,}853{,}480$

$P(7) = [_{10}C_7][_{30}C_5]/[_{40}C_{12}] = 120 \cdot 142506/5586853480 = 17{,}100{,}720/5{,}586{,}853{,}480$

$P(8) = [_{10}C_8][_{30}C_4]/[_{40}C_{12}] = 45 \cdot 27405/5586853480 = 1{,}233{,}225/5{,}586{,}853{,}480$

$P(9) = [_{10}C_9][_{30}C_3]/[_{40}C_{12}] = 10 \cdot 4060/5586853480 = 40{,}600/5{,}586{,}853{,}480$

$P(10) = [_{10}C_{10}][_{30}C_2]/[_{40}C_{12}] = 1 \cdot 435/5586853480 = 435/5{,}586{,}853{,}480$

Total: $5{,}586{,}853{,}440/5{,}586{,}853{,}480$

There are $_{40}C_{12}$ = 5,586,853,440 total ways that a group of 12 persons can be selected from a class of 40. The number of females in each group can range from x=0 to x=10. The probability of each of those x values is given above – in fraction form, to indicate the exact numbers of ways involved for each type of group.

The following table, with the probabilities converted to decimal form, follows the methods of this chapter to calculate the mean and standard deviation of the probability distribution. The given table is a probability distribution since $0 \le P(x) \le 1$ for each x and $\Sigma P(x) = 1$.

x	P(x)	x·P(x)	x^2	$x^2 \cdot P(x)$
0	0.0154815	0	0	0
1	0.0977782	0.0977783	1	0.0977783
2	0.2420012	0.4840026	4	0.9680051
3	0.3073032	0.9219096	9	2.7657289
4	0.2200011	0.8800046	16	3.5200186
5	0.0918266	0.4591329	25	2.2956643
6	0.0223190	0.1339137	36	0.8034825
7	0.0030608	0.0214262	49	0.1499834
8	0.0002207	0.0017659	64	0.0141272
9	0.0000073	0.0000654	81	0.0005886
10	0.0000001	0.0000008	100	0.0000078
	1.0000000	3.0000000		10.6153846

$\mu = \Sigma[x \cdot P(x)]$
$= 3$

$\sigma^2 = \Sigma[x^2 \cdot P(x)] - \mu^2$
$= 10.6153846 - 3^2$
$= 1.6153846$

$\sigma = 1.2709$, rounded to 1.3

Statistical Literacy and Critical Thinking

1. A random variable is a variable that has a single value, determined by chance, for each outcome of a procedure. Yes, a discrete random variable can have an infinite number of possible values – e.g., the number of coin tosses required to get the first head can be 1,2,3,4,...

2. A discrete random variable has a countable number of possible values – i.e., its possible values are either finite, or they can by placed in a well-defined 1-to-1 correspondence with the positive integers. A continuous random variable has an infinite number of possible values that are associated with measurements on a continuous scale – i.e., a scale in which there are generally no gaps or interruptions between the possible values.

3. In a binomial probability distribution, there are exactly two possible outcomes: p is the probability of one of the outcomes (usually called "success"), and q is the probability of the other outcome (usually called "failure"). Since the two outcomes are mutually exclusive and exhaustive, $p+q = 1$. This means that $q = 1-p$ and that $p = 1-q$.

4. No. Other named discrete probability distributions identified in the exercises are the geometric distribution (Section 5-3, Exercise 46), the hypergeometric distribution (Section 5-3, Exercise 47), and the Poisson distribution (section 5-3, Exercise 49). There are also general unnamed discrete probability distributions (e.g., Section 5-2, Exercise 11.)

Chapter Quick Quiz

1. No. Since $P(x=0)+P(x=1) = 0.8+0.8 = 1.6$ violates the requirement that $\Sigma P(x) = 1$, the given scenario does not define a probability distribution.

x	P(x)	x·P(x)
0	0.3	0.0
1	0.7	0.7
	1.0	0.7

2. As shown by the table at the right, $\mu = \Sigma[x \cdot P(x)] = 0.7$.

3. The scenario describes a binomial probability distribution with $n = 400$ and $p = 0.5$.
   $\mu = np = (400)(0.5) = 200$

4. The scenario describes a binomial probability distribution with $n = 400$ and $p = 0.5$.
   $\sigma^2 = npq = (400)(0.5)(0.5) = 100$
   $\sigma = 10$

5. Yes. Any value more than two standard deviations from the mean is generally considered unusual, and 35 is $(35\text{-}20.0)/4.0 = 3.75$ standard deviations from the mean.

6. Yes. The scenario describes a binomial probability distribution with $n = 5$ and $p = 0.2$
   A check of the given P(x) values indicates they were generated by the binomial formula
   $$P(x) = \frac{n!}{(n-x)!x!} p^x q^{n-x}.$$

7. $P(x \geq 1) = 1 - P(x=0)$
   $= 1 - 0.4096$
   $= 0.5904$

8. $P(x \leq 4) = P(x=0) + P(x=1) + P(x=2) + P(x=3) + P(x=4)$
   $= 0.4096 + 0.4096 + 0.1536 + 0.0256 + 0.0016$
   $= 0.9984$

   $P(x=5) = 1 - P(x \leq 4)$
   $= 1 - 0.9984$
   $= 0.0016$

9. $P(x=2 \text{ or } x=3) = P(x=2) + P(x=3)$
   $= 0.1536 + 0.0256$
   $= 0.1792$

10. Yes. Since $P(x=5) = 0.0016 \leq 0.05$, getting all five answers correct by random guessing would be an unusual event.

Review Exercises

1. Let x = the number of deaths in the week before Thanksgiving.
The scenario describes a binomial probability distribution
with n = 8 and p = 0.5.
The table at the right was constructed using Table A-1.
It could also be constructed using the binomial formula $P(x) = \frac{n!}{(n-x)!x!} p^x q^{n-x}$.

x	P(x)
0	0.004
1	0.031
2	0.109
3	0.219
4	0.273
5	0.219
6	0.109
7	0.031
8	0.004
	1.000

2. Since the random variable follows a binomial probability distribution.
$\mu = np = (8)(0.5) = 4.o$
$\sigma^2 = npq = (8)(0.5)(0.5) = 2$
$\sigma = 1.414$, rounded to 1.4
Unusual values are those outside $\mu \pm 2\sigma$
$4.0 \pm 2(1.414)$
4.0 ± 2.8
1.2 to 6.8
Yes. Since 8 is outside the above limits, it would be unusual for all 8 of the deaths to occur during the week before Thanksgiving. Also, P(8) = 0.004 < 0.05 implies an unusual result.
NOTE: All 8 deaths occurring in the week before Thanksgiving does not support the theory about postponing death. That theory suggests that there should be a larger than usual number of deaths during the week after the holiday.

3. Let x = the number of deaths in the week before Thanksgiving.
The scenario describes a binomial probability distribution with n = 20 and p = 0.5.
a. $P(x) = \frac{n!}{(n-x)!x!} p^x q^{n-x}$.
$P(x{=}14) = [20!/(6!14!)](0.5)^{14}(0.5)^6$
$= [38760](0.5)^{14}(0.5)^6$
$= 0.0370$
b. Yes. Since $0.0370 \leq 0.05$, having exactly 14 of the deaths during the week before Thanksgiving would be considered unusual.
c. No. In situations involving multiple ordered outcomes, the unusualness of a particular outcome is generally determined by the probability of getting that outcome or a more extreme outcome. In this instance the appropriate probability is $P(x \geq 14)$ – or, alternatively, one could use the $\mu \pm 2\sigma$ guideline to determine whether 14 was outside the range of usual values.

4. The following table describes the situation.
Based on expectations, since 315,075 > 193,000, she should continue.
$E = \Sigma[x \cdot P(x)] = \$315{,}075.0$

x	P(x)	x·P(x)
75	0.2	15.0
300	0.2	60.0
75,000	0.2	15,000.0
500,000	0.2	100,000.0
1,000,000	0.2	200,000.0
	1.0	315,075.0

5. The table at the right describes the situation.
 a. $E = \Sigma[x \cdot P(x)] = 0.012474746$
 $= 1.25¢$
 b. expected loss = price of stamp – 1.25¢
 = 42¢ – 1.25¢
 = 40.75¢ [as of fall 2008]

x	P(x)	x·P(x)
1,000,000	0.000000011	0.011111111
100,000	0.000000009	0.000909091
25,000	0.000000009	0.000227273
5,000	0.000000027	0.000136362
2,500	0.000000036	0.000090909
0	0.999999908	0
	1.000000000	0.012474746

 No. From a purely financial point of view, it is not worth entering this contest.
 NOTE: The biggest winner in the sweepstakes appears to be the USPS. Based on the probabilities of winning, it appears that the company expects about 100,000,000 entries. At 42¢ per entry, the sweepstakes would generate 42 million dollars of entry postage alone!

6. No. Since $P(x=0)+P(x=1)+P(x=2)+P(x=3)+P(x=4) = 0.0016+0.0250+0.1432+0.3892+0.4096 = 0.9686$ violates the requirement that $\Sigma P(x) = 1$, the given scenario does not describe a probability distribution.

7. a. There are 10 possible values [0,1,2,3,4,5,6,7,8,9] for each digit.
 There are $10 \cdot 10 \cdot 10 \cdot 10 = 10{,}000$ ways to select the 4 digits.
 Since only one selection wins, $P(W) = 1/10{,}000 = 0.0001$
 b. Refer to the table in part (e). The first two columns give the desired probability distribution.
 c. The expected number of wins in 365 tries is $365(.0001) = 0.0365$.
 d. The scenario describes a binomial probability distribution with $n = 365$ and $p = 0.0001$.

$$P(x) = \frac{n!}{(n-x)!x!} p^x q^{n-x}.$$

$$P(x=1) = [365!/(364!1!)](0.0001)^1(0.9999)^{364}$$
$$= [365](0.0001)^1(0.9999)^{364}$$
$$= 0.0352$$

 e. Refer to the table. $E = \Sigma[x \cdot P(x)] = -0.5000 = -50¢$.

x	P(x)	x·P(x)
−1	0.9999	−0.9999
4999	0.0001	0.4999
	1.0000	−0.5000

8. a. Let x = the number fired for not getting along with others.
 binomial problem: $n = 5$ and $p = 0.17$, use the binomial formula

$$P(x) = \frac{n!}{(n-x)!x!} p^x q^{n-x}$$

$$P(x \geq 4) = P(x=4) + P(x=5)$$
$$= [5!/(1!4!)](0.17)^4(0.83)^1 + [5!/(0!5!)](0.17)^5(0.83)^0$$
$$= [5](0.17)^4(0.83)^1 + [1](0.17)^5(0.83)^0$$
$$= 0.003466 + 0.000142$$
$$= 0.003608, \text{ rounded to } 0.00361$$

 b. Yes. Since $0.00361 \leq 0.05$, such an event would be unusual for a typical company operating with $p = 0.17$. Either a rare event has occurred or, more likely, this company is different.

9. Let x = the number of checks with leading digit 1.
 binomial problem: $n = 784$ and $p = 0.301$, use the binomial formula
 a. The expected number is the mean, $\mu = np = (784)(0.301) = 236.0$
 b. For the binomial distribution,
 $\mu = np = (784)(0.301) = 236.0$
 $\sigma^2 = npq = (784)(0.301)(0.699) = 164.9528$
 $\sigma = 12.843$, rounded to 12.8

c. Usual values are those within $\mu \pm 2\sigma$

$236.0 \pm 2(12.843)$

236.0 ± 25.7

210.3 to 261.7 [using rounded values, 210.4 to 261.6]

d. Yes. Since 0 is so far outside the above limits, that would be very strong evidence that the checks are different (and likely fraudulent).

Cumulative Review Exercises

1. Arranged in order, the values are: 115.00 134.83 142.94 188.00 217.60

summary statistics: $n = 5$ $\Sigma x = 798.37$ $\Sigma x^2 = 134529.7325$

a. $\bar{x} = (\Sigma x)/n = 798.37/5 = \$159.674$

b. $\tilde{x} = x_3 = \$142.94$

c. $R = x_n - x_1 = 217.60 - 115.00 = \$102.60$

d. s = \$41.985 [the square root of the answer given in part e]

e. $s^2 = [n(\Sigma x^2) - (\Sigma x)^2]/[n(n-1)]$

$= [5(134529.7325) - (798.37)^2]/[5(4)]$

$= 35254.0056/20$

$= 1762.700 \text{ dollars}^2$

f. Usual values are those within $\bar{x} \pm 2s$

$159.674 \pm 2(41.9845)$

159.674 ± 83.969

75.705 to 243.643 [using rounded values, 75.704 to 243.644]

g. No. Since all of the sample values are within the above limits, none of them is unusual.

h. Ratio, since differences are meaningful and there is a meaningful zero.

i. Discrete, since checks must be written in whole numbers of cents.

j. Convenience, since they were selected because they happened to be close at hand.

k. Since the sample was a random sample, the sample mean of the n items observed should be a good estimate of the population mean of all N items. The estimate for the population total is $N\bar{x} = (134)(159.674) = \$21,396.316$.

2. Let x = the number of companies that test employees for substance abuse.

binomial problem: n = 10 and p = 0.80, use Table A-1

a. $P(x=5) = 0.026$

b. $P(x \geq 5) = P(x=5) + P(x=6) + P(x=7) + P(x=8) + P(x=9) + P(x=10)$

$= 0.026 + 0.088 + 0.201 + 0.302 + 0.268 + 0.107$

$= 0.992$

c. For the binomial distribution,

$\mu = np = (10)(0.80) = 8.0$

$\sigma^2 = npq = (10)(0.80)(0.20) = 1.60$

$\sigma = 1.2649$, rounded to 1.3

d. Usual values are those within $\mu \pm 2\sigma$

$8.0 \pm 2(1.2649)$

8.0 ± 2.5

5.5 to 10.5 [using rounded values, 5.4 to 10.6]

3. Let x = the number of HIV cases.
binomial problem: n = 150 and p = 0.10, use the binomial formula

$$P(x) = \frac{n!}{(n-x)!x!} p^x q^{n-x}.$$

a. $\mu = np = (150)(0.10) = 15.0$
$\sigma^2 = npq = (150)(0.10)(0.90) = 13.5$
$\sigma = 3.674$, rounded to 3.7

b. Usual values are those within $\mu \pm 2\sigma$
$15.0 \pm 2(3.674)$
15.0 ± 7.3
7.7 to 22.3 [using rounded values, 7.6 to 22.4]

No. Since 12 is within the above limits, it is not an unusually low result. There is not sufficient evidence to suggest that the program is effective in lowering the 10% rate.

4. Let S = the selected passenger was one who survived.

a. $P(S) = 706/2223 = 0.318$

b. $P(S_1 \text{ and } S_2) = P(S_1) \cdot P(S_2|S_1)$
$= (706/2223)(705/2222)$
$= 0.101$

c. $P(\overline{S}) = 1 - P(S)$
$= 1 - 706/2223$
$= 1517/2223$

$P(\overline{S}_1 \text{ and } \overline{S}_2) = P(\overline{S}_1) \cdot P(\overline{S}_2 | \overline{S}_1)$
$= (1517/2223)(1516/2222)$
$= 0.466$

5. No. The correct value is the weighted mean of the 50 state means, with the population of each state as the weight. If the means are not weighted, then small population states count equal with large population states – and a small population state with unusually high or low per capita consumption has undue influence and incorrectly affects the national mean as much as a large population state.

Chapter 6

Normal Probability Distributions

6-2 The Standard Normal Distribution

1. The word "normal" as used when referring to a normal distribution does carry with it some of the meaning the word has in ordinary language. Normal distributions occur in nature and describe the normal, or natural, state of many common phenomena. But in statistics the term "normal" has a specific and well-defined meaning in addition to its generic connotations of being "typical" – it refers to a specific bell-shaped distribution generated by a particular mathematical formula.

3. A normal distribution can be centered about any value and have any level of spread. A *standard* normal distribution has a center (as measured by the mean) of 0 and has a spread (as measured by the standard deviation) of 1.

5. The height of the rectangle is 0.5. Probability corresponds to area, and the area of a rectangle is (width)·(height).
$P(x>124.0) = (\text{width})\cdot(\text{height})$
$= (125.0-124.0)(0.5)$
$= (1.0)(0.5)$
$= 0.50$

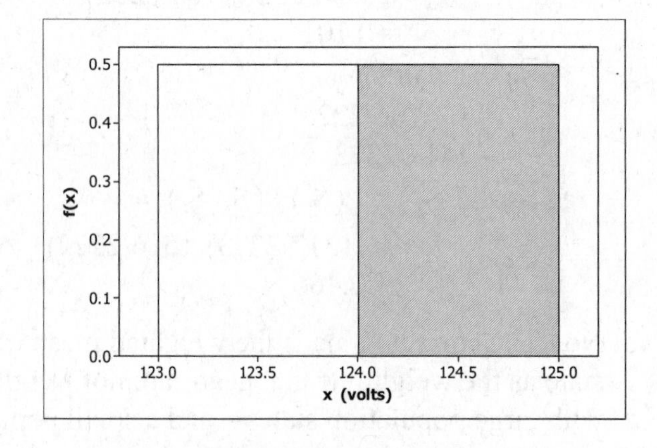

7. The height of the rectangle is 0.5. Probability corresponds to area, and the area of a rectangle is (width)·(height).
$P(123.2<x<124.7) = (\text{width})\cdot(\text{height})$
$= (124.7-123.2)(0.5)$
$= (1.7)(0.5)$
$= 0.75$

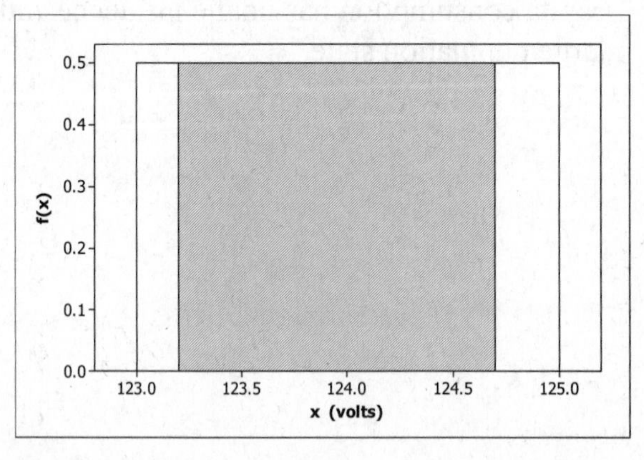

NOTE: For problems 9-16, the answers are re-expressed (when necessary) in terms of items that can be read directly from Table A-2. In general, this step is omitted in subsequent exercises and the reader is referred to the accompanying sketches to see how the indicated probabilities and z scores relate to Table A-2. "A" is used to denote the tabled value of the area to the left of the given z score. As a crude check, always verify that

$A>0.5000$ corresponds to a positive z score and $z>0$ corresponds to an $A>0.5000$
$A<0.5000$ corresponds to a negative z score and $z<0$ corresponds to an $A<0.5000$

9. $P(z<0.75) = 0.7734$

11. $P(-0.60<z<1.20) = P(z<1.20) - P(z<-0.60)$
$= 0.8849 - 0.2743$
$= 0.6106$

13. For $A = 0.9798$, $z = 2.05$.

15. If the area to the right of z is 0.1075, $A = 1 - 0.1075 = 0.8925$.
For $A = 0.8925$, $z = 1.24$.

NOTE: The sketch is the key to Exercises 17-36. It tells whether to subtract two Table A-2 probabilities, to subtract a Table A-2 probability from 1, etc. For the remainder of chapter 6, THE ACCOMPANYING SKETCHES ARE NOT TO SCALE and are intended only as aids to help the reader understand how to use the table values to answer the questions. In addition, the probability of any single point in a continuous distribution is zero – i.e., $P(x=a) = 0$ for any single point a. For normal distributions, therefore, this manual ignores $P(x=a)$ and uses $P(x>a) = 1 - P(x<a)$.

17. $P(z<-1.50) = 0.0668$

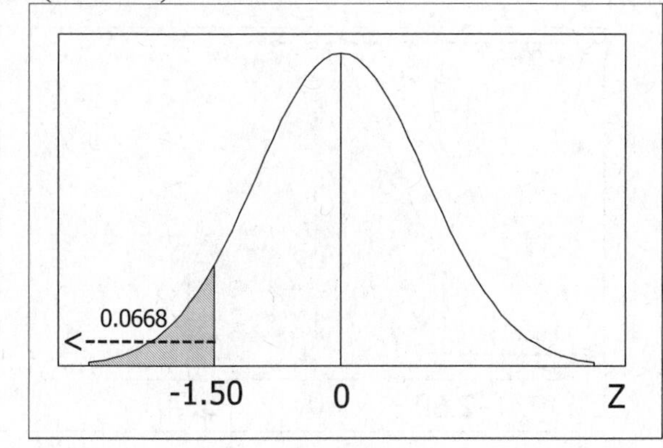

19. $P(z<1.23) = 0.8907$

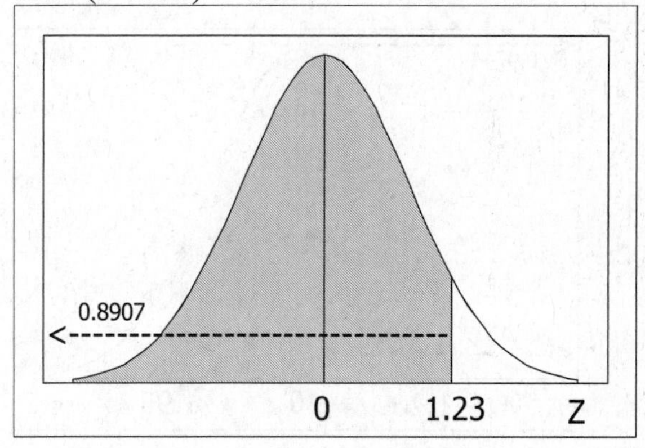

21. $P(z>2.22)$
$= 1 - 0.9868$
$= 0.0132$

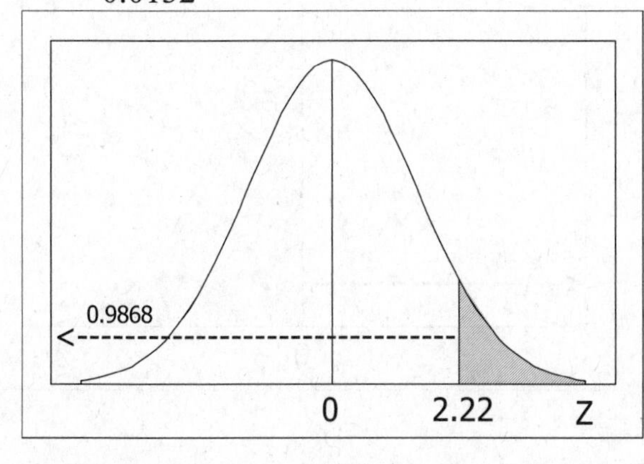

23. $P(z>-1.75)$
$= 1 - 0.0401$
$= 0.9599$

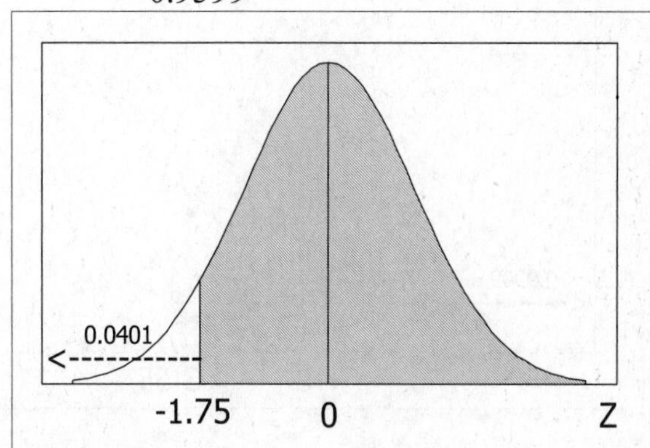

25. $P(0.50<z<1.00)$
$= 0.8413 - 0.6915$
$= 0.1498$

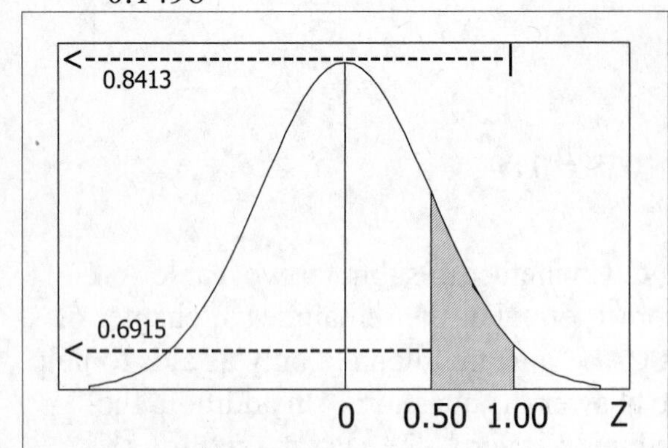

27. $P(-3.00<z<-1.00)$
$= 0.1587 - 0.0013$
$= 0.1574$

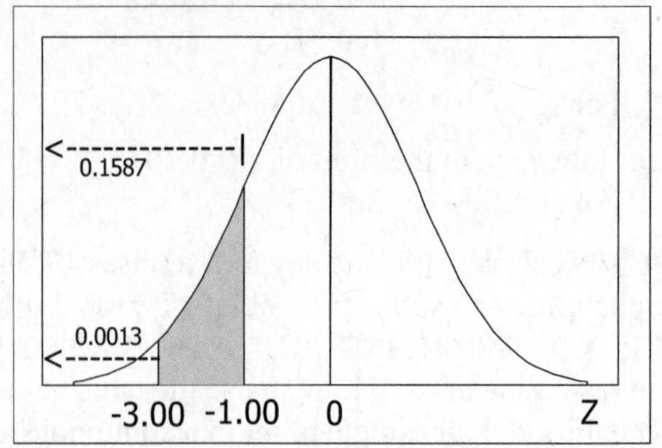

29. $P(-1.20<z<1.95)$
$= 0.9744 - 0.1151$
$= 0.8593$

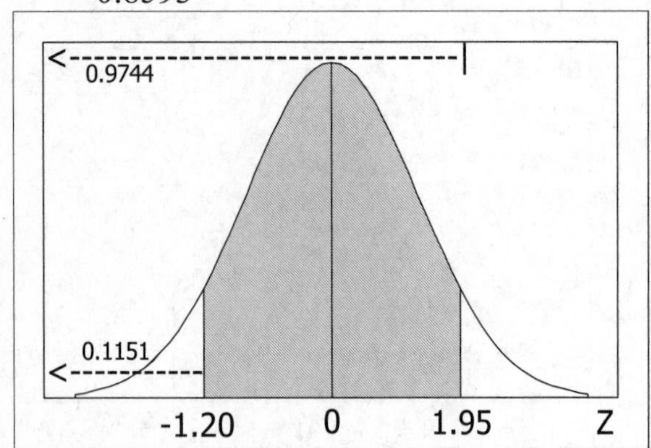

31. $P(-2.50<z<5.00)$
$= 0.9999 - 0.0062$
$= 0.9937$

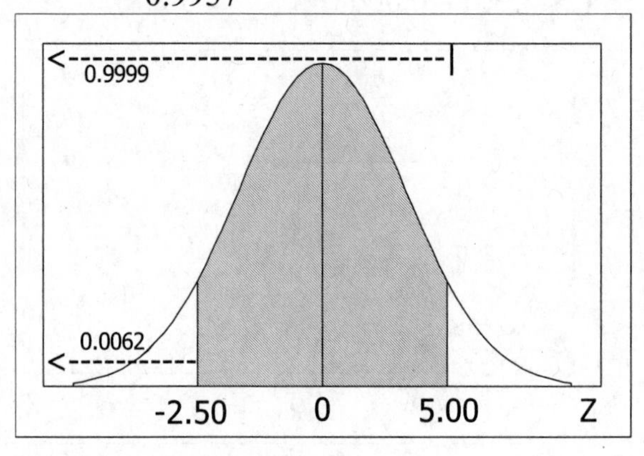

33. $P(z<3.55) = 0.9999$

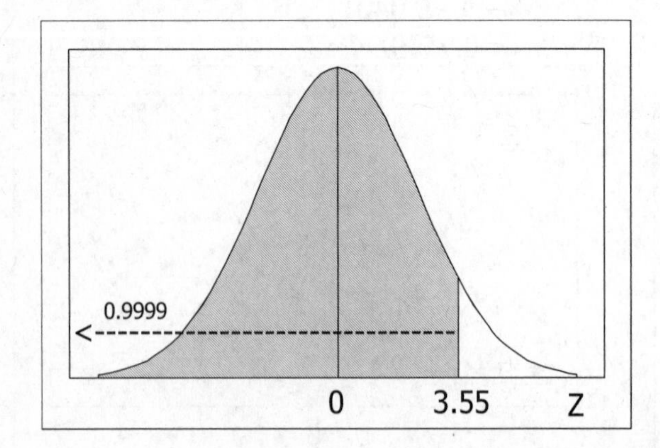

35. $P(z>0.00)$
$= 1 - 0.5000$
$= 0.5000$

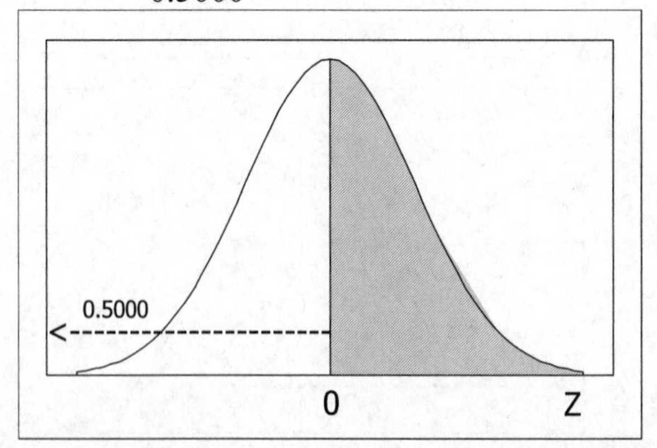

37. $P(-1<z<1) = 0.8413 - 0.1587 = 0.6826$
about 68%

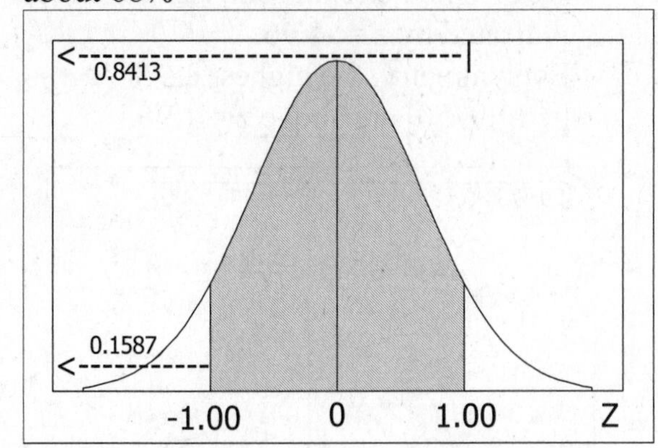

39. $P(-3<z<3) = 0.9987 - 0.0013 = 0.9974$
About 99.7%

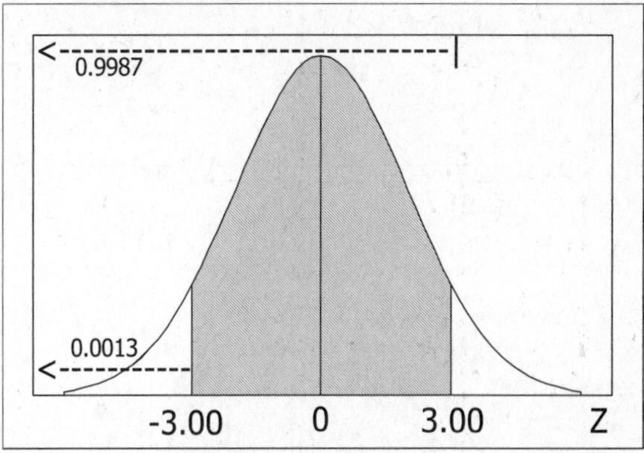

41. For $z_{0.05}$, the cumulative area is 0.9500.
The closest entry is 0.9500,
for which $z = 1.645$

43. For $z_{0.10}$, the cumulative area is 0.9000.
The closest entry is 0.8997,
for which $z = 1.28$.

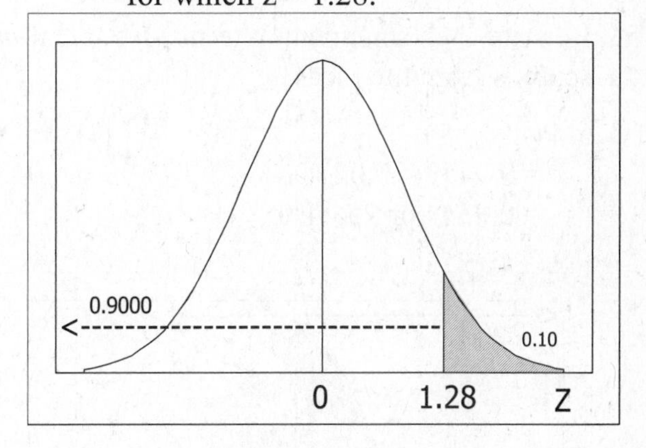

45. $P(-1.96<z<1.96)$
$= 0.9750 - 0.0250$
$= 0.9500$

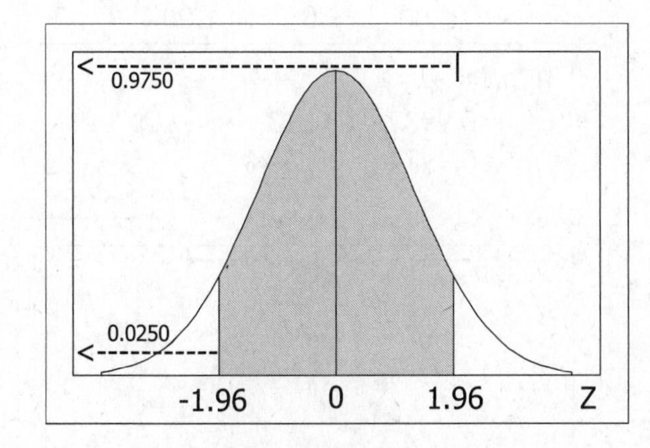

47. $P(z<-2.575 \text{ or } z>2.575)$
$= P(z<-2.575) + P(z>2.575)$
$= 0.0050 + (1 - 0.9950)$
$= 0.0050 + 0.0050$
$= 0.0100$

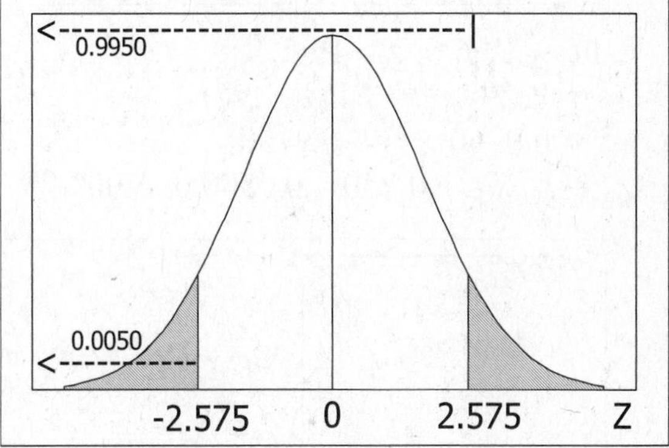

49. For P_{95}, the cumulative area is 0.9500.
The closest entry is 0.9500,
for which z = 1.645

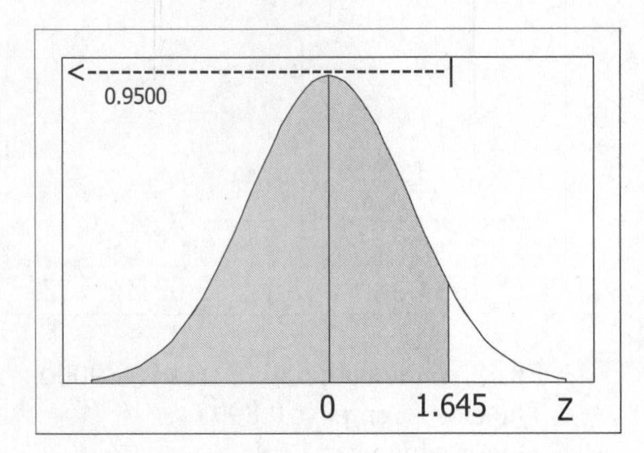

51. For the lowest 2.5%, the cumulative area is 0.0250 (exact entry in the table), indicated by z = -1.96.
By symmetry, the highest 2.5% [= 1–.9750] are above z = 1.96.

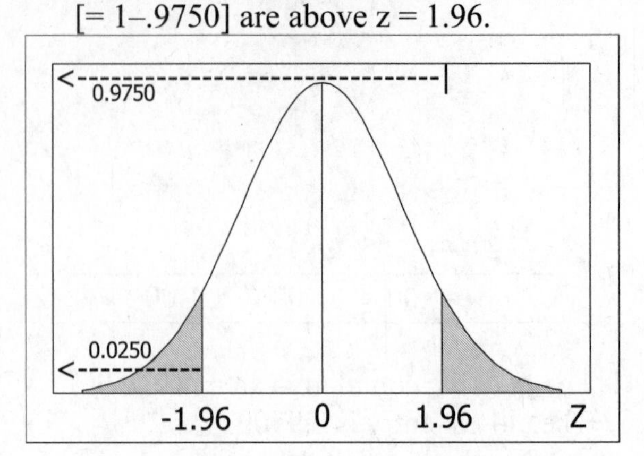

53. Rewrite each statement in terms of z, recalling that z is the number of standard deviations a score is from the mean.

a. $P(-2<z<2)$
= 0.9772 – 0.0228
= 0.9544 or 95.44%

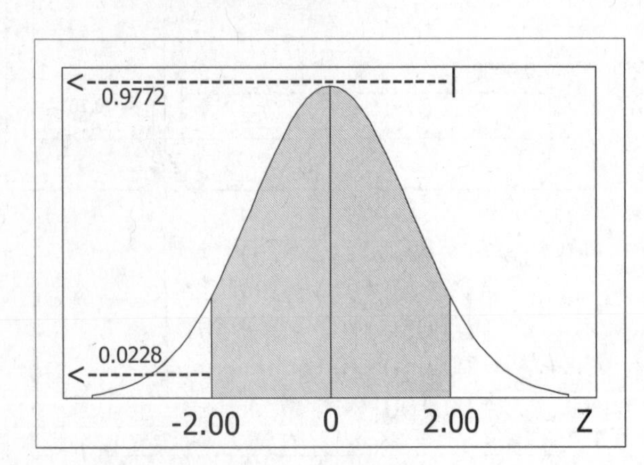

b. $P(z<-1 \text{ or } z>1)$
= $P(z<-1) + P(z>1)$
= 0.1587 + (1 – 0.8413)
= 0.1587 + 0.1587 = 0.3174 or 31.74%

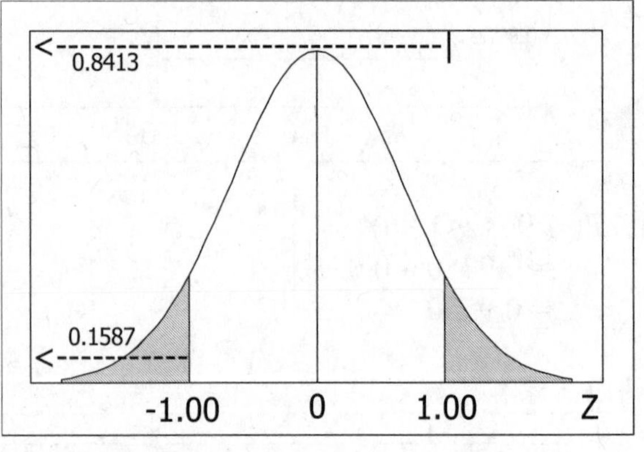

c. $P(z<-1.96 \text{ or } z>1.96)$
= $P(z<-1.96) + P(z>1.96)$
= 0.0250 + (1 – 0.9750)
= 0.250 + 0.250 = 0.0500 or 5.00%

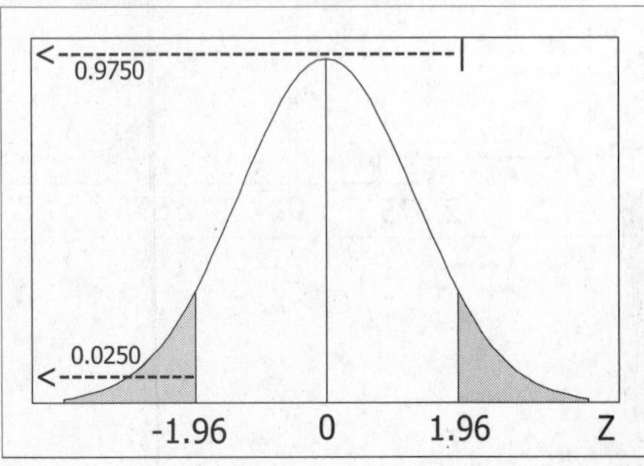

d. $P(-3<z<3)$
= 0.9987 – 0.0013
= 0.9974 or 99.74%

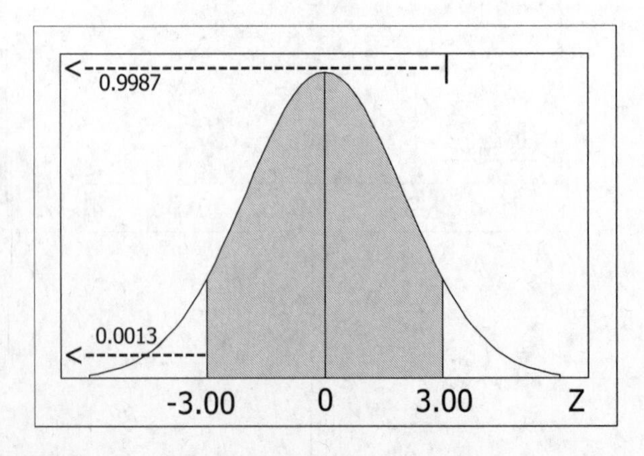

e. $P(z<-3 \text{ or } z>3)$
$= P(z<-3) + P(z>3)$
$= 0.0013 + (1 - 0.9987)$
$= 0.0013 + 0.0013 = 0.0026 = 0.26\%$

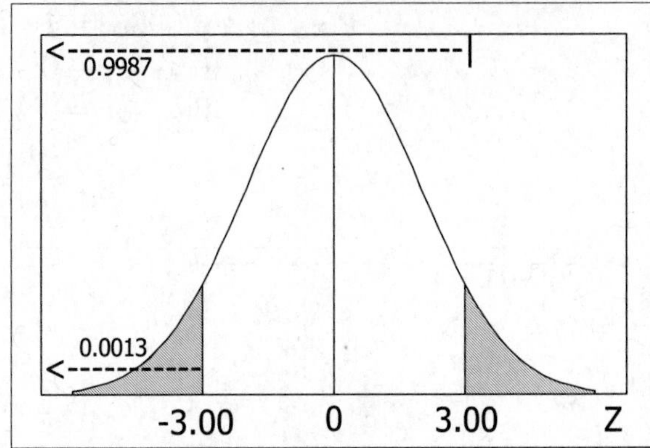

55. The sketches are the key. They tell what probability (i.e., cumulative area) to look up when reading Table A-2 "backwards." They also provide a check against gross errors by indicating whether a score is above or below zero.

a. $P(z<a) = 0.9599$
$a = 1.75$
(see the sketch at the right)

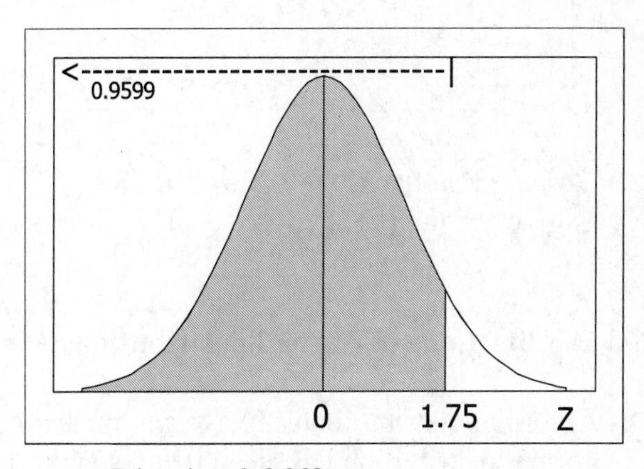

b. $P(z>b) = 0.9772$
$P(z<b) = 1 - 0.9772$
$= 0.0228$
$b = -2.00$

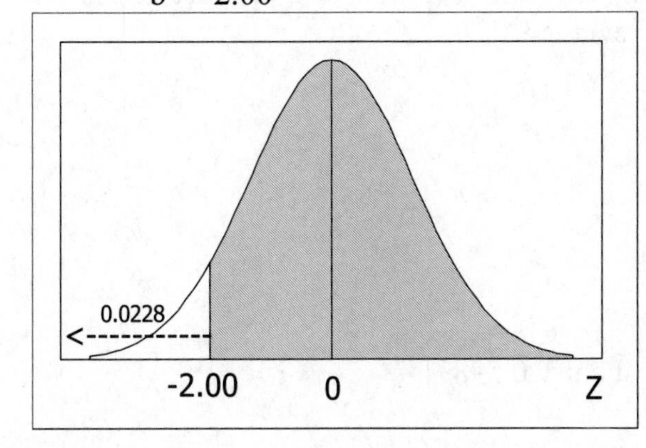

c. $P(z>c) = 0.0668$
$P(z<c) = 1 - 0.0668$
$= 0.9332$
$c = 1.50$

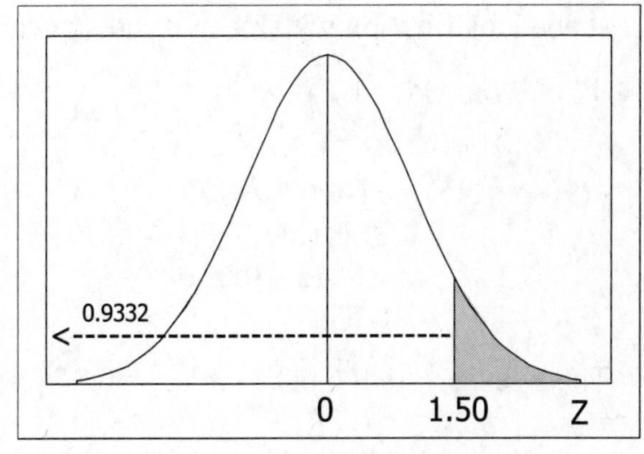

d. Since $P(z<-d) = P(z>d)$ by symmetry and $\Sigma P(z) = 1$,

$$\begin{aligned} P(z<-d) + P(-d<z<d) + P(z>d) &= 1 \\ P(z<-d) + 0.5878 + P(z<-d) &= 1 \\ 2 \cdot P(z<-d) + 0.5878 &= 1 \\ 2 \cdot P(z<-d) &= 0.4122 \\ P(z<-d) &= 0.2061 \\ -d &= -0.82 \\ d &= 0.82 \end{aligned}$$

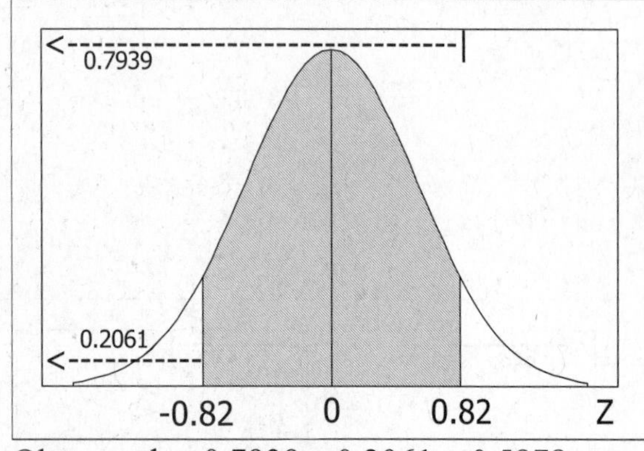

Observe that 0.7939 – 0.2061 = 0.5878, as given in the problem.

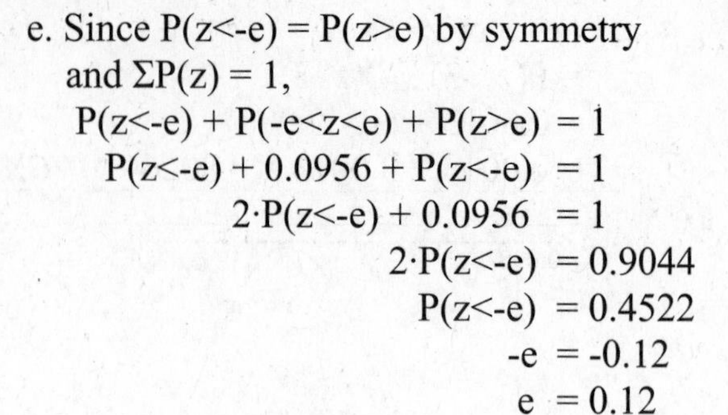

e. Since $P(z<-e) = P(z>e)$ by symmetry and $\Sigma P(z) = 1$,

$$\begin{aligned} P(z<-e) + P(-e<z<e) + P(z>e) &= 1 \\ P(z<-e) + 0.0956 + P(z<-e) &= 1 \\ 2 \cdot P(z<-e) + 0.0956 &= 1 \\ 2 \cdot P(z<-e) &= 0.9044 \\ P(z<-e) &= 0.4522 \\ -e &= -0.12 \\ e &= 0.12 \end{aligned}$$

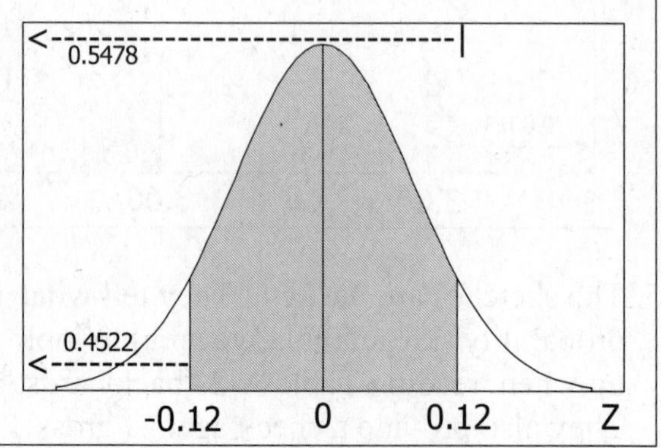

Observe that 0.5478 – 0.4522 = 0.0956, as given in the problem.

6-3 Applications of Normal Distributions

1. A normal distribution can have any mean and any positive standard deviation. A standard normal distribution has mean 0 and standard deviation 1 – and it follows a "nice" bell-shaped curve. Non-standard normal distributions can follow bell-shaped curves that are tall and thin, or short and fat.

3. For any distribution, converting to z scores using the formula $z = (x-\mu)/\sigma$ produces a same-shaped distribution with mean 0 and standard deviation 1.

5. $P(x<120) = P(z<1.33)$
   $= 0.9082$

7. $P(90<x<115) = P(-0.67<z<1.00)$
   $= P(z<1.00) - P(z<-0.67)$
   $= 0.8413 - 0.2514$
   $= 0.5899$

9. The z score with 0.6 below it is z = 0.25 [closest entry 0.5987].
   $x = \mu + z\sigma$
   $= 100 + (0.25)(15)$
   $= 100 + 3.75 = 103.75$, rounded to 103.8

11. The z score with 0.95 above is the z score with 0.05 below it; z = -1.645 [bottom of table].
   $x = \mu + z\sigma$
   $= 100 + (-1.645)(15)$
   $= 100 - 24.675 = 75.325$, rounded to 75.3

13. normal distribution: $\mu = 100$ and $\sigma = 15$
$P(x<115)$
$= P(z<1.00)$
$= 0.8413$

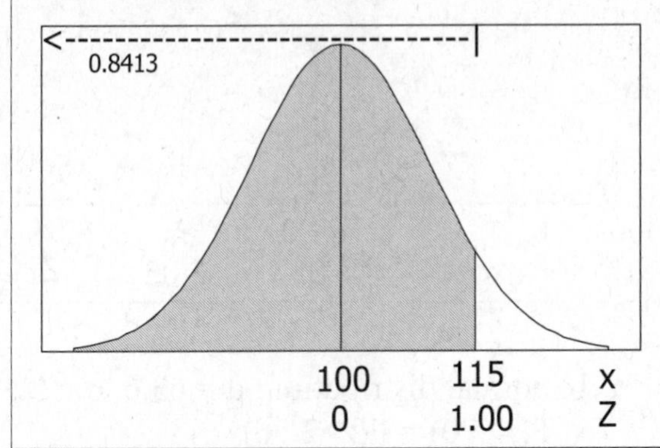

15. normal distribution: $\mu = 100$ and $\sigma = 15$
$P(90<x<110) = P(-0.67<z<0.67)$
$= 0.7486 - 0.2514$
$= 0.4972$

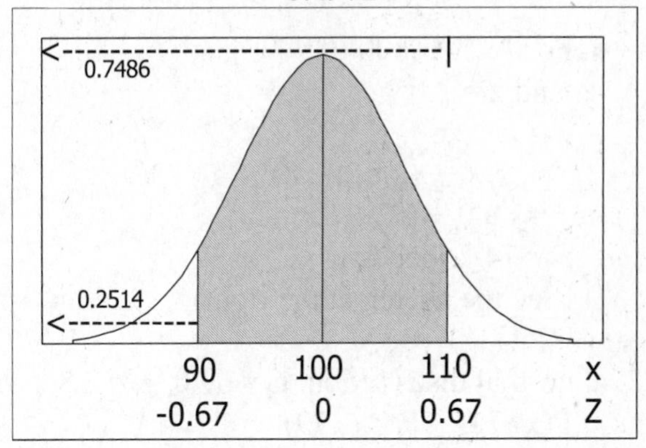

17. normal distribution: $\mu = 100$ and $\sigma = 15$
For P_{30}, $A = 0.3000$ [0.3015] and $z = -0.52$.
and $z = -0.52$.
$x = \mu + z\sigma$
$= 100 + (-0.52)(15)$
$= 100 - 7.8 = 92.2$

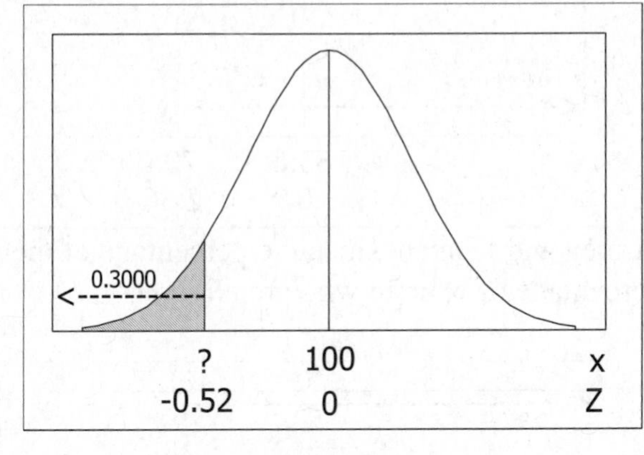

19. normal distribution: $\mu = 100$ and $\sigma = 15$
For Q_3, $A = 0.7500$ [0.7486]
and $z = 0.67$.
$x = \mu + z\sigma$
$= 100 + (0.67)(15)$
$= 100 + 10.05 = 110.05$, rounded to 110.1

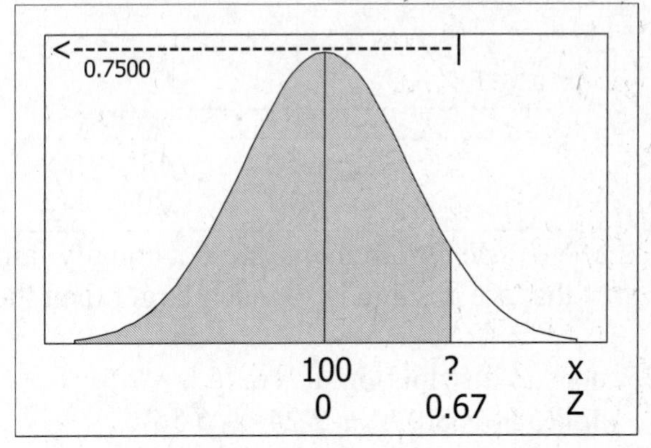

21. a. normal distribution: $\mu = 69.0$, $\sigma = 2.8$
$P(x<72) = P(z<1.07)$
$= 0.8577$ or 85.77%

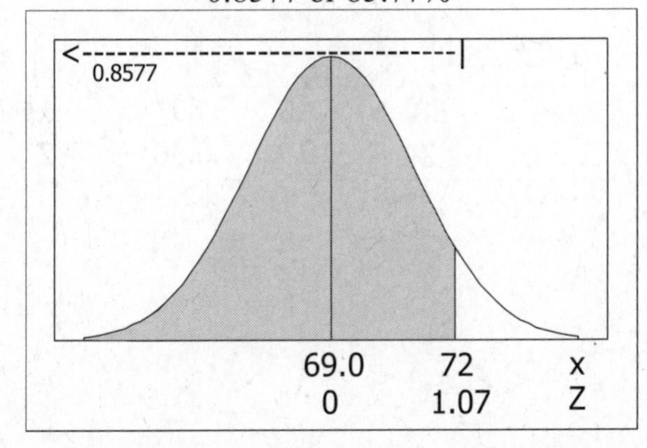

b. normal distribution: $\mu = 63.6$, $\sigma = 2.5$
$P(x<72) = P(z<3.36)$
$= 0.9996$ or 99.96%

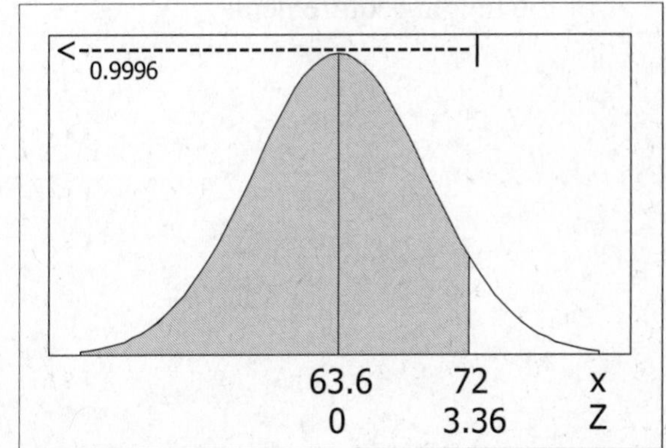

c. No. It is not adequate in that 14% of the men need to bend to enter, and they may be in danger of injuring themselves if they fail to recognize the necessity to bend.

d. For A = 0.9800 [0.9798], and z = 2.05.
$x = \mu + z\sigma$
$= 69.0 + (2.05)(2.8)$
$= 69.0 + 5.7$
$= 74.7$ inches
(See the sketch at the right)

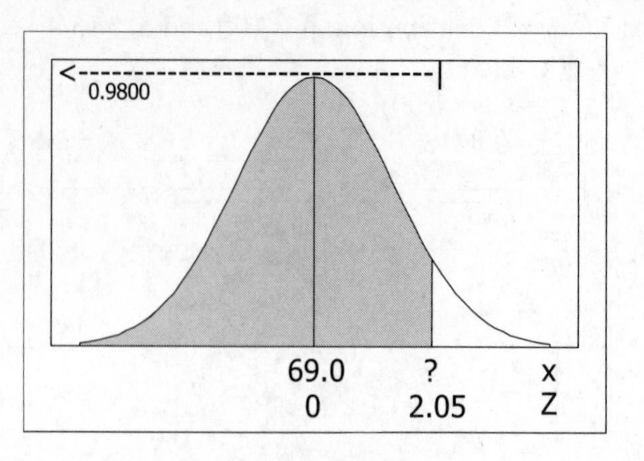

23. a. normal distribution: $\mu = 69.0$, $\sigma = 2.8$
$P(x>74) = P(z>1.79)$
$= 1 - 0.9633$
$= 0.0367$ or 3.67%

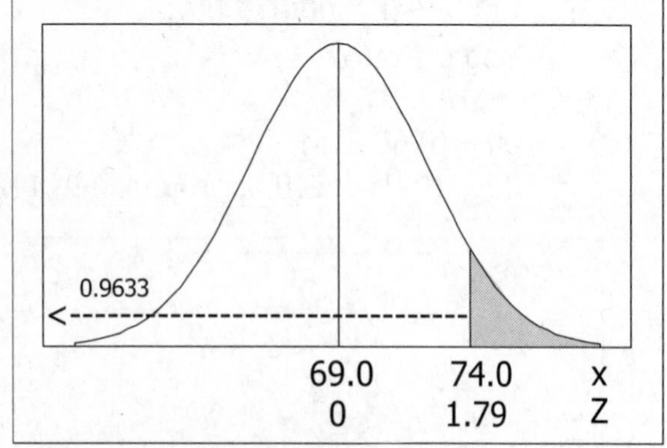

b. normal distribution: $\mu = 63.6$, $\sigma = 2.5$
$P(x>70) = P(z>2.56)$
$= 1 - 0.9948$
$= 0.0052$ or 0.52%

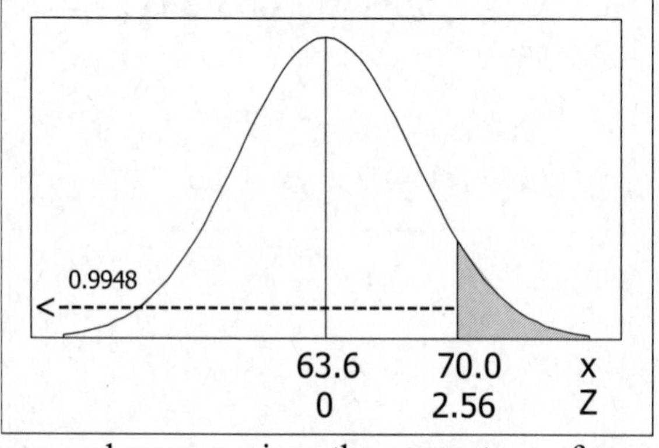

c. No. The requirements are not equally fair for men and women, since the percentage of men that are eligible is so much larger than the percentage of women who are eligible.

25. normal distribution: $\mu = 63.6$ $\sigma = 2.5$
a. $P(58<x<80) = P(-2.24<z<6.56)$
$= 0.9999 - 0.0125$
$= 0.9874$ or 98.74%
No. Only $1 - 0.9874 = 0.0126 = 1.26\%$ of the women are not eligible because of the height requirements.

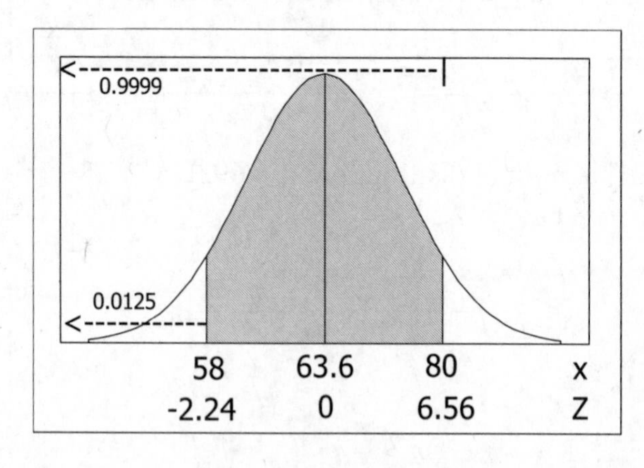

b. For the shortest 1%, A = 0.0100 [0.0099] and z = -2.33.

$x = \mu + z\sigma$
$= 63.6 + (-2.33)(2.5)$
$= 63.6 - 5.8$
$= 57.8$ inches

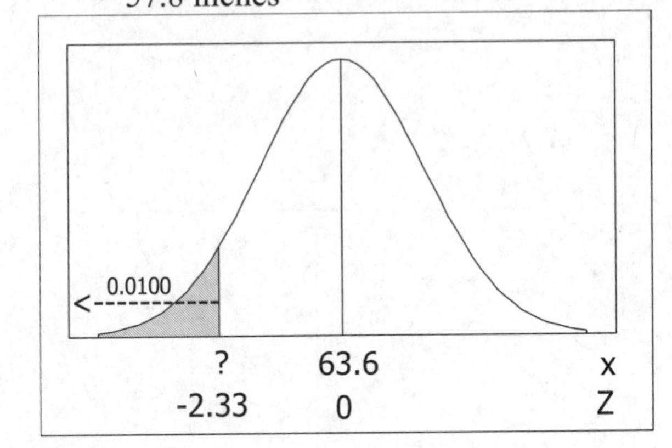

For the tallest 2%, A = 0.9800 [0.9798] and z = 2.05.

$x = \mu + z\sigma$
$= 63.6 + (2.05)(2.5)$
$= 63.6 + 5.1$
$= 68.7$ inches

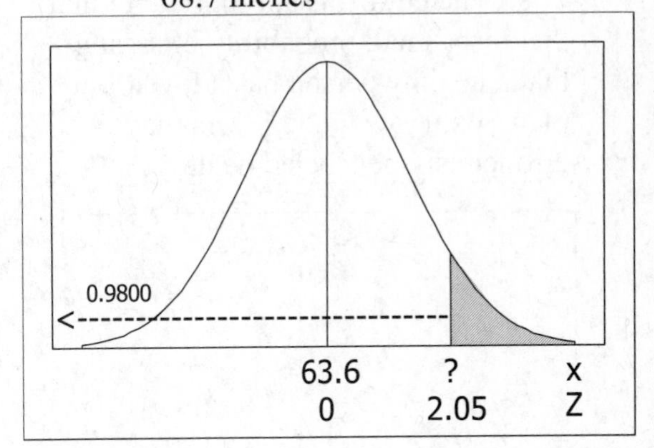

27. normal distribution: $\mu = 3570$ and $\sigma = 500$

a. $P(x<2700)$
$= P(z<-1.74)$
$= 0.409$ or 4.09%

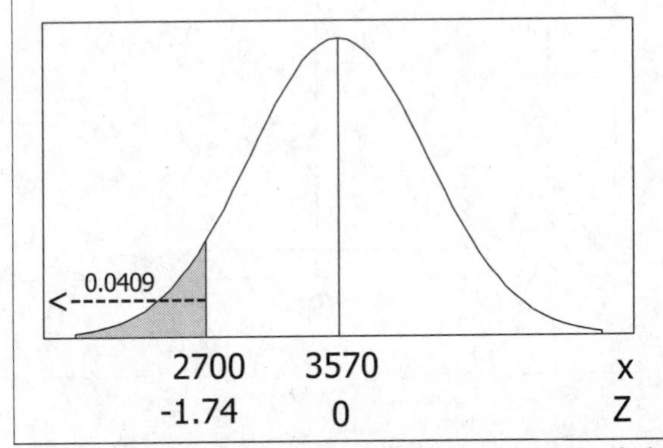

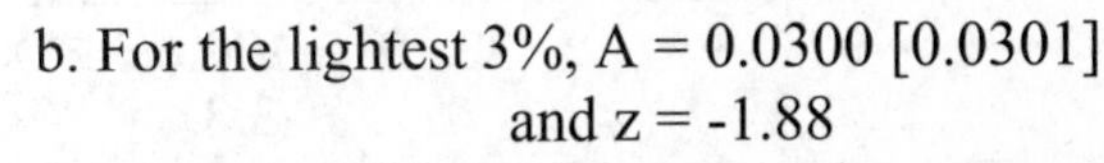
b. For the lightest 3%, A = 0.0300 [0.0301] and z = -1.88

$x = \mu + z\sigma$
$= 3570 + (-1.88)(500)$
$= 3570 - 940 = 2630$ g

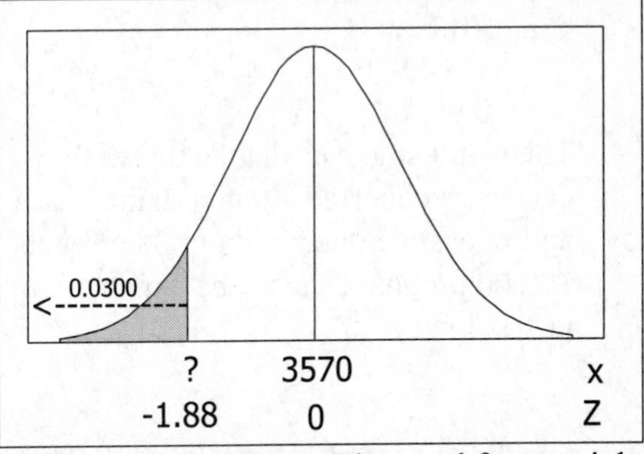

c. Not all babies below a certain birth weight require special treatment. The need for special treatment is determined at least as much by developmental considerations as by weight alone. Also, the birth weight identifying the bottom 3% is not a static figure and would have to be updated periodically – perhaps creating unnecessary uncertainty and inconsistency.

29. normal distribution: $\mu = 98.20$ and $\sigma = 0.62$

a. $P(x>100.6)$
$= P(z>3.87)$
$= 1 - 0.9999$
$= 0.0001$

Yes. The cut-off is appropriate in that there is a small probability of saying that a healthy person has a fever, but many with low grade fevers may erroneously be labeled healthy.

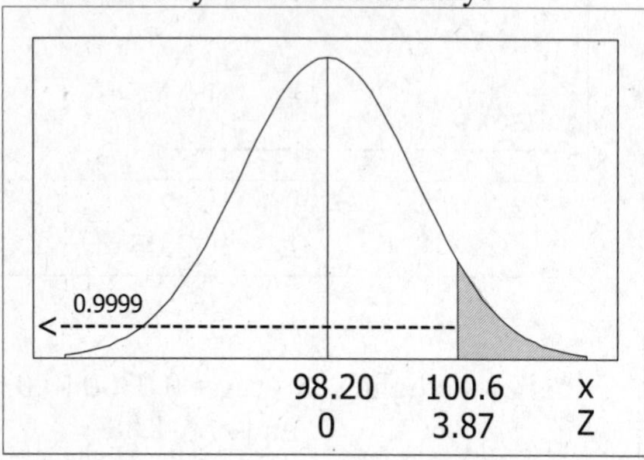

b. For the highest 5%, A = 0.9500 and z = 1.645.

$x = \mu + z\sigma$
$= 98.20 + (1.645)(0.62)$
$= 98.20 + 1.02$
$= 99.22$ °F

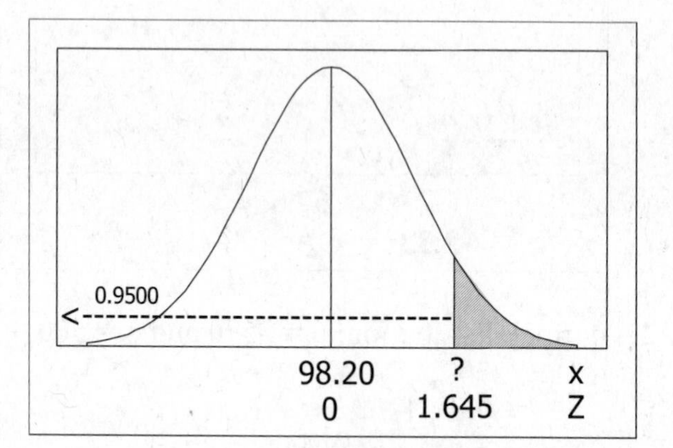

31. normal distribution: $\mu = 268$ and $\sigma = 15$

a. $P(x>308)$
$= P(z>2.67)$
$= 1 - 0.9962$
$= 0.0038$

The result suggests that an unusual event has occurred – but certainly not an impossible one, as about 38 of every 10,000 pregnancies can be expected to last as long.

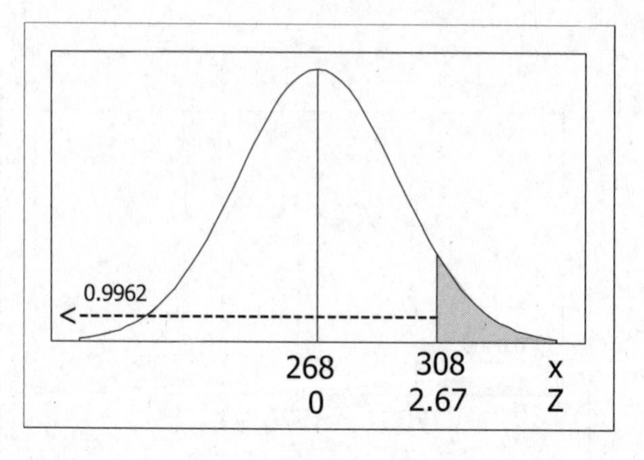

b. For the lowest 4%, A = 0.0400 [0.0401] and z = -1.75.

$x = \mu + z\sigma$
$= 268 + (-1.75)(15)$
$= 268 - 26$
$= 242$ days

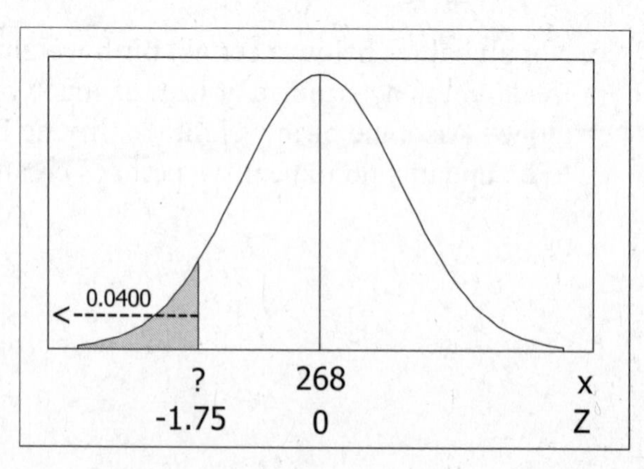

33\. a. The distribution and summary statistics, obtained using statistical software, are given below. The data appear to have a distribution that is approximately normal.

$n = 40$
$\overline{x} = 118.9$
$s = 10.46$

pressure	frequency
90- 99	1
100-109	4
110-119	17
120-129	12
130-139	5
140-149	0
150-159	1
	40

NOTE: Using rounded σ=10.5 in part (b) gives $x_5 = 101.6$ and $x_{95} = 136.2$.

b. normal distribution: $\mu = 118.9$, $\sigma = 10.46$
For P_5, A = 0.0500 and z = -1.645.
For P_{95}, A = 0.9500 and z = 1.645.
$x_5 = \mu + z\sigma = 118.9 + (-1.645)(10.46)$
$= 118.9 - 17.2 = 101.7$ mm
$x_{95} = \mu + z\sigma = 118.9 + (1.645)(10.46)$
$= 118.9 + 17.2 = 136.1$ mm

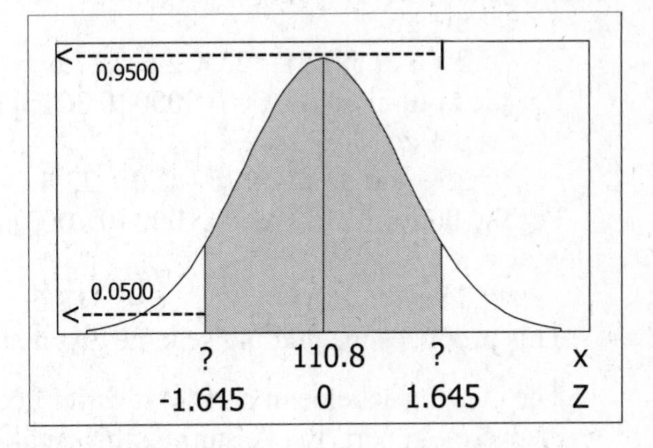

35\. a. The z scores are always unit free. Because the numerator and the denominator of the fraction $z = (x-\mu)/\sigma$ have the same units, the units will divide out.

b. For a population of size N, $\mu = \Sigma x/N$ and $\sigma^2 = \Sigma(x-\mu)^2/N$.
As shown below, $\mu = 0$ and $\sigma = 1$ will be true for *any* set of z scores – regardless of the shape of the original distribution.

$\Sigma z = \Sigma(x-\mu)/\sigma = (1/\sigma)[\Sigma(x-\mu)]$
$= (1/\sigma)[\Sigma x - \Sigma\mu]$
$= (1/\sigma)[N\mu - N\mu] = (1/\sigma)[0] = 0$
$\Sigma z^2 = \Sigma[(x-\mu)/\sigma]^2 = (1/\sigma)^2[\Sigma(x-\mu)^2]$
$= (1/\sigma^2)[N\sigma^2] = N$
$\mu_Z = (\Sigma z)/N = 0/N = 0$
$\sigma^2_Z = \Sigma(z-\mu_Z)^2/N = \Sigma(z-0)^2/N = \Sigma z^2/N = N/N = 1$ and $\sigma_Z = 1$

The re-scaling from x scores to z scores will not affect the basic shape of the distribution – and so in this case the z scores will be normal, as was the original distribution.

37\. normal distribution: $\mu = 25$ and $\sigma = 5$

a. For a population of size N, $\mu = \Sigma x/N$ and $\sigma^2 = \Sigma(x-\mu)^2/N$.
As shown below, adding a constant to each score increases the mean by that amount but does not affect the standard deviation. In non-statistical terms, shifting everything by k units does not affect the spread of the scores. This is true for *any* set of scores – regardless of the shape of the original distribution. Let y = x + k.

$\mu_Y = [\Sigma(x+k)]/N$
$= [\Sigma x + \Sigma k]/N$
$= [\Sigma x + Nk]/N$
$= (\Sigma x)/N + (Nk)/N$
$= \mu_X + k$

$\sigma^2_Y = \Sigma[y - \mu_Y]^2/N$
$= \Sigma[(x+k) - (\mu_X + k)]^2/N$
$= \Sigma[x - \mu_X]^2/N$
$= \sigma^2_X$ and so $\sigma_Y = \sigma_X$

If the teacher adds 50 to each grade,
new mean = 25 + 50 = 75
new standard deviation = 5 (same as before).

b. No. Curving should consider the variation. Had the test been more appropriately constructed, it is not likely that every student would score exactly 50 points higher. If the typical student score increased by 50, we would expect the better students to increase by more than 50 and the poorer students to increase by less than 50. This would make the scores more spread out and would increase the standard deviation.

c. For the top 10%, A = 0.9000 [0.8997] and z = 1.28.
$x = \mu + z\sigma$
$= 25 + (1.28)(5) = 25 + 6.4 = 31.4$
For the bottom 70%, A = 0.7000 [0.6985] and z = 0.52.
$x = \mu + z\sigma$
$= 25 + (0.52)(5) = 25 + 2.6 = 27.6$
For the bottom 30%, A = 0.3000 [0.3015] and z = -0.52.
$x = \mu + z\sigma$
$= 25 + (-0.52)(5) = 25 - 2.6 = 22.4$
For the bottom 10%, A = 0.1000 [0.1003] and z = -1.28.
$x = \mu + z\sigma$
$= 25 + (-1.28)(5) = 25 - 6.4 = 18.6$
This produces the grading scheme given at the right.

A: higher than 31.4
B: 27.6 to 31.4
C: 22.4 4o 27.6
D: 18.6 to 22.4
E: less than 18.6

d. The curving scheme in part (c) is fairer because it takes into account the variation as discussed in part (b). Assuming the usual 90-80-70-60 letter grade cut-offs, for example, the percentage of A's under the scheme in part (a) with $\mu = 75$ and $\sigma = 5$ is
$P(x>90) = 1 - P(x<90)$
$= 1 - P(z<3.00)$
$= 1 - 0.9987$
$= 0.0013$ or 0.13%
This is considerably less than the 10% A's under the scheme in part (c) and reflects the fact that the variation in part (a) is unrealistically small.

39. Work with z scores to answer the question using a standard normal distribution. This is a four-step process as follows.
Step 1. Find Q_1 and Q_3.
For Q_1, A = 0.2500 [0.2514] and z = -0.67.
For Q_3, A = 0.7500 [0.7486] and z = 0.67.
Step 2. Find the $IQR = Q_3 - Q_1$, and then determine (1.5)(IQR)
$IQR = Q_3 - Q_1 = 0.67 - (-0.67) = 1.34$
$(1.5)(IQR) = (1.5)(1.34) = 2.01$
Step 3. Find the lower and upper limits L and U beyond which outliers occur.
$L = Q_1 - (1.5)(IQR) = -0.67 - 2.01 = -2.68$
$U = Q_3 + (1.5)(IQR) = 0.67 + 2.01 = 2.68$
Step 4. Find P(outlier) = P(z<L) + P(z>U).
$P(outlier) = P(z<-2.68 \text{ or } z>2.68)$
$= P(z<-2.68) + P(z>2.68)$
$= 0.0037 + (1 - 0.9963)$
$= 0.0037 + 0.0037$
$= 0.0074$

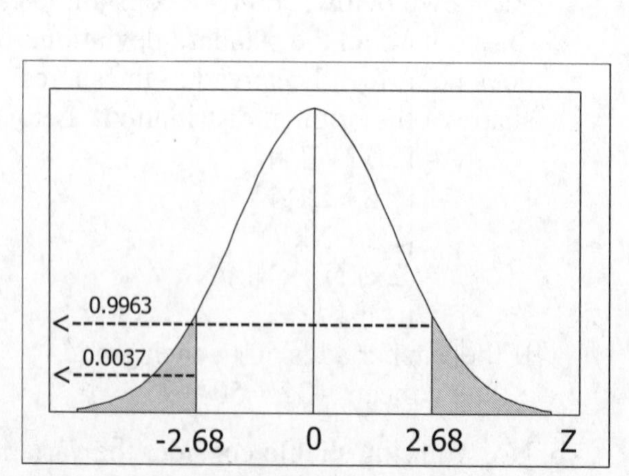

6-4 Sampling Distributions and Estimators

1. Given a population distribution of scores, one can take a sample of size n and calculate any of several statistics. A sampling distribution is the distribution of all possible values of such a particular statistic.

3. An unbiased estimator is one whose expected value is the true value of the parameter which it estimates. The sample mean is an unbiased estimator of the population mean because the expected value (or mean) of its sampling distribution is the population mean.

5. No. The students at New York University are not necessarily representative (by race, major, etc.) of the population of all U.S. college students.

7. The sample means will have a distribution that is approximately normal. They will tend to form a symmetric, unimodal and bell-shaped distribution around the value of the population mean.

9. a. The medians of the 9 samples are given in column 2 at the right. The sampling distribution of the median is given in columns 3 and 4 at the right.
   b. The population median is 3. The mean of the sample medians is $\Sigma\tilde{x}\cdot P(\tilde{x}) = 45/9 = 5$.
   c. In general the sample medians do not target the value of the population median. For this reason, sample median is not a good estimator of the population median.

sample	$\tilde{x}$	$\tilde{x}$	$P(\tilde{x})$	$\tilde{x}\cdot P(\tilde{x})$
2,2	2	2	1/9	2/9
2,3	2.5	2.5	2/9	5/9
2,10	6	3	1/9	3/9
3,2	2.5	6	2/9	12/9
3,3	3	6.5	2/9	13/9
3,10	6.5	10	1/9	10/9
10,2	6		9/9	45/9
10,3	6.5			
10,10	10			

NOTE: Section 5-2 defined the mean of a probability distribution of x's as $\mu_x = \Sigma[x\cdot P(x)]$. If the variable is designated by the symbol y, then the mean of a probability distribution of y's is $\mu_y = \Sigma[y\cdot P(y)]$. In this section, the variables are statistics – like $\overline{x}$ and $\hat{p}$. In such cases, the formula for the mean may be adjusted – to $\mu_{\overline{x}} = \Sigma[\overline{x}\cdot P(\overline{x})]$ and $\mu_{\hat{p}} = \Sigma[\hat{p}\cdot P(\hat{p})]$. In a similar manner, the formula for the variance of a probability distribution may also be adjusted to match the variable being considered.

11. a. The variances of the 9 samples are given in column 2 at the right. The sampling distribution of the variance is given in columns 3 and 4 at the right.
   b. Since the values 2,3,10 are considered a population, the population variance is $\sigma^2 = \Sigma(x-\mu)^2/N = (3^2 + 2^2 + 5^2)/3 = 38/3$. The mean of the sample variances is $\Sigma s^2\cdot P(s^2) = 114/9 = 38/3$.
   c. The sample variance always targets the value of the population variance. For this reason, the sample variance is a good estimator of the population variance.

sample	s^2	s^2	$P(s^2)$	$s^2\cdot P(s^2)$
2,2	0	0	3/9	0/9
2,3	0.5	0.5	2/9	1/9
2,10	32	24.5	2/9	49/9
3,2	0.5	32	2/9	64/9
3,3	0		9/9	114/9
3,10	24.5			
10,2	32			
10,3	24.5			
10,10	0			

The information below and the box at the right apply to Exercises 13-16.

original population of scores in order: 46 49 56 58
summary statistics: N= 4 $\Sigma x = 209$ $\Sigma x^2 = 11017$
$\mu = \Sigma x/N$
$= 209/4 = 52.25$
population $\tilde{x} = (x_2+x_3)/2$
$= (49+56)/2 = 52.5$
population $R = x_n - x_1$
$= 58 - 46 = 12$
$\sigma^2 = [\Sigma(x-\mu)^2]/N$
$= [(-6.25)^2 + (-3.25)^2 + (3.75)^2 + (5.75)^2]/4$
$= [39.0625 + 10.5625 + 14.0625 + 33.0625]/4$
$= 96.75/4 = 24.1875$

sample	$\bar{x}$	$\tilde{x}$	R	s^2
46,46	46.0	46.0	0	0.0
46,49	47.5	47.5	3	4.5
46,56	51.0	51.0	10	50.0
46,58	52.0	52.0	12	72.0
49,46	47.5	47.5	3	4.5
49,49	49.0	49.0	0	0.0
49,56	52.5	52.5	7	24.5
49,58	53.5	53.5	9	40.5
56,46	51.0	51.0	10	50.0
56,49	52.5	52.5	7	24.5
56,56	56.0	56.0	0	0.0
56,58	57.0	57.0	2	2.0
58,46	52.0	52.0	12	72.0
58,49	53.5	53.5	9	40.5
58,56	57.0	57.0	2	2.0
58,58	58.0	58.0	0	0.0

13. a. The sixteen possible samples are given in the "samples" column of the box preceding this exercise
b. The sixteen possible means are given in column 2 of the box preceding this exercise. The sampling distribution of the mean is given in the first two columns at the right.
c. The population mean is 52.25. The mean of the sample means is $\Sigma\bar{x}\cdot P(\bar{x}) = 836.0/16 = 52.25$. They are the same.
d. Yes. The sample mean always targets the value of the population mean. For this reason, the sample mean is a good estimator of the population mean.

$\bar{x}$	$P(\bar{x})$	$\bar{x}\cdot P(\bar{x})$
46.0	1/16	46.0/16
47.5	2/16	95.0/16
49.0	1/16	49.0/16
51.0	2/16	102.0/16
52.0	2/16	104.0/16
52.5	2/16	105.0/16
53.5	2/16	107.0/16
56.0	1/16	56.0/16
57.0	2/16	114.0/16
58.0	1/16	58.0/16
	16/16	836.0/16

15. a. The sixteen possible samples are given in the "samples" column of the box preceding Exercise 13.
b. The sixteen possible ranges are given in column 4 of the box preceding Exercise 13. The sampling distribution of the range is given in the first two columns at the right.
c. The population range is 12. The mean of the sample ranges is $\Sigma R\cdot P(R) = 5.375$. They are not the same.
d. No. The sample ranges do not always target the value of the population range. For this reason, the sample range is not a good estimator of the population range.

R	P(R)	R·P(R)
0	4/16	0/16
2	2/16	4/16
3	2/16	6/16
7	2/16	14/16
9	2/16	18/16
10	2/16	20/16
12	2/16	24/16
	16/16	86/16

17. Let $\hat{p}$ be the symbol for the sample proportion. The 9 possible sample proportions are given in column 2 at the right. The sampling distribution of the proportion is given in columns 3 and 4 at the right. The population proportion of odd numbers is 1/3. The mean of the sample proportions is $\Sigma\hat{p}\cdot P(\hat{p})$ = 3.0/9 = 1/3. They are the same. The sample proportion always targets the value of the population proportion. For this reason, the sample proportion is a good estimator of the population proportion.

sample	$\hat{p}$	$\hat{p}$	$P(\hat{p})$	$\hat{p}\cdot P(\hat{p})$
2,2	0.0	0.0	4/9	0.0/9
2,3	0.5	0.5	4/9	2.0/9
2,10	0.0	1.0	1/9	1.0/9
3,2	0.5		9/9	3.0/9
3,3	1.0			
3,10	0.5			
10,2	0.0			
10,3	0.5			
10,10	0.0			

19. a. The sampling distribution of the proportion is given in columns 5 and 6 of the table at the right.
    b. The mean of the sampling distribution is $\Sigma\hat{p}\cdot P(\hat{p})$ = 12.0/16 = 0.75.
    c. The population proportion of females is 3/4 = 0.75. Yes, the two values are the same. The sample proportion always targets the value of the population proportion. For this reason, the sample proportion is a good estimator of the population proportion.

pair	$\hat{p}$	pair	$\hat{p}$	$\hat{p}$	$P(\hat{p})$	$\hat{p}\cdot P(\hat{p})$
mm	0.0	bm	0.5	0.0	1/16	0.0/16
ma	0.5	ba	1.0	0.5	6/16	3.0/16
mb	0.5	bb	1.0	1.0	9/16	9.0/16
mc	0.5	bc	1.0		16/16	12.0/16
am	0.5	cm	0.5			
aa	1.0	ca	1.0			
ab	1.0	cb	1.0			
ac	1.0	cc	1.0			

21. Use of the formula for the given values is shown below. The resulting distribution agrees with, and therefore describes, the sampling distribution for the proportion of girls in a family of size 2.

$$P(x) = 1/[2(2-2x)!(2x)!]$$
$$P(x=0) = 1/[2(2)!(0)!] = 1/[2\cdot 2\cdot 1] = 1/4$$
$$P(x=0.5) = 1/[2(1)!(1)!] = 1/[2\cdot 1\cdot 1] = 1/2$$
$$P(x=1) = 1/[2(0)!(2)!] = 1/[2\cdot 1\cdot 2] = 1/4$$

6-5 The Central Limit Theorem

1. The standard error of a statistic is the standard deviation of its sampling distribution. For any given sample size n, the standard error of the mean is the standard deviation of the distribution of all possible sample means of size n. Its symbol and numerical value are $\sigma_{\bar{x}} = \sigma/\sqrt{n}$.

3. The subscript $\bar{x}$ is used to distinguish the mean and standard deviation of the sample means from the mean and standard deviation of the original population. This produces the notation

$$\mu_{\bar{x}} = \mu \qquad \sigma_{\bar{x}} = \sigma/\sqrt{n}$$

5. a. normal distribution
$\mu = 1518$
$\sigma = 325$
$P(x<1500)$
$= P(z<-0.06)$
$= 0.4761$

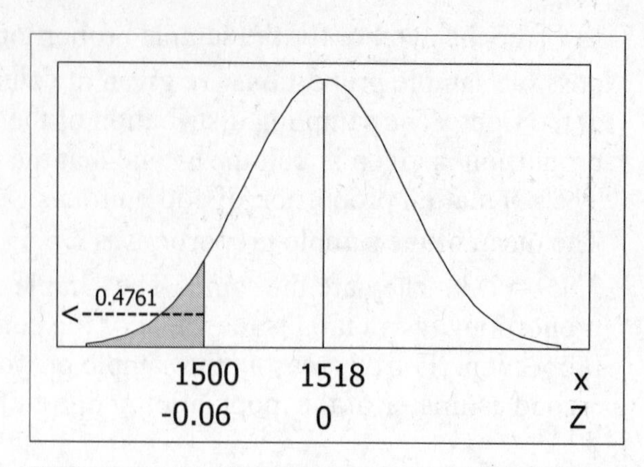

b. normal distribution,
since the original distribution is so
$\mu_{\bar{x}} = \mu = 1518$
$\sigma_{\bar{x}} = \sigma/\sqrt{n} = 325/\sqrt{100} = 32.5$
$P(\bar{x}<1500)$
$= P(z<-0.55)$
$= 0.2912$

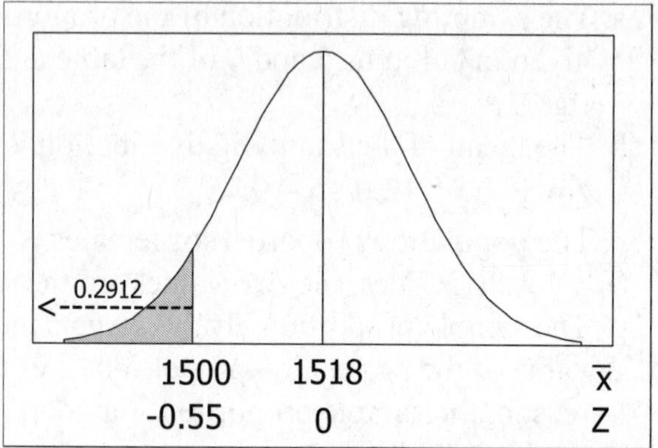

7. a. normal distribution
$\mu = 1518$
$\sigma = 325$
$P(1550<x<1575)$
$= P(0.10<z<0.18)$
$= 0.5714 - 0.5398$
$= 0.0316$

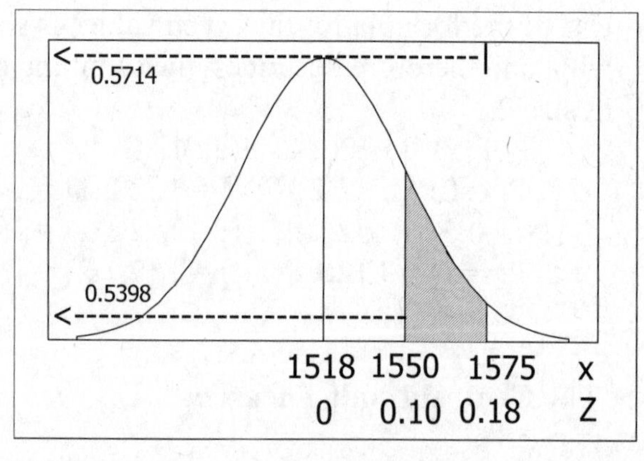

b. normal distribution,
since the original distribution is so
$\mu_{\bar{x}} = \mu = 1518$
$\sigma_{\bar{x}} = \sigma/\sqrt{n} = 325/\sqrt{25} = 65$
$P(1550<\bar{x}<1575)$
$= P(0.49<z<0.88)$
$= 0.8106 - 0.6879$
$= 0.1227$

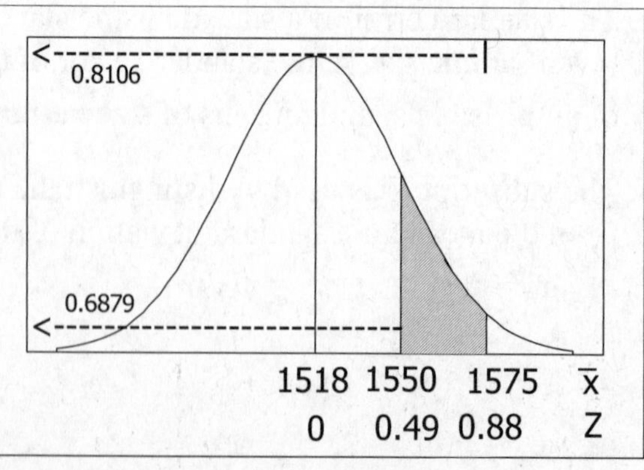

c. Since the original distribution is normal, the Central Limit Theorem can be used in part (b) even though the sample size does not exceed 30.

9. a. normal distribution
   $\mu = 172$
   $\sigma = 29$
   $P(x>180)$
   $= P(z>0.28)$
   $= 1 - 0.6103$
   $= 0.3897$

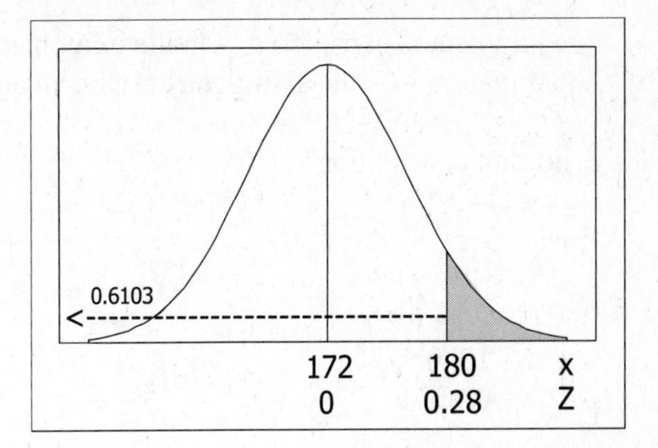

b. normal distribution,
   since the original distribution is so
   $\mu_{\bar{x}} = \mu = 172$
   $\sigma_{\bar{x}} = \sigma/\sqrt{n} = 29/\sqrt{20} = 6.48$
   $P(\bar{x}>180)$
   $= P(z>1.23)$
   $= 1 - 0.8907$
   $= 0.1093$

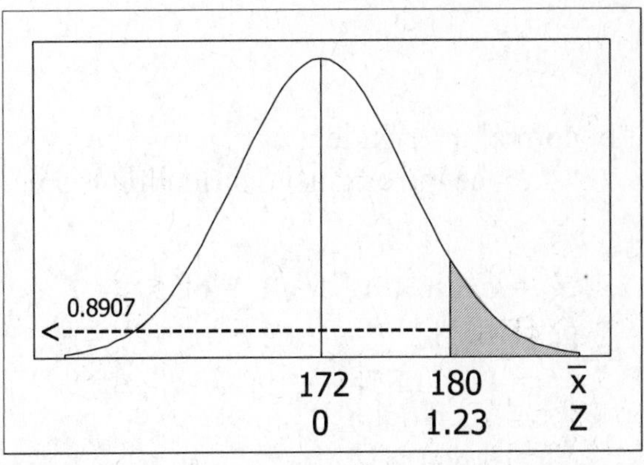

c. Yes. A capacity of 20 is not appropriate when the passengers are all adult men, since a 10.93% probability of overloading is too much of a risk.

11. a. normal distribution
   $\mu = 172$
   $\sigma = 29$
   $P(x>167)$
   $= P(z>-0.17)$
   $= 1 - 0.4325$
   $= 0.5675$

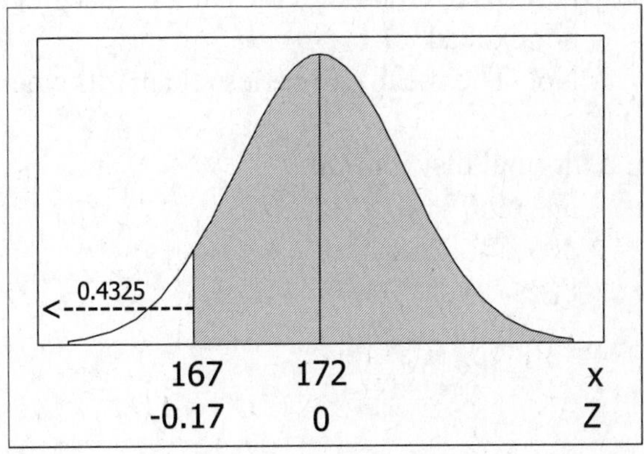

b. normal distribution,
   since the original distribution is so
   $\mu_{\bar{x}} = \mu = 172$
   $\sigma_{\bar{x}} = \sigma/\sqrt{n} = 29/\sqrt{12} = 8.372$
   $P(\bar{x}>167)$
   $= P(z>-0.60)$
   $= 1 - 0.2743$
   $= 0.7257$

c. No. It appears that the 12 person capacity could easily exceed the 2004 lbs – especially when the weight of clothes and

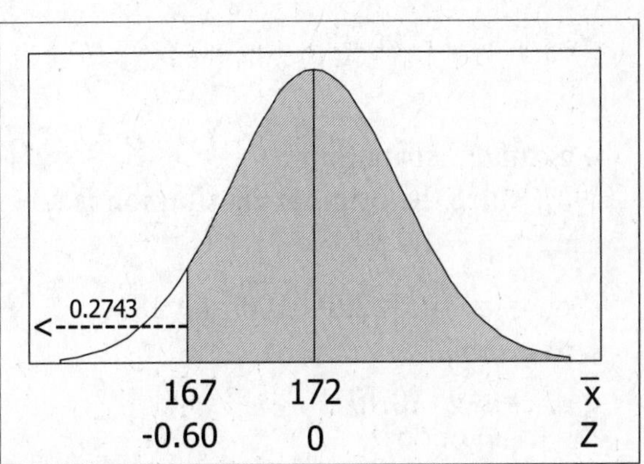

equipment is considered. On the other hand, skiers may be lighter than the general population – as the skiing may not be an activity that attracts heavier persons.

13. a. normal distribution
$\mu = 114.8$
$\sigma = 13.1$
$P(x>140)$
$= P(z>1.92)$
$= 1 - 0.9726$
$= 0.0274$

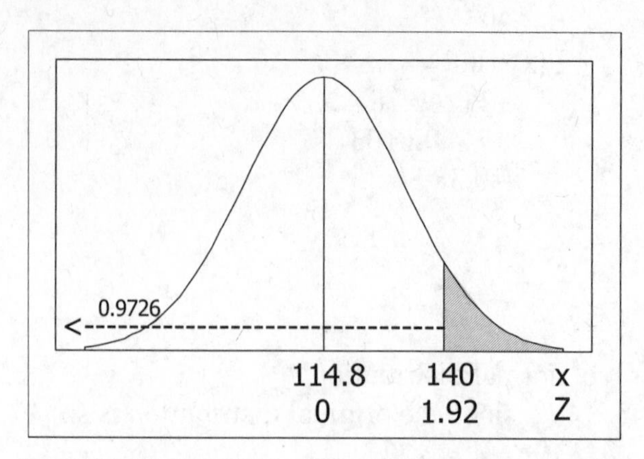

b. normal distribution,
since the original distribution is so
$\mu_{\bar{x}} = \mu = 114.8$
$\sigma_{\bar{x}} = \sigma/\sqrt{n} = 13.1/\sqrt{4} = 6.55$
$P(\bar{x}>140)$
$= P(z>3.85)$
$= 1 - 0.9999$
$= 0.0001$

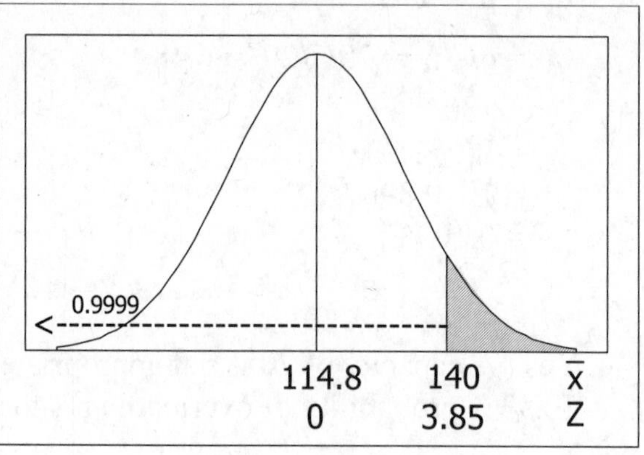

c. Since the original distribution is normal, the Central Limit Theorem can be used in part (b) even though the sample size does not exceed 30.

d. No. The mean can be less than 140 when one or more of the values is greater than 140.

15. a. normal distribution
$\mu = 69.0$
$\sigma = 2.8$
$P(x<72)$
$= P(z<1.07)$
$= 0.8577$

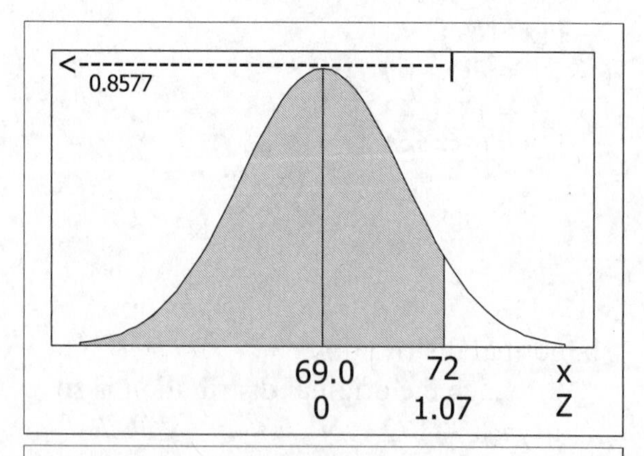

b. normal distribution,
since the original distribution is so
$\mu_{\bar{x}} = \mu = 69.0$
$\sigma_{\bar{x}} = \sigma/\sqrt{n} = 2.8/\sqrt{100} = 0.28$
$P(\bar{x}<72)$
$= P(z<10.17)$
$= 0.9999$

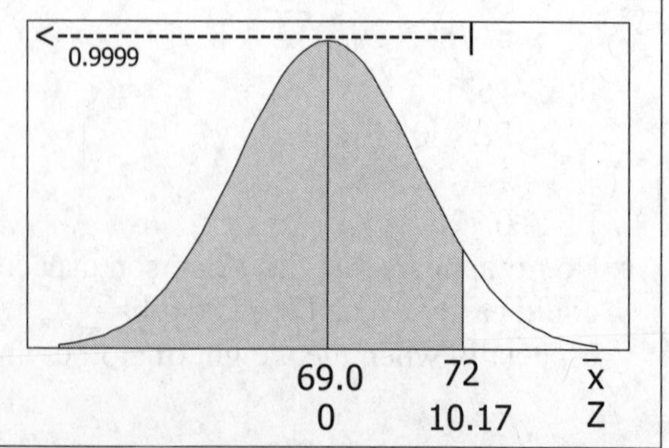

c. The probability in part (a) is more relevant. Part (a) deals with individual passengers, and these are the persons whose safety and comfort need to be considered. Part (b) deals with group means – and it is possible for statistics that apply "on the average" to actually describe only a small portion of the population of interest.

d. Women are generally smaller than men. Any design considerations that accommodate larger men will automatically accommodate larger women.

17. a. normal distribution
$\mu = 143$
$\sigma = 29$
$P(140<x<211)$
$= P(-0.10<z<2.34)$
$= 0.9904 - 0.4602$
$= 0.5302$

b. normal distribution,
since the original distribution is so
$\mu_{\bar{x}} = \mu = 143$
$\sigma_{\bar{x}} = \sigma/\sqrt{n} = 29/\sqrt{36} = 4.833$
$P(140<\bar{x}<211)$
$= P(-0.62<z<14.07)$
$= 0.9999 - 0.2676$
$= 0.7323$

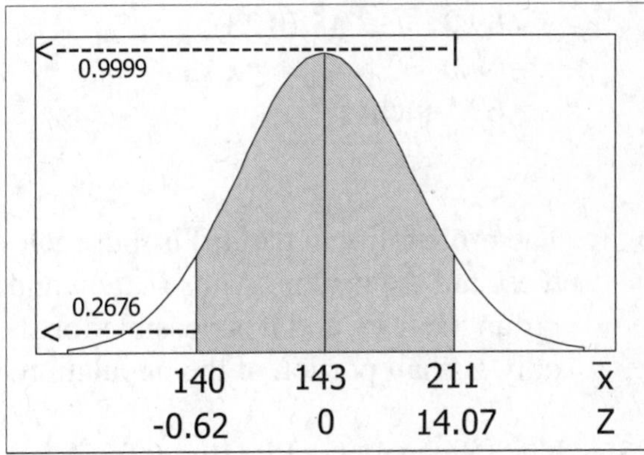

c. The information from part (a) is more relevant, since the seats will be occupied by one woman at a time.

19. normal distribution,
by the Central Limit Theorem
$\mu_{\bar{x}} = \mu = 12.00$
$\sigma_{\bar{x}} = \sigma/\sqrt{n} = 0.09/\sqrt{36} = 0.015$
$P(\bar{x}>12.29)$
$= P(z>19.33)$
$= 1 - 0.9999$
$= 0.0001$

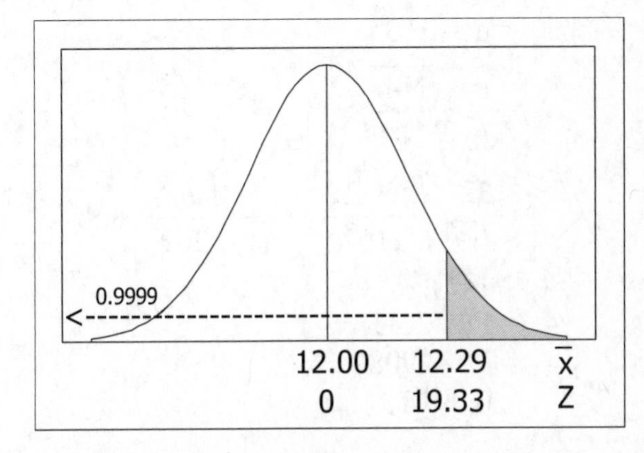

Yes. The results suggest that the Pepsi cans are being filled with more than 12.00 oz of product. This is undoubtedly done on purpose, to minimize probability of producing cans with less than the stated 12.00 oz of product.

21. a. normal distribution
$\mu = 69.0$
$\sigma = 2.8$
The z score with 0.9500 below it is 1.645.
$x = \mu + z\sigma$
$= 69.0 + (1.645)(2.8)$
$= 69.0 + 4.6$
$= 73.6$ inches

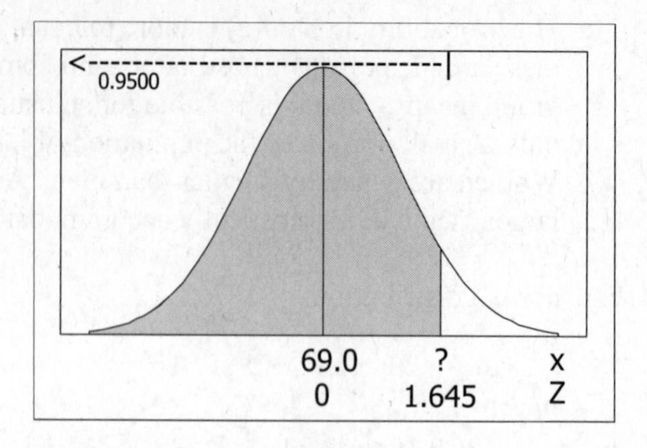

b. normal distribution,
since the original distribution is so
$\mu_{\bar{x}} = \mu = 69.0$
$\sigma_{\bar{x}} = \sigma/\sqrt{n} = 2.8/\sqrt{100} = 0.28$
The z score with 0.9500 below it is 1.645
$\bar{x} = \mu_{\bar{x}} + z\sigma_{\bar{x}}$
$= 69.0 + (1.645)(0.28)$
$= 69.0 + 0.5$
$= 69.5$ inches

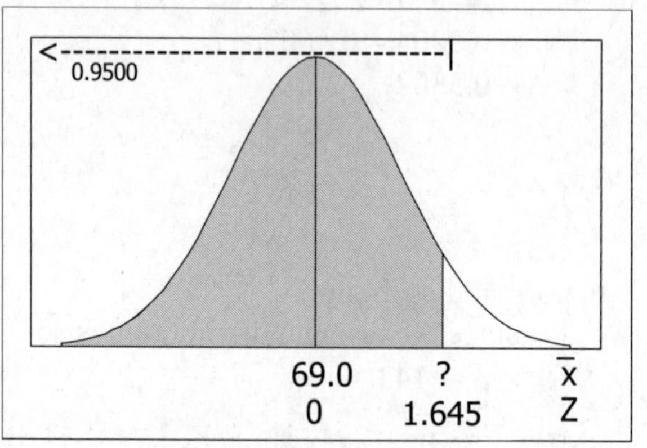

c. The probability in part (a) is more relevant. Part (a) deals with individual passengers, and these are the persons whose safety and comfort need to be considered. Part (b) deals with group means – and it is possible for statistics that apply "on the average" to actually describe only a small portion of the population of interest.

23. a. Yes. Since n/N = 12/210 = 0.0571 > 0.05, use the finite population correction factor.
b. For 12 passengers, the 2100 lb limit implies a mean weight of 2100/12 = 175 lbs.
c. normal distribution,
since the original distribution is normal
$\mu_{\bar{x}} = \mu = 163$

$$\sigma_{\bar{x}} = \frac{\sigma}{\sqrt{n}}\sqrt{\frac{N-n}{N-1}}$$

$$= \frac{32}{\sqrt{12}}\sqrt{\frac{210-12}{210-1}} = \frac{32}{\sqrt{12}}\sqrt{\frac{198}{209}} = 8.991$$

$P(\bar{x} > 175)$
$= P(z > 1.33)$
$= 1 - 0.9082$
$= 0.0918$

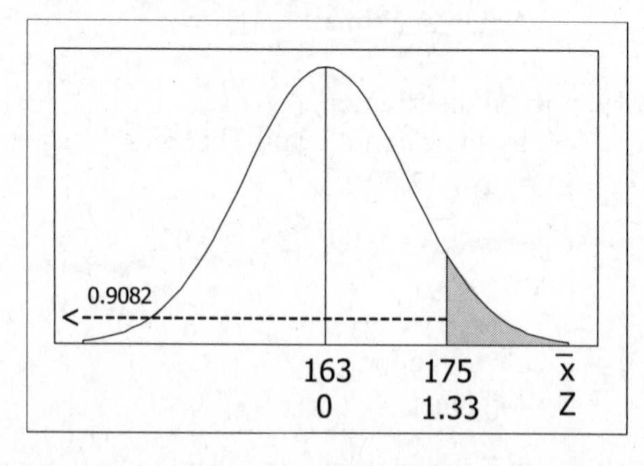

d. The best approach is systematic trial and error. The above calculations indicate that for
$n = 12$, $P(\Sigma x<2100) = P(\bar{x}<2100/n) = P(\bar{x}< 175) = 0.9082$.
Repeat the above calculations for
$n = 11, 10$, etc. until $P(\Sigma x<2100) = P(\bar{x}<2100/n)$ reaches 0.9990.

*For $n = 11$, $\bar{x} = 2100/11 = 190.91$
normal distribution,
since the original distribution is normal

$\mu_{\bar{x}} = \mu = 163$

$$\sigma_{\bar{x}} = \frac{\sigma}{\sqrt{n}}\sqrt{\frac{N-n}{N-1}}$$

$$= \frac{32}{\sqrt{11}}\sqrt{\frac{210-11}{210-1}} = \frac{32}{\sqrt{11}}\sqrt{\frac{199}{209}} = 9.415$$

$P(\bar{x}<190.91)$
$= P(z<2.96)$
$= 0.9985$
Since $0.9985 < 0.999$, try $n = 10$.

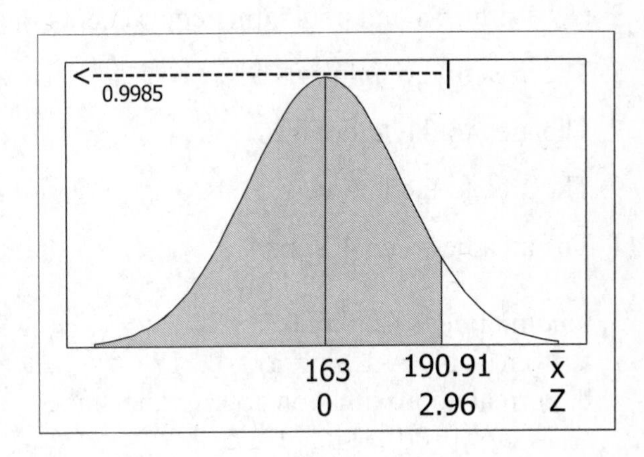

*For $n = 10$, $\bar{x} = 2100/10 = 210$
normal distribution,
since the original distribution is normal

$\mu_{\bar{x}} = \mu = 163$

$$\sigma_{\bar{x}} = \frac{\sigma}{\sqrt{n}}\sqrt{\frac{N-n}{N-1}}$$

$$= \frac{32}{\sqrt{10}}\sqrt{\frac{210-10}{210-1}} = \frac{32}{\sqrt{10}}\sqrt{\frac{200}{209}} = 9.899$$

$P(\bar{x}<210)$
$= P(z<4.75)$
$= 0.9999$
Since $0.9999 > 0.999$, set $n = 10$.

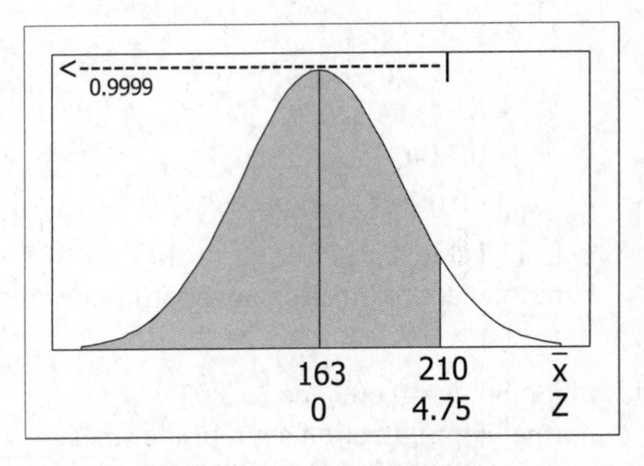

NOTE: The z score with 0.999 below it is 3.10. To find n exactly, solve the following equation to get $n = 10.9206$. Since n must be a whole number, round down to set $n = 10$.

$$z = \frac{\bar{x} - \mu_{\bar{x}}}{\sigma_{\bar{x}}}$$

$$3.10 = \frac{2100/n - 163}{\frac{32}{\sqrt{n}}\sqrt{\frac{210-n}{210-1}}}$$

6-6 Normal as Approximation to Binomial

1. Assuming the true proportion of households watching *60 Minutes* remains relative constantly during the two years of sampling, the two years of sample proportions represent samples from the same population. Since the sampling distribution of the proportion approximates a normal distribution, the histogram depicting the two years of sample proportions should be approximately normal – e.g., bell-shaped.

3. No. With n=4 and p=0.5, the requirements that $np \geq 5$ and $nq \geq 5$ are not met.

5. The area to the right of 8.5. In symbols, $P(x>8) = P_C(x>8.5)$.

7. The area to the left of 4.5. In symbols, $P(x<5) = P_C(x<4.5)$.

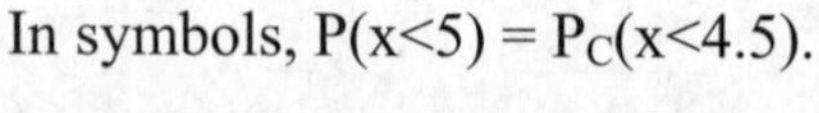

9. The area to the left of 15.5. In symbols, $P(x \leq 15) = P_C(x<15.5)$.

11. The area between 4.5 and 9.5. In symbols, $P(4 \leq x \leq 9) = P_C(4.5<x<9.5)$.

13. binomial: n=10 and p=0.5
    a. from Table A-1, $P(x=3) = 0.117$
    b. normal approximation appropriate since
        $np = 10(0.5) = 5.0 \geq 5$
        $nq = 10(0.5) = 5.0 \geq 5$
    $\mu = np = 10(0.5) = 5.0$
    $\sigma = \sqrt{npq} = \sqrt{10(0.5)(0.5)} = 1.581$
    $P(x=3)$
        $= P(2.5<x<3.5)$
        $= P(-1.58<z<-0.95)$
        $= 0.1711 - 0.0571$
        $= 0.1140$

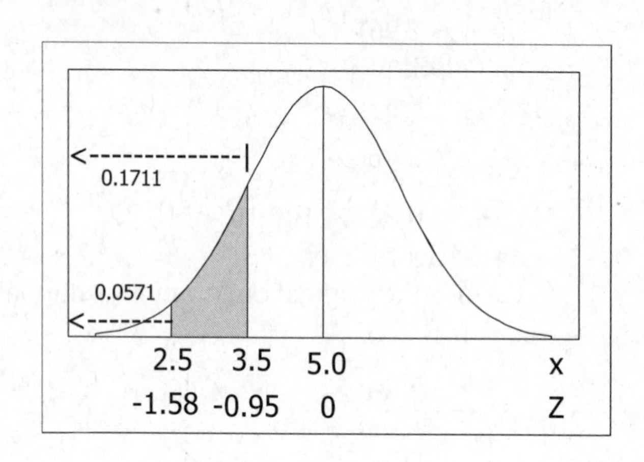

15. binomial: n=8 and p=0.9
    a. from Table A-1, $P(x \geq 6) = 0.149 + 0.383 + 0.430 = 0.962$
    b. normal approximation not appropriate since
        $nq = 8(0.1) = 0.8 < 5$

17. binomial: n=40,000 and p=0.03
    normal approximation appropriate since
        $np = 40{,}000(0.03) = 1200 \geq 5$
        $nq = 40{,}000(0.97) = 3880 \geq 5$
    $\mu = np = 40{,}000(0.03) = 1200$
    $\sigma = \sqrt{npq} = \sqrt{40000(0.03)(0.97)} = 34.117$
    $P(x \geq 1300)$
        $= P(x>1299.5)$
        $= P(z>2.92)$
        $= 1 - 0.9982$
        $= 0.0018$
    No. It is not likely that the goal of at least 1300 will be reached.

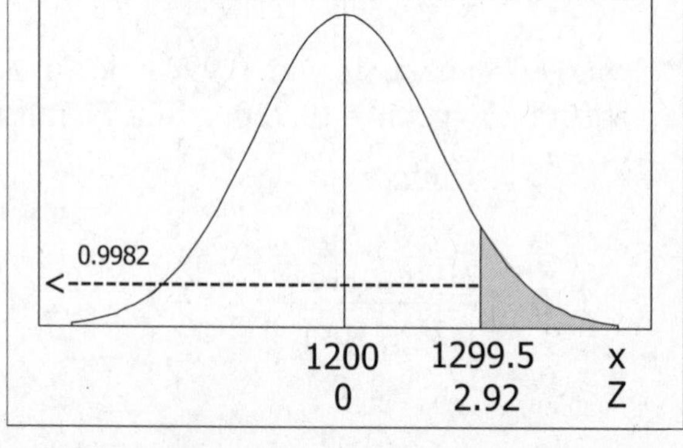

19. binomial: n=574 and p=0.50
normal approximation appropriate since
$np = 574(0.50) = 287 \geq 5$
$nq = 574(0.50) = 287 \geq 5$
$\mu = np = 574(0.50) = 287$
$\sigma = \sqrt{npq} = \sqrt{574(0.50)(0.50)} = 11.979$
$P(x \geq 525)$
$= P(x>524.5)$
$= P(z>19.83)$
$= 1 - 0.9999$
$= 0.0001$

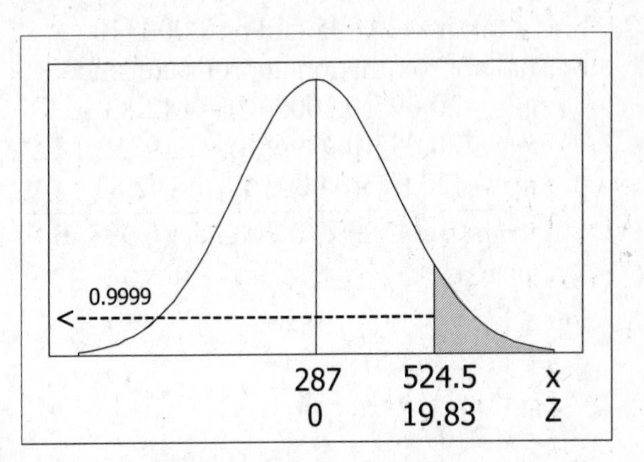

Yes. Since the probability of getting at least 525 girls by chance alone is so small, it appears that the method is effective and that the genders were not being determined by chance alone.

21. binomial: n=580 and p=0.25
normal approximation appropriate since
$np = 580(0.25) = 145 \geq 5$
$nq = 580(0.75) = 435 \geq 5$
$\mu = np = 580(0.25) = 145$
$\sigma = \sqrt{npq} = \sqrt{580(0.25)(0.75)} = 10.428$

a. $P(x=152)$
$= P(151.5<x<152.5)$
$= P(0.62<z<0.72)$
$= 0.7642 - 0.7324$
$= 0.0318$

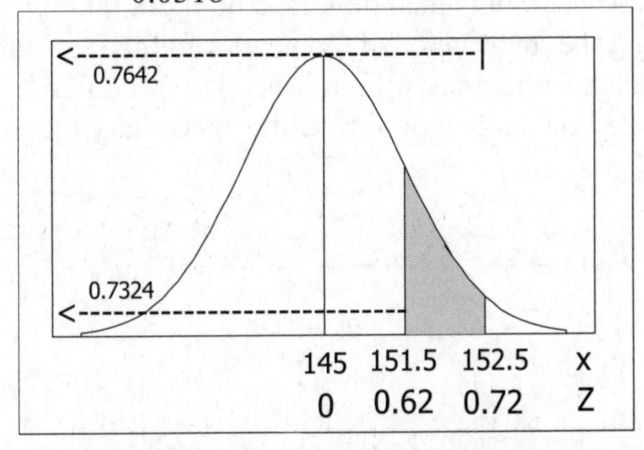

b. $P(x \geq 152)$
$= P(x>151.5)$
$= P(z>0.62)$
$= 1 - 0.7324$
$= 0.2676$

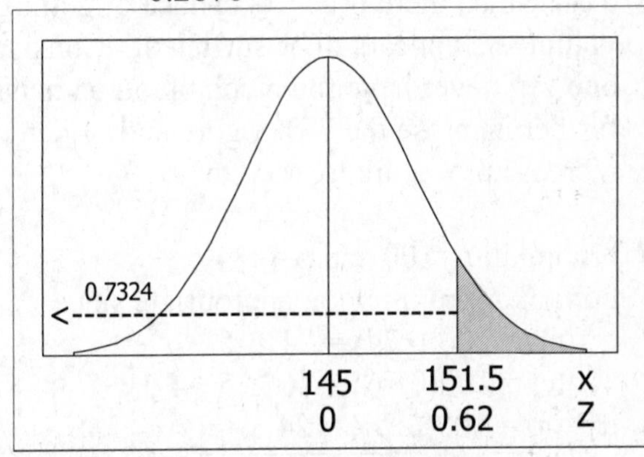

c. The part (b) answer is the useful probability. In situations involving multiple ordered outcomes, the unusualness of a particular outcome is generally determined by the probability of getting that outcome or a more extreme outcome.

d. No. Since 0,2676 > 0.05, Mendel's result could easily occur by chance alone if the true rate were really 0.25.

23. binomial: n=420,095 and p=0.000340
normal approximation appropriate since
$np = 420,095(0.000340) = 142.83 \geq 5$
$nq = 420,095(0.999660) = 419952.17 \geq 5$
$\mu = np = 420,095(0.000340) = 142.83$
$\sigma = \sqrt{npq} = \sqrt{420095(0.000340)(0.999660)}$
$= 11.949$
$P(x \leq 135)$
$= P(x < 135.5)$
$= P(z < -0.61)$
$= 0.2709$

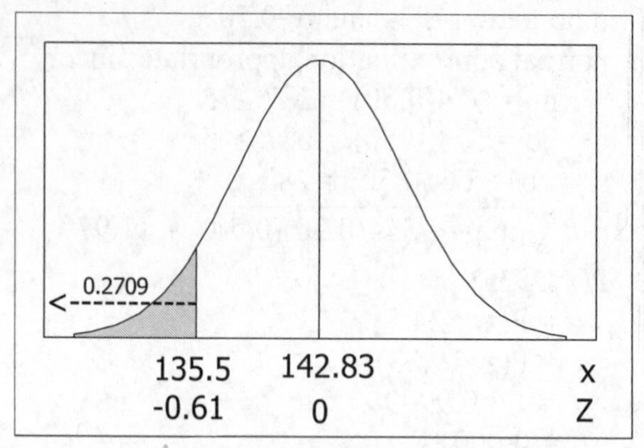

To conclude that cell phones increase the likelihood of experiencing such cancers requires $x > 142.83$. Since $135 < 142.83$, these results definitely do not support the media reports.

25. binomial: n=200 and p=0.06
normal approximation appropriate since
$np = 200(0.06) = 12 \geq 5$
$nq = 200(0.94) = 188 \geq 5$
$\mu = np = 200(0.06) = 12$
$\sigma = \sqrt{npq} = \sqrt{200(0.06)(0.94)} = 3.359$
$P(x \geq 10)$
$= P(x > 9.5)$
$= P(z > -0.74)$
$= 1 - 0.2296$
$= 0.7704$

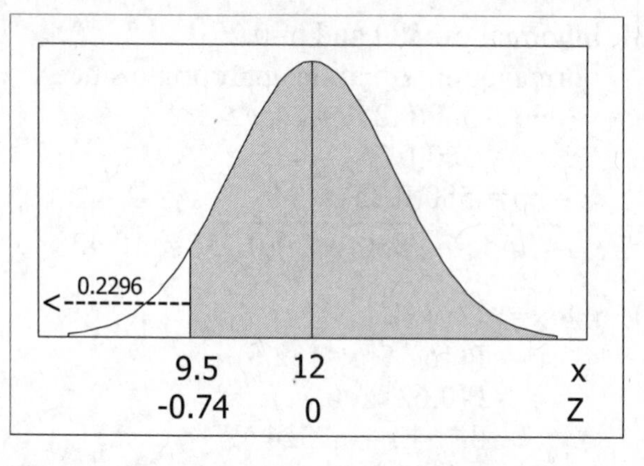

Yes. Since there is a 77% chance of getting at least 10 universal donors, a pool of 200 volunteers appears to be sufficient. Considering the importance of the need, and the fact that one can never have too much blood on hand, the hospital may want to use a larger pool to further increase the 77% figure and/or determine how large a pool would be necessary to increase the figure to, say, 95%.

27. binomial: n=100 and p=0.24
normal approximation appropriate since
$np = 100(0.24) = 24 \geq 5$
$nq = 100(0.76) = 76 \geq 5$
$\mu = np = 100(0.24) = 24$
$\sigma = \sqrt{npq} = \sqrt{100(0.24)(0.76)} = 4.271$
$P(x \geq 27)$
$= P(x > 26.5)$
$= P(z > 0.59)$
$= 1 - 0.7224$
$= 0.2776$

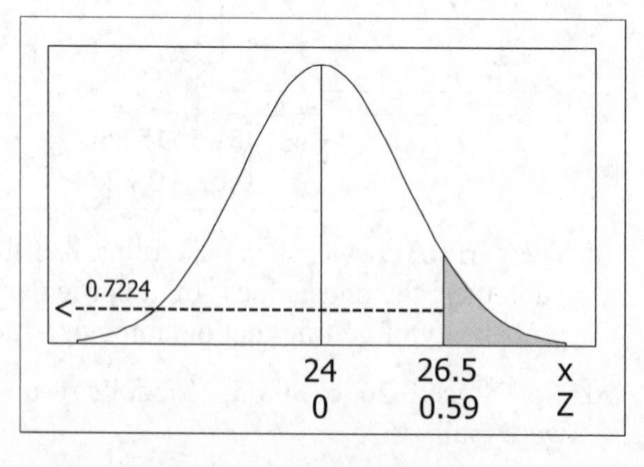

No. Since $0.2776 > 0.05$, 27 is not an unusually high number of blue M&M's.

29. binomial: n=863 and p=0.019
normal approximation appropriate since
$np = 863(0.019) = 16.397 \geq 5$
$nq = 863(0.981) = 846.603 \geq 5$
$\mu = np = 863(0.019) = 16.397$
$\sigma = \sqrt{npq} = \sqrt{863(0.019)(0.981)} = 4.011$
$P(x \geq 19)$
$= P(x>18.5)$
$= P(z>0.52)$
$= 1 - 0.6985$
$= 0.3015$

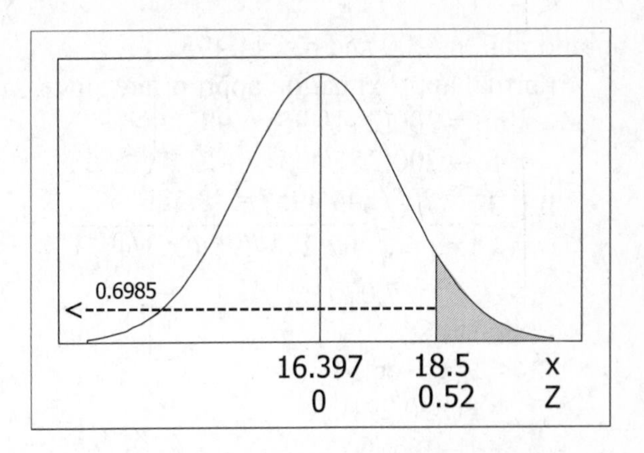

Since the 0.3015 > 0.05, 19 or more persons experiencing flu symptoms is not an unusual occurrence for a normal population. There is no evidence to suggest that flu symptoms are an adverse reaction to the drug.

31. Let x = the number that show up.
binomial: n=236 and p=0.9005
normal approximation appropriate since
$np = 236(0.9005) = 212.518 \geq 5$
$nq = 236(0.0995) = 23.482 \geq 5$
$\mu = np = 236(0.9005) = 212.518$
$\sigma = \sqrt{npq} = \sqrt{236(0.9005)(0.0995)} = 4.598$
$P(x>213)$
$= P(x>213.5)$
$= P(z>0.21)$
$= 1 - 0.5832$
$= 0.4168$

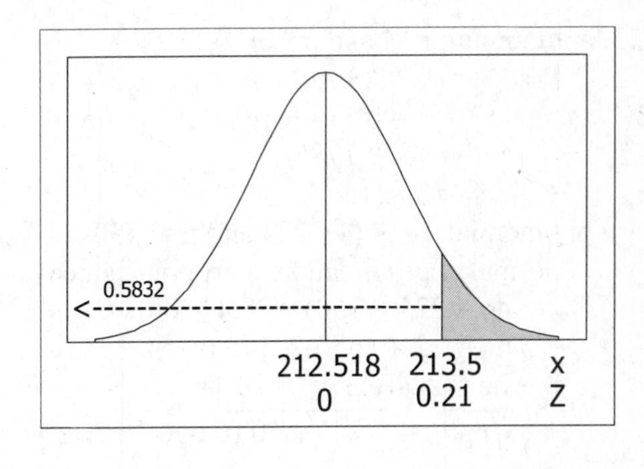

No. Since the 0.4168 > 0.05, overbooking is not an unusual event. It is a real concern for both the airline and the passengers.

33. Marc is placing 200(5) = \$1000 in bets. To make a profit his return must be more than \$1000. In part (a) each win returns the original 5 plus 35(5), for a total of 5 + 175 = \$180.
For a profit, he must win more than 1000/180 = 5.55 times – i.e., he needs at least 6 wins.
In part (b) each win returns the original 5 plus 1(5), for a total of 5 + 5 = \$10.
For a profit, he must win more than 1000/10 = 100 times – i.e., he needs at least 101 wins.

a. binomial: n=200 and p=1/38
normal approximation appropriate since
$np = 200(1/38) = 5.26 \geq 5$
$nq = 200(37/38) = 194.74 \geq 5$
$\mu = np = 200(1/38) = 5.263$
$\sigma = \sqrt{npq} = \sqrt{200(1/38)(37/38)} = 2.263$
$P(x \geq 6)$
$= P(x>5.5)$
$= P(z>0.10)$
$= 1 - 0.5398$
$= 0.4602$

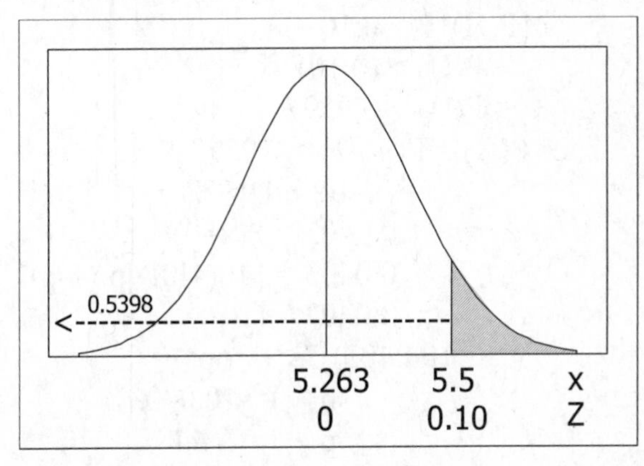

b. binomial: n=200 and p=244/495

normal approximation appropriate since

$np = 200(244/495) = 98.59 \geq 5$

$nq = 200(251/495) = 101.41 \geq 5$

$\mu = np = 200(244/495) = 98.586$

$\sigma = \sqrt{npq} = \sqrt{200(244/495)(251/495)}$

$= 7.070$

$P(x \geq 101)$

$= P(x>100.5)$

$= P(z>0.27)$

$= 1 - 0.6064$

$= 0.3936$

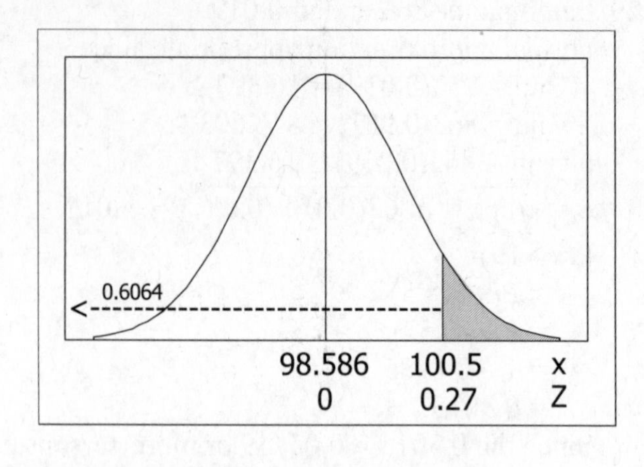

c. Since 0.4602 > 0.3936, the roulette game is the better "investment" – but since both probabilities are less than 0.5000, he would do better not to play at all.

35. a. binomial: n=4 and p=0.350

$P(x \geq 1) = 1 - P(x=0)$

$= 1 - [4!/(4!0!)](0.350)^0(0.650)^4$

$= 1 - 0.1785$

$= 0.8215$

b. binomial: n=56(4)=224 and p=0.350

normal approximation appropriate since

$np = 224(0.350) = 78.4 \geq 5$

$nq = 224(0.650) = 145.6 \geq 5$

$\mu = np = 224(0.350) = 78.4$

$\sigma = \sqrt{npq} = \sqrt{224(0.350)(0.650)} = 7.139$

$P(x \geq 56)$

$= P(x>55.5)$

$= P(z>-3.20)$

$= 1 - 0.0007$

$= 0.9993$

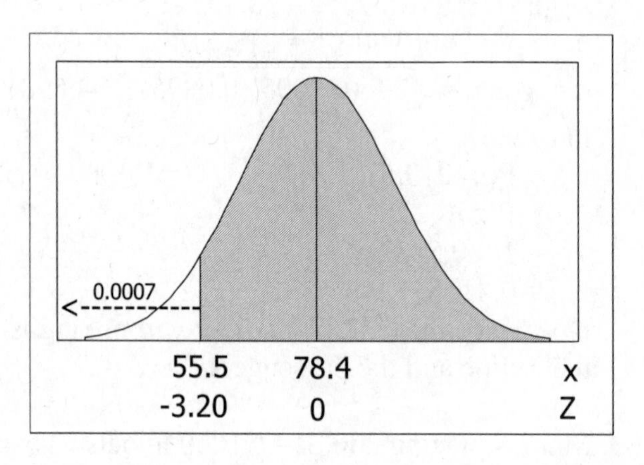

c. Let H = getting at least one hit in 4 at bats.

P(H) = 0.8215 [from part (a)]

For 56 consecutive games, $[P(H)]^{56} = [0.8125]^{56} = 0.0000165$

d. The solution below employs the techniques and notation of parts (a) and (c).

for $[P(H)]^{56} > 0.10$

$P(H) > (0.10)^{1/56}$

$P(H) > 0.9597$

for $P(H) = P(x \geq 1) > 0.9597$

$1 - P(x=0) > 0.9597$

$0.0403 > P(x=0)$

$0.0403 > [4!/(4!0!)]p^0(1-p)^4$

$0.0403 > (1-p)^4$

$(0.0403)^{1/4} > 1 - p$

$p > 1 - (0.0403)^{1/4}$

$p > 1 - 0.448$

$p > 0.552$

6-7 Assessing Normality

1. A normal quantile plot can be used to determine whether sample data come form a normal distribution. In theory, it compares the z scores for the sample data with the z scores for normally distributed data with the same cumulative relative frequencies as the sample data. In practice, it uses the sample data directly – since converting to z scores is a linear transformation that re-labels the scores but does not change their distribution.

3. Because the weights are likely to follow a normal distribution, on expects the points in a normal quantile plot to approximate a straight line.

5. No, the data do not appear to come from a population with a normal distribution. The points are not reasonably close to a straight line.

7. Yes, the data appear to come from a population with a normal distribution. The points are reasonably close to a straight line.

9. Yes, the data appear to come from a population with a normal distribution. The frequency distribution and histogram indicate that the data is approximately bell-shaped.

duration (hours)	frequency
0 – 49	1
50 – 99	8
100 – 149	18
150 – 199	23
200 – 249	25
250 – 299	19
300 – 349	11
350 – 399	8
400 – 449	2
	115

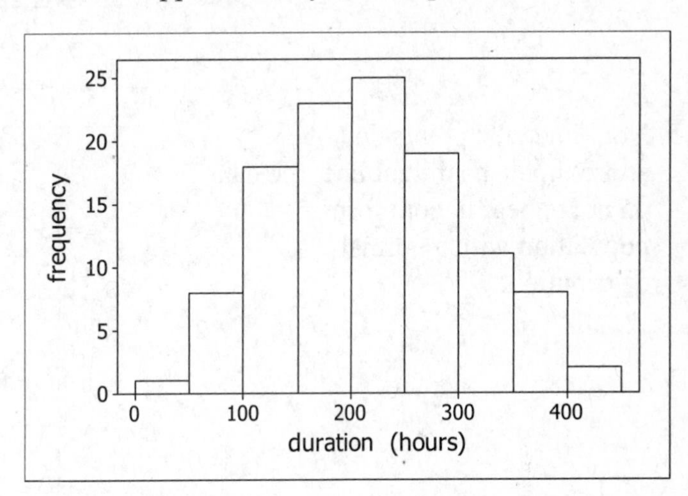

11. No, the data do not appear to come from a population with a normal distribution. The frequency distribution and the histogram indicate there is a concentration of data at the lower end.

degree days	frequency
0 – 499	20
500 – 999	5
1000 – 1499	8
1500 – 1999	9
2000 – 2499	5
2500 – 2599	1
	48

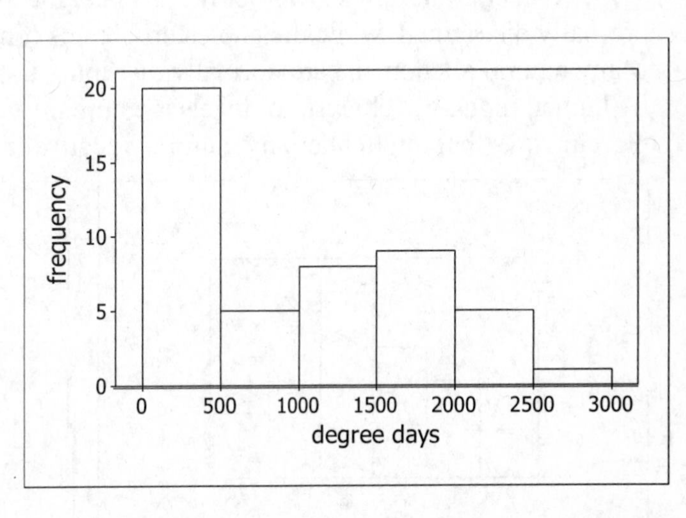

NOTE: The normal quantile plots for Exercises 13-16 may be obtained directly from many sources or constructed using Minitab for n scores in C1 using the commands at the right, where (2n-1) and (2n) are the actual values, and then plotting C1 on the x-axis and C4 on the y-axis. See Exercises 19 and 20 for more detail on the mechanics of this process, which is the five-step "manual construction" process given in the text.

```
MTB> Sort C1 C1
MTB> Set C2
DATA> 1:(2n-1)/2
DATA> end
MTB> Let C3 = C2/(2n)
MTB> INVCDF C3 C4
```

13. Yes. Since the points approximate a straight line, the data appear to come from a population with a normal distribution. The gaps/groupings in the durations may reflect the fact that the times have to reflect whole numbers of orbits or other physical constraints.

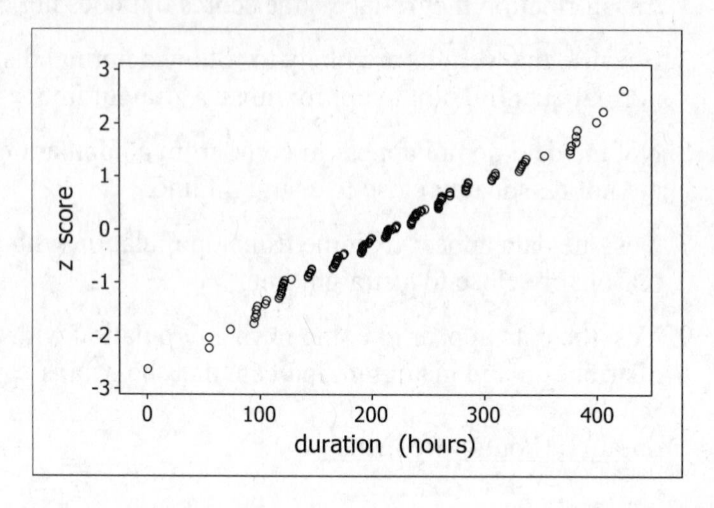

15. No. Since the points do not approximate a straight line, the data do not appear to com form a population with a normal distribution.

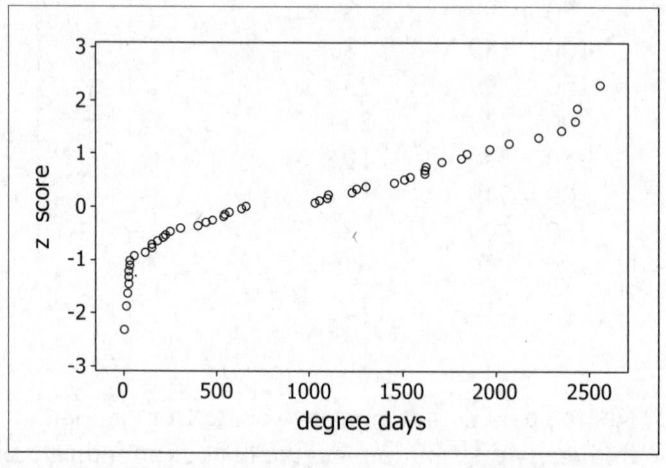

17. The two histograms are shown below. The heights (on the left) appear to be approximately normally distributed, while the cholesterol levels (on the right) appear to be positively skewed. Many natural phenomena are normally distributed. Height is a natural phenomenon unaffected by human input, but cholesterol levels are humanly influenced (by diet, exercise, medication, etc.) in ways that might alter any naturally occurring distribution.

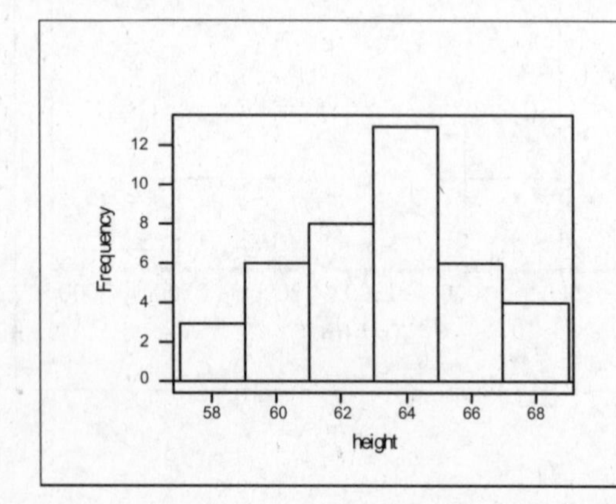

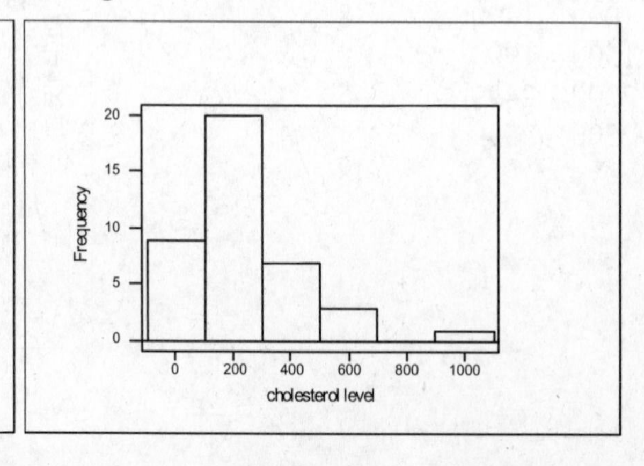

19. The corresponding z scores in the table below were determined following the five-step "manual construction" procedure of the text.
    (1) Arrange the n scores in order and place them in the x column.
    (2) For each x_i, calculate the cumulative probability using $cp_i = (2i-1)/2n$ for $i = 1,2,\ldots,n$.
    (3) For each cp_i, find the z_i for which $P(z<z_i) = cp_i$ for $i = 1,2,\ldots,n$.

    The resulting normal quantile plot indicates that the data appear to come from a population with a normal distribution.

i	x	cp	z
1	127	0.10	-1.28
2	129	0.30	-0.52
3	131	0.50	0
4	136	0.70	0.52
5	146	0.90	1.28

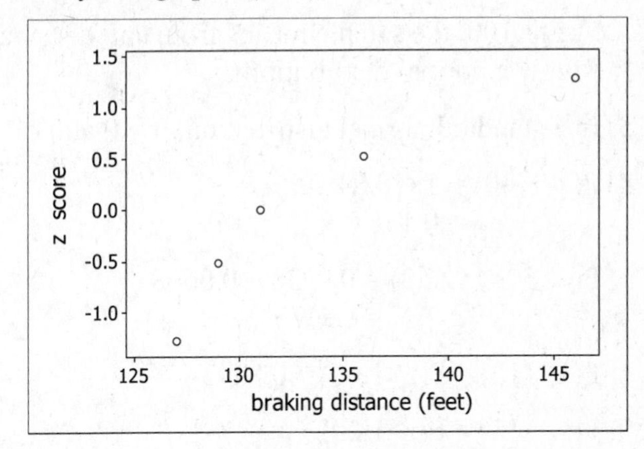

21. a. Yes. Adding two inches to each height shifts the entire distribution to the right, but it does not affect the shape of the distribution.
    b. Yes. Changing to a different unit measure re-labels the horizontal axis, but it does not change relationships between the data points or affect the shape of the distribution.
    c. No. Unlike a linear transformation of the form f(x) = ax + b, the log function f(x) = log(x) does not have the same effect on all segments of the horizontal axis.

Statistical Literacy and Critical Thinking

1. A normal distribution is one that is symmetric and bell-shaped. More technically, it is one that can be described by the following formula, where μ and σ are specified values,

$$f(x) = \frac{e^{-\frac{1}{2\sigma^2}(x-\mu)^2}}{\sigma\sqrt{2\pi}}.$$

   A standard normal distribution is a normal distribution that has a mean of 0 and a standard deviation of 1. More technically, it is the distribution that results when μ = 0 and σ = 1 in the above formula.

2. In statistics, a normal distribution is one that is symmetric and bell-shaped. This statistical use of the term "normal" is not to be confused with the common English word "normal" in the sense of "typical."

3. Assuming there are no trends over time in the lengths of movies, and that there is mean length that remains fairly constant from year to year, the sample means will follow a normal distribution centered around that enduring mean length.

4. Not necessarily. Depending on how the survey was conducted, the sample may be a convenience sample (if AOL simply polled its own customers) or a voluntary response sample (if the responders were permitted decide for themselves whether or not to participate).

Chapter Quick Quiz

1. The symbol $z_{0.03}$ represents the z score with 0.0300 below it. According to Table A-1 [closest value is 0.0301], this is -1.88.

2. For n=100, the sample means from any distribution with finite mean and standard deviation follow a normal distribution.

3. In a standard normal distribution, $\mu = 0$ and $\sigma = 1$.

4. $P(z>1.00) = 1 - 0.8413$
   $= 0.1587$

5. $P(-1.50<z<2.50) = 0.9938 - 0.0668$
   $= 0.9270$

6. $P(x<115) = P(z<1.00) = 0.8413$

7. $P(x>118) = P(z>1.20)$
   $= 1 - 0.8849$
   $= 0.1151$

8. $P(88<x<112) = P(-0.80<z<0.80)$
   $= 0.7881 - 0.2119$
   $= 0.5762$

9. For n=25, $\mu_{\bar{x}} = \mu = 100$ and $\sigma_{\bar{x}} = \sigma/\sqrt{n} = 15/\sqrt{25} = 3$.
   $P(\bar{x}<103) = P(z<1.00) = 0.8413$

10. For n=100, $\mu_{\bar{x}} = \mu = 100$ and $\sigma_{\bar{x}} = \sigma/\sqrt{n} = 15/\sqrt{100} = 1.5$.
    $P(\bar{x}>103) = P(z>2.00)$
    $= 1 - 0.9772$
    $= 0.0228$

Review Exercises

1. a. normal distribution: $\mu = 69.0$ and $\sigma = 2.8$
   $P(x>75)$
   $= P(z>2.14)$
   $= 1 - 0.9838$
   $= 0.0162$ or 1.62%

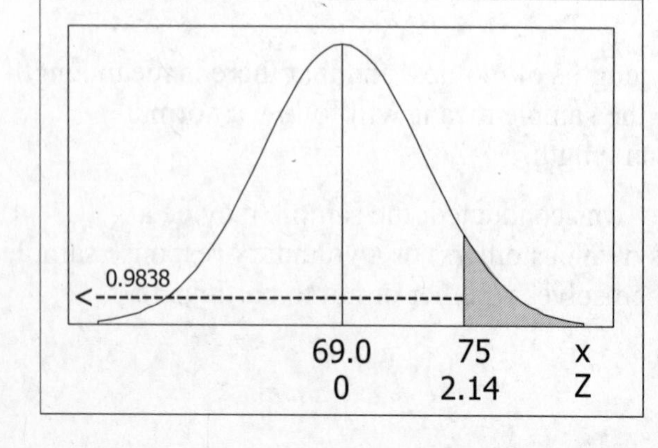

   b. normal distribution: $\mu = 63.6$ and $\sigma = 2.5$
   $P(x>75)$
   $= P(z>4.56)$
   $= 1 - 0.9999$
   $= 0.0001$ or 0.01%

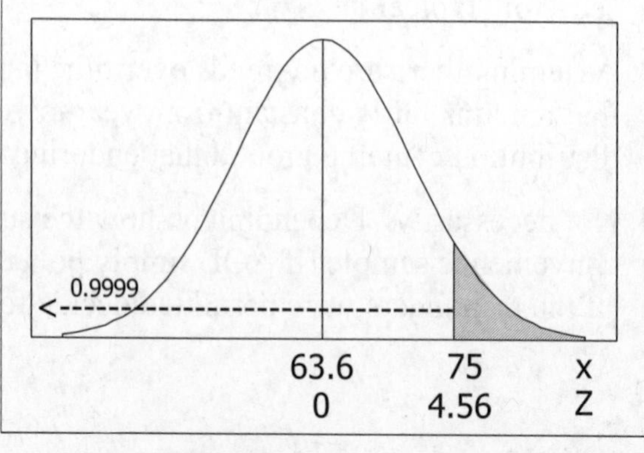

c. The length of a day bed appears to be adequate to meet the needs of all but the very tallest men and women.

2. For the tallest 5%, A = 0.9500 and z = 1.645

$x = \mu + z\sigma$
$= 69.0 + (1.645)(2.8)$
$= 69.0 + 4.6$
$= 73.6$ inches

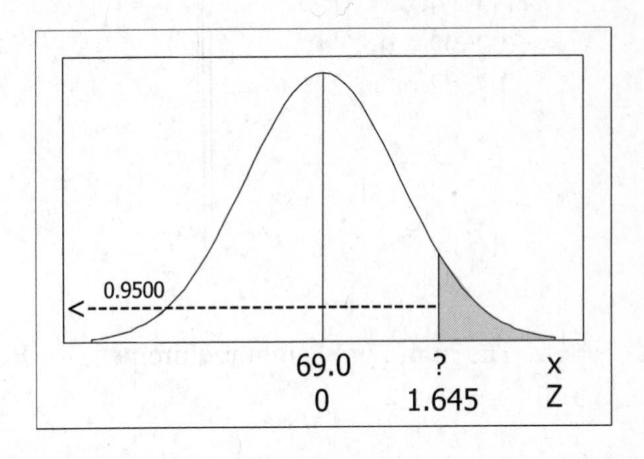

3. a. normal distribution: $\mu = 69.0$ and $\sigma = 2.8$

$P(x>78)$
$= P(z>3.21)$
$= 1 - 0.9993$
$= 0.0007$ or 0.07% of the men

normal distribution: $\mu = 63.6$ and $\sigma = 2.5$

$P(x>78)$
$= P(z>5.76)$
$= 1 - 0.9999$
$= 0.0001$ or 0.01% of the women

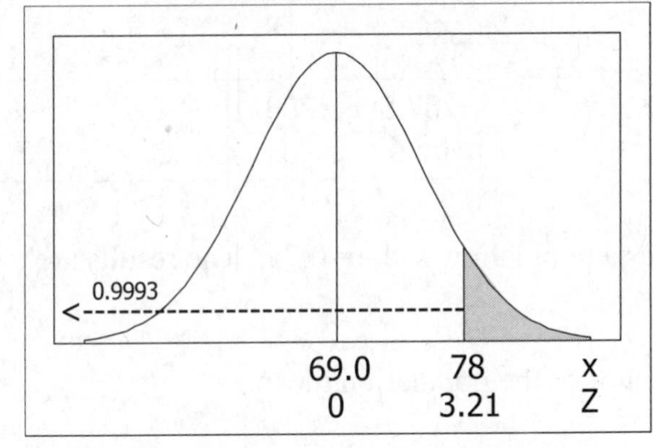

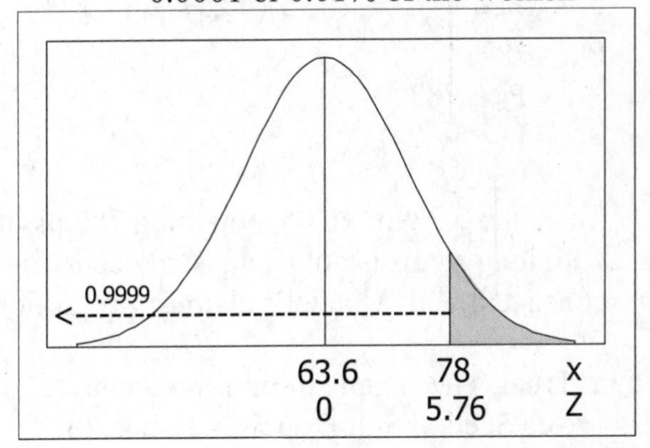

b. For the tallest 1%, A = 0.9900 [0.9901] and z = 2.33

$x = \mu + z\sigma$
$= 69.0 + (2.33)(2.8)$
$= 69.0 + 6.5$
$= 75.5$ inches

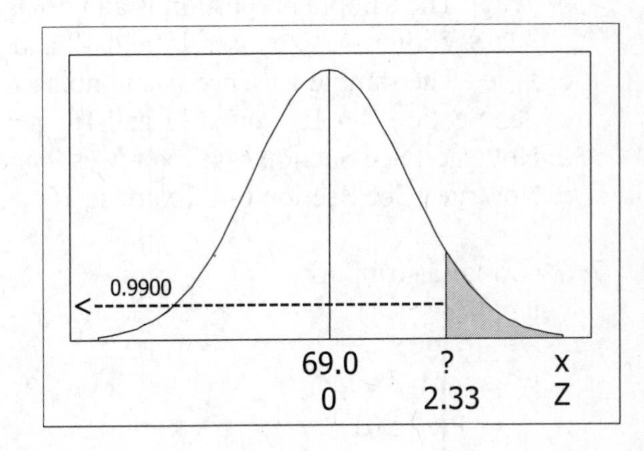

4. normal distribution: $\mu = 63.6$ and $\sigma = 2.5$
   $P(66.5<x<71.5)$
   $= P(1.16<z<3.16)$
   $= 0.9992 - 0.8770$
   $= 0.1222$ or 12.22%

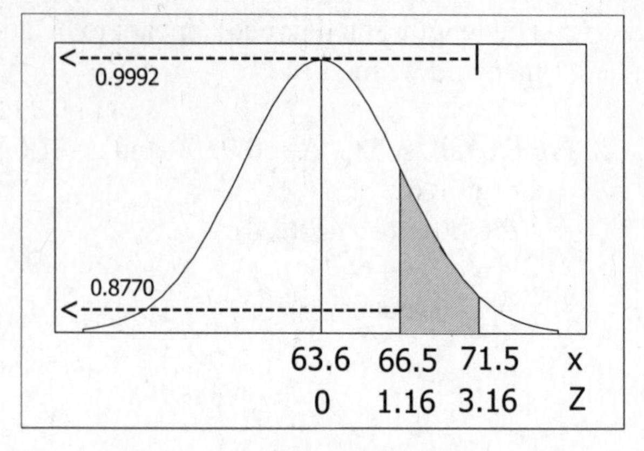

   Yes. The minimum height requirement for Rockettes is greater than the mean height of all women.

5. binomial: n=1064 and p=0.75
   normal approximation appropriate since
   $np = 1064(0.75) = 798 \geq 5$
   $nq = 1064(0.25) = 266 \geq 5$
   $\mu = np = 1064(0.75) = 798$
   $\sigma = \sqrt{npq} = \sqrt{1064(0.75)(0.25)} = 14.124$
   $P(x \leq 787)$
   $= P(x<787.5)$
   $= P(z<-0.74)$
   $= 0.2296$
   Since the $0.2296 > 0.05$, obtaining 787 plants with long stems is not an unusual occurrence for a population with p=0.75. The results are consistent with Mendel's claimed proportion.

6. a. True. The sample mean is an unbiased estimator of the population mean. See Section 6-4, Exercises 12 and 13.
   b. True. The sample proportion is an unbiased estimator of the population proportion. See Section 6-4, Exercises 17 and 18 and 19.
   c. True. The sample variance is an unbiased estimator of the population variance. See Section 6-4, Exercises 11 and 16.
   d. Not true. See Section 6-4, Exercises 9 and 14.
   e. Not true. See Section 6-4, Exercise 15.

7. a. normal distribution
   $\mu = 178.1$
   $\sigma = 40.7$
   $P(x>260)$
   $= P(z>2.01)$
   $= 1 - 0.9778$
   $= 0.0222$

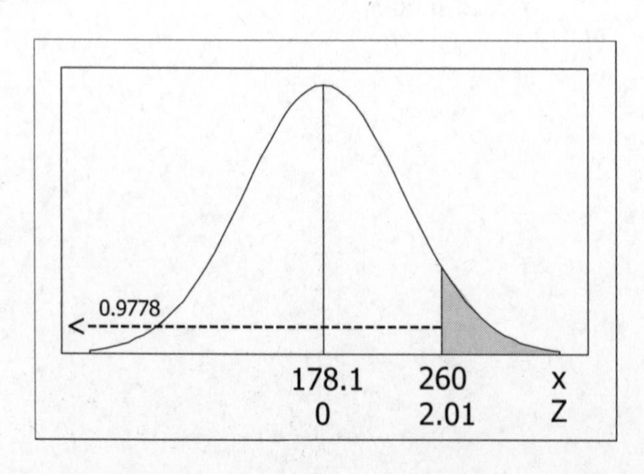

b. normal distribution
$\mu = 178.1$
$\sigma = 40.7$
$P(170<x<200)$
$= P(-0.20<z<0.54)$
$= 0.7054 - 0.4207$
$= 0.2847$

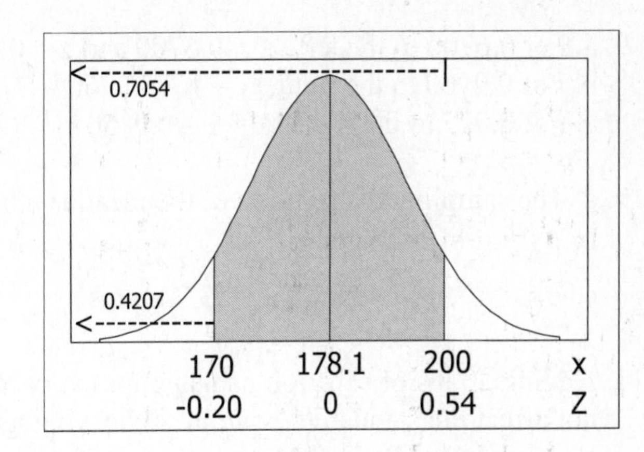

c. normal distribution,
since the original distribution is so
$\mu_{\bar{x}} = \mu = 178.1$
$\sigma_{\bar{x}} = \sigma/\sqrt{n} = 40.7/\sqrt{9} = 13.567$
$P(170<\bar{x}<200)$
$= P(-0.60<z<1.61)$
$= 0.9463 - 0.2743$
$= 0.6720$

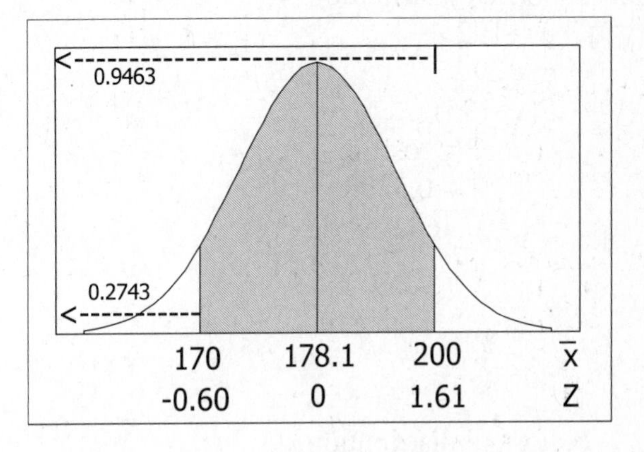

d. For the top 3%,
A = 0.9700 [0.9699] and z = 1.88
$x = \mu + z\sigma$
$= 178.1 + (1.88)(40.7)$
$= 178.1 + 76.5$
$= 254.6$ mg/100mL

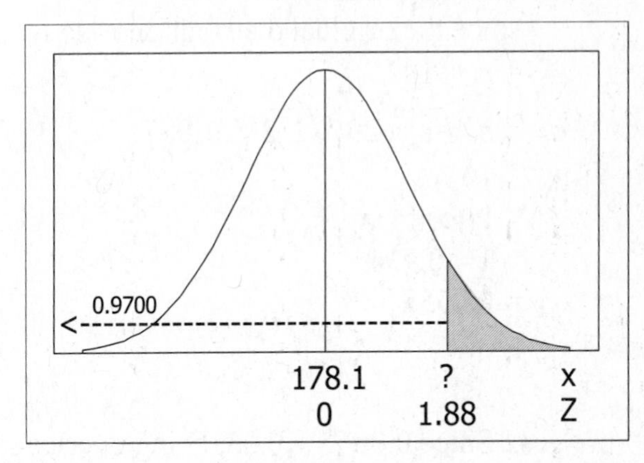

8. binomial: n=40 and p=0.50
normal approximation appropriate since
$np = 40(0.50) = 20 \geq 5$
$nq = 40(0.50) = 20 \geq 5$
$\mu = np = 40(0.50) = 20$
$\sigma = \sqrt{npq} = \sqrt{40(0.50)(0.50)} = 3.162$
$P(x \leq 15)$
$= P(x<15.5)$
$= P(z<-1.42)$
$= 0.0778$

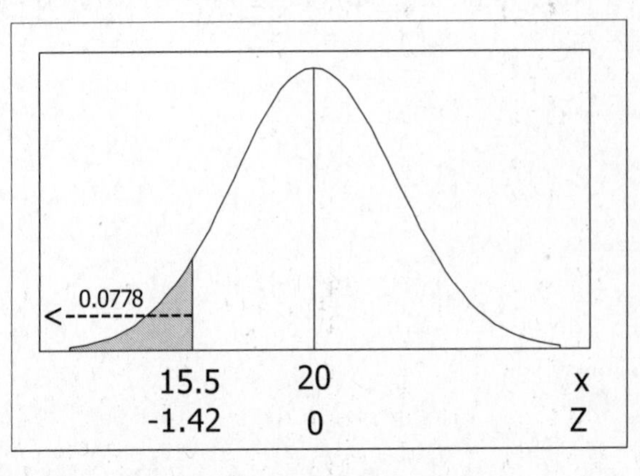

Since the 0.0778 > 0.05, obtaining 15 or fewer women is not an unusual occurrence for a population with p=0.75. No. There is not strong evidence to charge gender discrimination.

9. a. For 0.6700 to the left, A = 0.6700 and z = 0.44.
   b. For 0.9960 to the right, A = 1 – 0.9960 = 0.0040 and z = -2.65.
   c. For 0.025 to the right, A = 1 – 0.0250 = 0.9750 and z = 1.96.

10. a. The sampling distribution of the mean is a normal distribution.
   b. $\mu_{\bar{x}} = \mu = 3420$ grams
   c. $\sigma_{\bar{x}} = \sigma/\sqrt{n} = 495/\sqrt{85} = 53.7$ grams

11. Adding 20 lbs of carry-on baggage for every male will increase the mean weight by 20 but will not affect the standard deviation of the weights.
   a. normal distribution
   $\mu = 192$
   $\sigma = 29$
   $P(x>195)$
   $= P(z>0.10)$
   $= 1 - 0.5398$
   $= 0.4602$

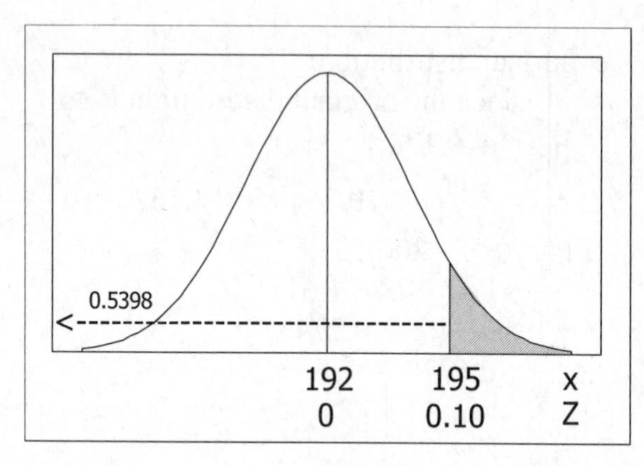

   b. normal distribution,
   since the original distribution is so
   $\mu_{\bar{x}} = \mu = 192$
   $\sigma_{\bar{x}} = \sigma/\sqrt{n} = 29/\sqrt{213} = 1.987$
   $P(\bar{x}>195)$
   $= P(z>1.51)$
   $= 1 - 0.9345$
   $= 0.0655$

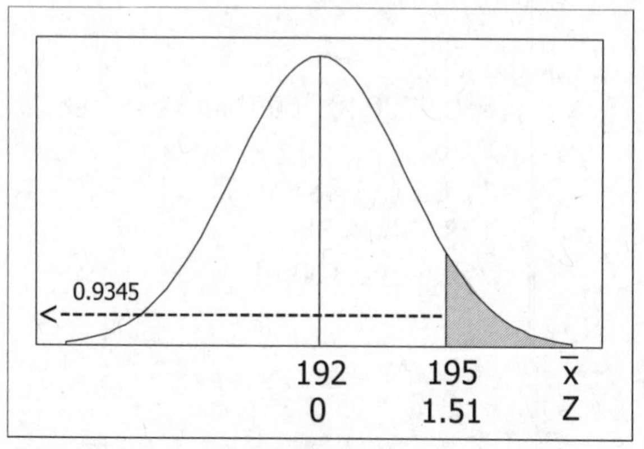

   Yes. Since 0.0655 > 0.05, being overloaded under those circumstances (i.e., all passengers are male and carrying 20 lbs of carry-on luggage) would not be unusual.

12. For review purposes, this exercise is worked in detail using a frequency distribution, a frequency histogram, and a normal quantile plot.

a. Using a frequency distribution and a frequency histogram.

weight (grams)	frequency
7.9500 – 7.9999	1
8.0000 – 8.0499	5
8.0500 – 8.0999	8
8.1000 – 8.1499	6
	20

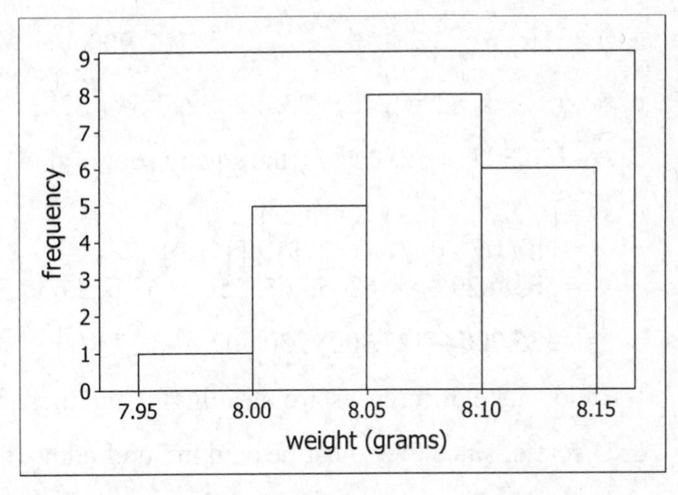

b. Using a normal quantile plot. Using the five-step "manual construction of a normal quantile plot" given in the text, the corresponding z scores in the table below were determined as follows.

(1) Arrange the n=20 scores in order and place them in the x column.

(2) For each x_i, calculate the cumulative probability using $cp_i = (2i-1)/2n$ for $i = 1,2,\ldots,n$.

(3) For each cp_i, find the z_i for which $P(z<z_i) = cp_i$ for $i = 1,2,\ldots,n$.

i	x	cp	z
1	7.9817	0.025	-1.96
2	8.0241	0.075	-1.44
3	8.0271	0.125	-1.15
4	8.0307	0.175	-0.93
5	8.0342	0.225	-0.76
6	8.0345	0.275	-0.60
7	8.0510	0.325	-0.45
8	8.0538	0.375	-0.32
9	8.0658	0.425	-0.19
10	8.0719	0.475	-0.06
11	8.0775	0.525	0.06
12	8.0813	0.575	0.19
13	8.0894	0.625	0.32
14	8.0954	0.675	0.45
15	8.1008	0.725	0.60
16	8.1041	0.775	0.76
17	8.1072	0.825	0.93
18	8.1238	0.875	1.15
19	8.1281	0.925	1.44
20	8.1384	0.975	1.96

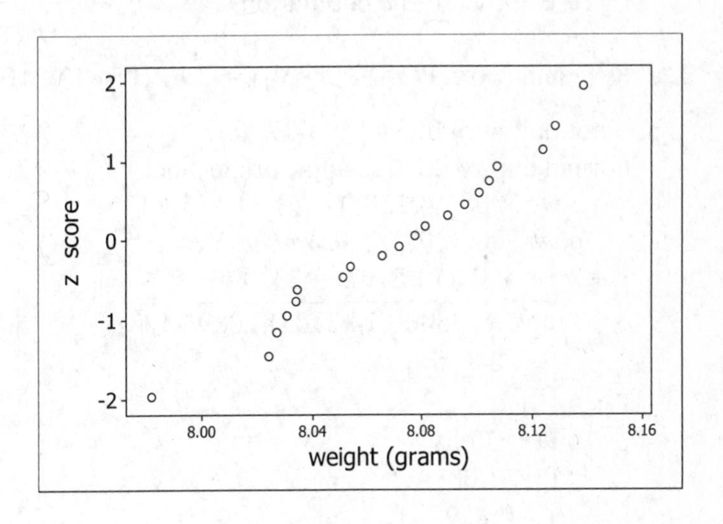

Yes. The weights appear to come from a population that has a normal distribution. The frequency distribution and the histogram suggest a bell-shaped distribution, and the points on the normal quantile plot are reasonably close to a straight line.

Cumulative Review Exercises

1. Arranged in order, the values are: 125 128 138 159 212 235 360 492 530 900
   summary statistics: $n = 10$ $\Sigma x = 3279$ $\Sigma x^2 = 1639067$

   a. $\bar{x} = (\Sigma x)/n = 3279/10 = 327.9$ or \$327,900

   b. $\tilde{x} = (x_5 + x_6)/2 = (212 + 235)/2 = 223.5$ or \$223,500

   c. $s = 250.307$ or \$250,307 [the square root of the answer given in part d]

   d. $s^2 = [n(\Sigma x^2) - (\Sigma x)^2]/[n(n-1)]$
   $= [10(1639067) - (3279)^2]/[10(9)]$
   $= 5638829/90 = 62653.655556$ or 62,653,655,556 dollars^2

   e. $z = (235{,}000 - 327{,}900)/250{,}307 = -0.37$

   f. Ratio, since differences are meaningful and there is a meaningful zero.

   g. Discrete, since they must be paid in hundredths of dollars (i.e., in whole cents).

2. a. A simple random sample of size n is a sample selected in such a way that every sample of size n has the same chance of being selected.

   b. A voluntary response sample is one for which the respondents themselves made the decision and effort to be included. Such samples are generally unsuited for statistical purposes because they are typically composed of persons with strong feelings on the topic and are not representative of the population.

3. a. $P(V_1 \text{ and } V_2) = P(V_1)\cdot P(V_2|V_1) = (14/2103)(13/2102) = 0.0000412$

   b. binomial: $n=5000$ and $p=14/2103$
   normal approximation appropriate since
   $np = 5000(14/2103) = 33.29 \geq 5$
   $nq = 5000(2089/2103) = 4966.71 \geq 5$
   $\mu = np = 5000(14/2103) = 33.286$
   $\sigma = \sqrt{npq} = \sqrt{5000(14/2103)(2089/2103)}$
   $= 5.750$
   $P(x \geq 40)$
   $= P(x > 39.5)$
   $= P(z > 1.08)$
   $= 1 - 0.8599$
   $= 0.1401$

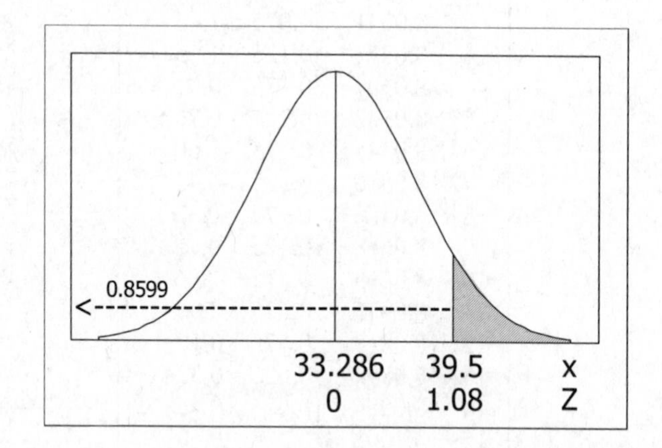

   c. No. Since $0.1401 > 0.05$, 40 is not an unusually high number of viral infections.

   d. No. To determine whether viral infections are an adverse reaction to using Nasonex, we would need to compare to $14/2103 = 0.00666$ rate for Nasonex users to the rate in the general population – and that information is not given.

4. No. By not beginning the vertical scale at zero, the graph exaggerates the differences.

5. a. Let L = a person is left-handed.
   $P(L) = 0.10$, for each random selection
   $P(L_1 \text{ and } L_2 \text{ and } L_3) = P(L_1)\cdot P(L_2)\cdot P(L_3)$
   $= (0.10)(0.10)(0.10)$
   $= 0.001$

   b. Let N = a person is not left-handed.
   $P(N) = 0.90$, for each random selection
   P(at least one left-hander) = 1 – P(no left-handers)
   $= 1 - P(N_1 \text{ and } N_2 \text{ and } N_3)$
   $= 1 - P(N_1)\cdot P(N_2)\cdot P(N_3)$
   $= 1 - (0.90)(0.90)(0.90)$
   $= 1 - 0.729$
   $= 0.271$

   c. binomial: n=3 and p=0.10
   normal approximation not appropriate since
   $np = 3(0.10) = 3 < 5$

   d. binomial: n=50 and p=0.10
   $\mu = np = 50(0.10) = 5$

   e. binomial problem: n=50 and p=0.10
   $\sigma = \sqrt{npq} = \sqrt{50(0.10)(0.90)} = 2.121$

   f. There are two previous approaches that may be used to answer this question.
   (1) An unusual score is one that is more than two standard deviations from the mean. Use the values for μ and σ from parts (d) and (e).
   $z = (x - \mu)/\sigma$
   $z_8 = (8 - 5)/2.121$
   $= 1.41$
   Since 8 is 1.41<2 standard deviations from the mean, it would not be an unusual result.
   (2) A score is unusual if the probability of getting that result or a more extreme result is less than or equal to 0.05.
   binomial: n = 50 and p = 0.10
   normal approximation appropriate since
   $np = 50(0.10) = 5 \geq 5$
   $nq = 50(0.90) = 45 \geq 5$
   Use the values for μ and σ from parts (d) and (e).
   $P(x \geq 8)$
   $= P(x > 7.5)$
   $= P(z > 1.18)$
   $= 1 - 0.8810$
   $= 0.1190$

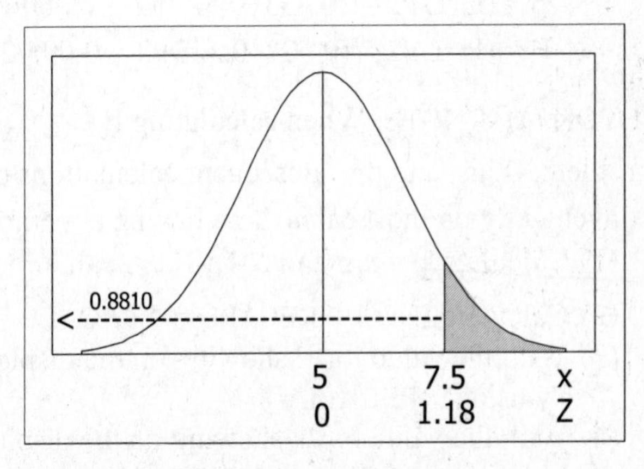

Since 0.1190 > 0.05, getting 8 left-handers in a group of 50 is not an unusual event.

Chapter 7

Estimates and Sample Sizes

7-2 Estimating a Population Proportion

1. The confidence level was not stated. The most common level of confidence is 95%, and sometimes that level is carelessly assumed without actually being stated.

3. By including a statement of the maximum likely error, a confidence interval provides information about the accuracy of an estimate.

5. For 99% confidence, $\alpha = 1–0.99 = 0.01$ and $\alpha/2 = 0.01/2 = 0.005$.
   For the upper 0.005, A = 0.9950 and z = 2.575.
   $z_{\alpha/2} = z_{0.005} = 2.575$

7. For $\alpha = 0.10$, $\alpha/2 = 0.10/2 = 0.05$.
   For the upper 0.05, A = 0.9500 and z = 1.645.
   $z_{\alpha/2} = z_{0.05} = 1.645$

9. Let L = the lower confidence limit; U = the upper confidence limit.
   $\hat{p} = (L+U)/2 = (0.200+0.500)/2 = 0.700/2 = 0.350$
   $E = (U–L)/2 = (0.500–0.200)/2 = 0.300/2 = 0.150$
   The interval can be expressed as 0.350 ± 0.150.

11. Let L = the lower confidence limit; U = the upper confidence limit.
    $\hat{p} = (L+U)/2 = (0.437+0.529)/2 = 0.966/2 = 0.483$
    $E = (U–L)/2 = (0.529–0.437)/2 = 0.092/2 = 0.046$
    The interval can be expressed as 0.483 ± 0.046.

13. Let L = the lower confidence limit; U = the upper confidence limit.
    $\hat{p} = (L+U)/2 = (0.320+0.420)/2 = 0.740/2 = 0.370$
    $E = (U–L)/2 = (0.420–0.320)/2 = 0.100/2 = 0.050$

15. Let L = the lower confidence limit; U = the upper confidence limit.
    $\hat{p} = (L+U)/2 = (0.433+0.527)/2 = 0.960/2 = 0.480$
    $E = (U–L)/2 = (0.527–0.433)/2 = 0.094/2 = 0.047$

IMPORTANT NOTE: When calculating $E = z_{\alpha/2}\sqrt{\hat{p}\hat{q}/n}$ do not round off in the middle of the problem. This, and the subsequent calculations of $U = \hat{p} + E$ and $L = \hat{p} - E$ may accomplished conveniently on most calculators having a memory as follows.

(1) Calculate $\hat{p} = x/n$ and STORE the value.
(2) Calculate E as 1 – RECALL = * RECALL = ÷ n = $\sqrt{\ }$ * $z_{\alpha/2}$ =
(3) With the value for E showing on the display, the upper confidence limit U can be calculated by using + RECALL =.
(4) With the value for U showing on the display, the lower confidence limit L can be calculated by using – RECALL ± + RECALL.

THE MANUAL USES THIS PROCEDURE, AND ROUNDS THE FINAL ANSWER TO 3 SIGNIFICANT DIGITS, EVEN THOUGH IT REPORTS INTERMEDIATE STEPS WITH A FINITE

NUMBER OF DECIMAL PLACES. If the above procedure does not work on your calculator, or to find out if some other procedure would be more efficient on your calculator, ask your instructor for assistance. You must become familiar with your own calculator – and be sure to do your homework on the same calculator you will use for the exams.

17. $\alpha = 0.05$ and $z_{\alpha/2} = z_{0.025} = 1.96$; $\hat{p} = x/n = 400/1000 = 0.40$

$E = z_{\alpha/2}\sqrt{\hat{p}\hat{q}/n} = 1.96\sqrt{(0.40)(0.60)/1000} = 0.0304$

19. $\alpha = 0.02$ and $z_{\alpha/2} = z_{0.01} = 2.33$; $\hat{p} = x/n = [492]/1230 = 0.40$

$E = z_{\alpha/2}\sqrt{\hat{p}\hat{q}/n} = 2.33\sqrt{(0.40)(0.60)/1230} = 0.0325$

NOTE: The value x=[492] was not given. In truth, any $486 \le x \le 498$ rounds to the given $\hat{p} = x/1230 = 40\%$. For want of a more precise value, $\hat{p} = 0.40$ is used in the calculation of E.

21. $\alpha = 0.05$ and $z_{\alpha/2} = z_{0.025} = 1.96$; $\hat{p} = x/n = 40/200 = 0.2000$

$\hat{p} \pm z_{\alpha/2}\sqrt{\hat{p}\hat{q}/n}$

$0.2000 \pm 1.96\sqrt{(0.2000)(0.8000)/200}$

0.2000 ± 0.0554

$0.145 < p < 0.255$

23. $\alpha = 0.01$ and $z_{\alpha/2} = z_{0.005} = 2.575$; $\hat{p} = x/n = 109/1236 = 0.0882$

$\hat{p} \pm z_{\alpha/2}\sqrt{\hat{p}\hat{q}/n}$

$0.0882 \pm 2.575\sqrt{(0.0882)(0.9118)/1236}$

0.0882 ± 0.0207

$0.0674 < p < 0.109$

25. $\alpha = 0.05$, $z_{\alpha/2} = z_{0.025} = 1.96$ and $E = 0.045$; $\hat{p}$ unknown, use $\hat{p} = 0.5$

$n = [(z_{\alpha/2})^2\hat{p}\hat{q}]/E^2$

$= [(1.96)^2(0.5)(0.5)]/(0.045)^2 = 474.27$, rounded up to 475

27. $\alpha = 0.01$, $z_{\alpha/2} = z_{0.005} = 2.575$ and $E = 0.02$; $\hat{p}$ estimated to be 0.14

$n = [(z_{\alpha/2})^2\hat{p}\hat{q}]/E^2$

$= [(2.575)^2(0.14)(0.86)]/(0.02)^2 = 1995.82$, rounded up to 1996

29. Let x = the number of girls born using the method.

a. $\hat{p} = x/n = 525/574 = 0.9146$, rounded to 0.915

b. $\alpha = 0.05$, $z_{\alpha/2} = z_{0.025} = 1.96$

$\hat{p} \pm z_{\alpha/2}\sqrt{\hat{p}\hat{q}/n}$

$0.9146 \pm 1.96\sqrt{(0.9146)(0.0854)/574}$

0.9146 ± 0.0229

$0.892 < p < 0.937$

c. Yes. Since 0.5 is not within the confidence interval, and below the interval, we can be 95% certain that the method is effective.

31. Let x = the number of deaths in the week before Thanksgiving.
  a. $\hat{p} = x/n = 6062/12000 = 0.5052$, rounded to 0.505
  b. $\alpha = 0.05$, $z_{\alpha/2} = z_{0.025} = 1.96$
  $$\hat{p} \pm z_{\alpha/2}\sqrt{\hat{p}\hat{q}/n}$$
  $$0.5052 \pm 1.96\sqrt{(0.5052)(0.4948)/12000}$$
  $$0.5052 \pm 0.0089$$
  $$0.496 < p < 0.514$$
  c. No. Since 0.5 is within the confidence interval, there is no evidence that people can temporarily postpone their death in such circumstances.

33. Let x = the number of yellow peas
  a. $\alpha = 0.05$, $z_{\alpha/2} = z_{0.025} = 1.96$ and $\hat{p} = x/n = 152/(428+152) = 152/580 = 0.2621$
  $$\hat{p} \pm z_{\alpha/2}\sqrt{\hat{p}\hat{q}/n}$$
  $$0.2621 \pm 1.96\sqrt{(0.2621)(0.7379)/580}$$
  $$0.2621 \pm 0.0358$$
  $$0.226 < p < 0.298 \text{ or } 22.6\% < p < 29.8\%$$
  b. No. Since 0.25 is within the confidence interval, it is a reasonable possibility for the true population proportion. The results do not contradict the theory.

35. Let x = the number that develop those types of cancer.
  a. $\alpha = 0.05$, $z_{\alpha/2} = z_{0.025} = 1.96$ and $\hat{p} = x/n = 135/420095 = 0.0003214$
  $$\hat{p} \pm z_{\alpha/2}\sqrt{\hat{p}\hat{q}/n}$$
  $$0.0003214 \pm 1.96\sqrt{(0.0003214)(0.9996786)/420095}$$
  $$0.0003214 \pm 0.0000542$$
  $$0.000267 < p < 0.000376 \text{ or } 0.0267\% < p < 0.0376\%$$
  b. No. Since 0.0340% = 0.000340 is within the confidence interval, it is a reasonable possibility for the true population value. The results do not provide evidence that cell phone users have a different cancer rate than the general population.

37. Let x = the number who say they use the Internet.
  $\alpha = 0.05$, $z_{\alpha/2} = z_{0.025} = 1.96$ and $\hat{p} = x/n = [2198]/3011 = 0.73$
  NOTE: The value x=[2198] was not given. In truth, any $2183 \leq x \leq 2213$ rounds to the given $\hat{p} = x/3011 = 73\%$. For want of a more precise value, $\hat{p} = 0.73$ is used in the calculations. Technically, this should limit the exercise to two significant digit accuracy
  $$\hat{p} \pm z_{\alpha/2}\sqrt{\hat{p}\hat{q}/n}$$
  $$0.73 \pm 1.96\sqrt{(0.73)(0.27)/3011}$$
  $$0.73 \pm 0.0159$$
  $$0.714 < p < 0.746$$
  No. Since 0.75 is not within the confidence interval, it is not likely to be the correct value of the population proportion and should not be reported as such. In this particular exercise, however, the above NOTE indicates that the third significant digit in the confidence interval endpoints is not reliable – and if $\hat{p}$ is really $2213/3011 = 0.73497$, for example, the confidence interval is $0.719 < p < 0.751$ and 75% is acceptable.

39. Let x = the number who indicate the outbreak would deter them from taking a cruise.
$\alpha = 0.05$, $z_{\alpha/2} = z_{0.025} = 1.96$ and $\hat{p} = x/n = [21302]/34358 = 0.62$
NOTE: The value x=[21302] was not given. In truth, any $21131 \le x \le 21473$ rounds to the given $\hat{p} = x/34358 = 62\%$. For want of a more precise value, $\hat{p} = 0.62$ is used in the calculations. Technically, this should limit the exercise to two significant digit accuracy.
$\hat{p} \pm z_{\alpha/2}\sqrt{\hat{p}\hat{q}/n}$
$0.62 \pm 1.96\sqrt{(0.62)(0.38)/34258}$
0.62 ± 0.0051
$0.615 < p < 0.625$
No. Since the sample is a voluntary response sample, the respondents are not likely to be representative of the population.

41. $\alpha = 0.01$, $z_{\alpha/2} = z_{0.005} = 2.575$ and $E = 0.02$
a. $\hat{p}$ unknown, use $\hat{p}$=0.5
$n = [(z_{\alpha/2})^2\hat{p}\hat{q}]/E^2$
$= [(2.575)^2(0.5)(0.5)]/(0.02)^2 = 4144.14$, rounded up to 4145
b. $\hat{p}$ estimated to be 0.73
$n = [(z_{\alpha/2})^2\hat{p}\hat{q}]/E^2$
$= [(2.575)^2(0.73)(0.27)]/(0.02)^2 = 3267.24$, rounded up to 3268

43. $\alpha = 0.05$, $z_{\alpha/2} = z_{0.025} = 1.96$ and $E = 0.03$; $\hat{p}$ unknown, use $\hat{p}$=0.5
$n = [(z_{\alpha/2})^2\hat{p}\hat{q}]/E^2$
$= [(1.96)^2(0.5)(0.5)]/(0.03)^2 = 1067.11$, rounded up to 1068

45. Let x = the number of green M&M's.
$\alpha = 0.05$, $z_{\alpha/2} = z_{0.025} = 1.96$ and $\hat{p} = x/n = 19/100 = 0.19$
$\hat{p} \pm z_{\alpha/2}\sqrt{\hat{p}\hat{q}/n}$
$0.1900 \pm 1.96\sqrt{(0.19)(0.81)/100}$
0.1900 ± 0.0769
$0.113 < p < 0.267$ or $11.3\% < p < 26.7\%$
Yes. Since 0.160 is within the confidence interval, this result is consistent with the claim that the true population proportion is 16%.

47. Let x = the number of days with precipitation.
$\alpha = 0.05$, $z_{\alpha/2} = z_{0.025} = 1.96$
Wednesdays: $\hat{p} = x/n = 16/53 = 0.3019$.
$\hat{p} \pm z_{\alpha/2}\sqrt{\hat{p}\hat{q}/n}$
$0.3019 \pm 1.96\sqrt{(0.3019)(0.6981)/53}$
0.3019 ± 0.1236
$0.178 < p < 0.425$

Sundays: $\hat{p} = x/n = 15/52 = 0.2885$
$\hat{p} \pm z_{\alpha/2}\sqrt{\hat{p}\hat{q}/n}$
$0.2885 \pm 1.96\sqrt{(0.2885)(0.7115)/52}$
0.2885 ± 0.1231
$0.165 < p < 0.412$

The confidence intervals are similar. It does not appear to rain more on either day.

49. $\alpha = 0.05$, $z_{\alpha/2} = z_{0.025} = 1.96$ and $E = 0.03$; $\hat{p}$ unknown, use $\hat{p} = 0.5$

$$n = \frac{N\hat{p}\hat{q}[z_{\alpha/2}]^2}{\hat{p}\hat{q}[z_{\alpha/2}]^2 + (N-1)E^2}$$
$$= \frac{(12784)(0.5)(0.5)[1.96]^2}{(0.5)(0.5)[1.96]^2 + (12783)(0.03)^2} = \frac{12277.8}{12.4651} = 984.97\text{, rounded up to } 985$$

No. The sample size is not too much lower than the n=1068 required for a population of millions of people.

51. $\alpha = 0.05$, $z_{\alpha/2} = z_{0.025} = 1.96$ and $\hat{p} = x/n = 3/8 = 0.3750$

$$\hat{p} \pm z_{\alpha/2}\sqrt{\hat{p}\hat{q}/n}$$
$$0.3750 \pm 1.96\sqrt{(0.3750)(0.6250)/8}$$
$$0.3750 \pm 0.3355$$
$$0.0395 < p < 0.710$$

Yes. The results are "reasonably close" – being shifted down 4.5% from the correct interval $0.085 < p < 0.755$. But depending on the context, such an error could be serious.

53. a. If $\hat{p} = x/n = 0/n = 0$, then

(1) $np \approx 0 < 5$, and the normal approximation to the binomial does not apply.

(2) $E = z_{\alpha/2}\sqrt{\hat{p}\hat{q}/n} = 0$, and there is no meaningful interval.

b. Since $\hat{p} = x/n = 0/20 = 0$, use $3/n = 3/20 = 0.15$ as the 95% upper bound for p.

NOTE: The corresponding interval would be $0 \le p < 0.15$. Do not use $0 < p < 0.15$, because the failure to observe any successes in the sample does not rule out p=0 as the true population proportion.

7-3 Estimating a Population Mean: σ Known

1. A point estimate is a single value used to estimate a population parameter. If the parameter in question is the mean of a population, the best point estimate is the mean of a random sample from that population.

3. It is estimated that the mean height of U.S. women is 63.195 inches. This result comes from the Third National Health and Nutrition Examination Survey of the U.S. Department of Health and Human Services. It is based on an in-depth study of 40 women and assumes a population standard deviation of 2.5 inches. The estimate has a margin of error of 0.775 inches with a 95% level of confidence. In other words, 95% of all such studies can be expected to produce estimates that are within 0.775 inches of the true population mean height of all U.S. women.

5. For 90% confidence, $\alpha = 1-0.90 = 0.10$ and $\alpha/2 = 0.10/2 = 0.05$.
For the upper 0.05, A = 0.9500 and z = 1.645.
$z_{\alpha/2} = z_{0.05} = 1.645$

7. For $\alpha = 0.20$, $\alpha/2 = 0.20/2 = 0.10$.
For the upper 0.10, A = 0.9000 and z = 1.28.
$z_{\alpha/2} = z_{0.10} = 1.28$

9. Since σ is known and $n>30$, the methods of this section may be used.
   $\alpha = 0.05$, $z_{\alpha/2} = z_{0.025} = 1.96$
   $E = z_{\alpha/2}\sigma/\sqrt{n} = 1.96(68)/\sqrt{50} = 18.8$ FICO units
   $\bar{x} \pm E$
   677.0 ± 18.8
   $658.2 < \mu < 695.8$ (FICO units)
   NOTE: The above interval assumes $\bar{x} = 677.0$. Technically, the failure to report $\bar{x}$ to tenths limits the endpoints of the confidence interval to whole number accuracy.

11. Since $n<30$ and the population is far from normal, the methods of this section may not be used.

13. $\alpha = 0.05$, $z_{\alpha/2} = z_{0.025} = 1.96$
   $n = [z_{\alpha/2}\cdot\sigma/E]^2$
   $= [(1.96)(68)/(3)]^2 = 1973.73$, rounded up to 1974

15. $\alpha = 0.01$, $z_{\alpha/2} = z_{0.005} = 2.575$
   $n = [z_{\alpha/2}\cdot\sigma/E]^2$
   $= [(2.575)(0.212)/(0.010)]^2 = 2980.07$, rounded up to 2981

17. $\bar{x} = 21.12$ mg

19. $E = (U - L)/2 = (22.387 - 19.853)/2 = 1.267$
   21.12 ± 1.267 (mg)

21. a. $\bar{x} = 146.22$ lbs
   b. $\alpha = 0.05$, $z_{\alpha/2} = z_{0.025} = 1.96$
   $\bar{x} \pm z_{\alpha/2}\cdot\sigma/\sqrt{n}$
   $146.22 \pm 1.96(30.86)/\sqrt{40}$
   146.22 ± 9.56
   $136.66 < \mu < 155.78$ (lbs)

23. a. $\bar{x} = 58.3$ seconds
   b. $\alpha = 0.05$, $z_{\alpha/2} = z_{0.025} = 1.96$
   $\bar{x} \pm z_{\alpha/2}\cdot\sigma/\sqrt{n}$
   $58.3 \pm 1.96(9.5)/\sqrt{40}$
   58.3 ± 2.9
   $55.4 < \mu < 61.2$ (seconds)
   c. Yes. Since the confidence interval contains 60 seconds, it is reasonable to assume that the sample mean was reasonably close to 60 seconds – and it was, in fact, 58.3 seconds.

25. a. $\alpha = 0.05$, $z_{\alpha/2} = z_{0.025} = 1.96$
   $\bar{x} \pm z_{\alpha/2}\cdot\sigma/\sqrt{n}$
   $1522 \pm 1.96(333)/\sqrt{125}$
   1522 ± 58
   $1464 < \mu < 1580$
   b. $\alpha = 0.01$, $z_{\alpha/2} = z_{0.005} = 2.575$
   $\bar{x} \pm z_{\alpha/2}\cdot\sigma/\sqrt{n}$
   $1522 \pm 2.575(333)/\sqrt{125}$
   1522 ± 77
   $1445 < \mu < 1599$
   c. The 99% confidence interval in part (b) is wider than the 95% confidence interval in part (a). For an interval to have more confidence associated with it, it must be wider to allow for more possibilities.

27. summary statistics: $n = 14$ $\Sigma x = 1875$ $\bar{x} = 133.93$
$\alpha = 0.05$, $z_{\alpha/2} = z_{0.025} = 1.96$
$\bar{x} \pm z_{\alpha/2} \cdot \sigma/\sqrt{n}$
$133.93 \pm 1.96(10)/\sqrt{14}$
133.93 ± 5.24
$128.7 < \mu < 139.2$ (mmHg)
Ideally, there is a sense in which all the measurements should be the same – and in that case there would be no need for a confidence interval. It is unclear what the given $\sigma = 10$ represents in this situation. Is it the true standard deviation in the values of all people in the population (in which case it would not be appropriate in this context where only a single person is involved)? Is it the true standard deviation in momentary readings on a single person (due to constant biological fluctuations)? Is it the true standard deviation in readings from evaluator to evaluator (when they are supposedly evaluating the same thing)? Using the methods of this section and assuming $\sigma = 10$, the confidence interval would be $128.7 < \mu < 139.2$ as given above even if all the readings were the same.

29. summary statistics: $n = 35$ $\Sigma x = 4305$ $\bar{x} = 123.00$
$\alpha = 0.05$, $z_{\alpha/2} = z_{0.025} = 1.96$
$\bar{x} \pm z_{\alpha/2} \cdot \sigma/\sqrt{n}$
$123.00 \pm 1.96(100)/\sqrt{35}$
123.00 ± 33.13
$89.9 < \mu < 156.1$ (million dollars)

31. $\alpha = 0.05$, $z_{\alpha/2} = z_{0.025} = 1.96$
$n = [z_{\alpha/2} \cdot \sigma/E]^2$
$= [(1.96)(15)/(5)]^2 = 34.57$, rounded up to 35

33. $\alpha = 0.05$, $z_{\alpha/2} = z_{0.025} = 1.96$
$n = [z_{\alpha/2} \cdot \sigma/E]^2$
$= [(1.96)(10.6)/(0.25)]^2 = 6906.27$, rounded up to 6907
The sample size is too large to be practical.

35. $\alpha = 0.05$, $z_{\alpha/2} = z_{0.025} = 1.96$
Using the range rule of thumb: $R = 40{,}000 - 0 = 40{,}000$, and $\sigma \approx R/4 = 40{,}000/4 = 10{,}000$.
$n = [z_{\alpha/2} \cdot \sigma/E]^2$
$= [(1.96)(10{,}000)/(100)]^2 = 38416$

37. Since $n/N = 125/200 = 0.625 > 0.05$, use the finite population correction factor.
$\alpha = 0.05$, $z_{\alpha/2} = z_{0.025} = 1.96$
$\bar{x} \pm [z_{\alpha/2}\sigma/\sqrt{n}] \cdot \sqrt{(N-n)/(N-1)}$
$1522 \pm [1.96(333)/\sqrt{125}] \cdot \sqrt{(200-125)/(200-1)}$
$1522 \pm [58.3774] \cdot [0.6139]$
1522 ± 36
$1486 < \mu < 1558$
The confidence interval becomes narrower because the sample is a larger portion of the population. As n approaches N, the length of the confidence interval shrinks to 0 – because when $n=N$ the true mean μ can be determined with certainty.

7-4 Estimating a Population Mean: σ Not Known

1. According to the point estimate ("average"), the parameter of interest is a population mean. But according to the margin of error ("percentage points"), the parameter of interest is a population proportion. It is possible that the margin of error the paper intended to communicate was 1% of $483 (or $4.83, which in a 95% confidence interval would correspond to a sample standard deviation of $226.57) – but the proper units for the margin of error in a situation like this are "dollars" and not "percentage points."

3. No; the estimate will not be good for at least two reasons. First, the sample is a convenience sample using the state of California, and California residents may not be representative of then entire country. Secondly, any survey that involves self-reporting (especially of financial information) is suspect because people tend to report favorable rather than accurate data.

5. σ unknown, normal population, n=23: use t with df =22
   $\alpha = 0.05$, $t_{df,\alpha/2} = t_{22,0.025} = 2.074$

IMPORTANT NOTE: This manual uses the following conventions.
(1) The designation "df" stands for "degrees of freedom."
(2) Since the t value depends on the degrees of freedom, a subscript may be used to clarify which t distribution is being used. For df =15 and $\alpha/2 = 0.025$, for example, one may indicate $t_{15,\alpha/2} = 2.132$. As with the z distribution, it is also acceptable to use the actual numerical value within the subscript and indicate $t_{15,.025} = 2.132$.
(3) Always use the closest entry in Table A-3. When the desired df is exactly halfway between the two nearest tabled values, be conservative and choose the one with the lower df.
(4) As the degrees of freedom increase, the t distribution approaches the standard normal distribution – and the "large" row of the t table actually gives z values. Consequently the z score for certain "popular" α and $\alpha/2$ values may be found by reading Table A-3 "frontwards" instead of Table A-2 "backwards." This is not only easier but also more accurate – since Table A-3 includes one more decimal place. Note the following examples.
   For "large" df and $\alpha/2 = 0.05$, $t_{\alpha/2} = 1.645 = z_{\alpha/2}$ (as found in the z table).
   For "large" df and $\alpha/2 = 0.01$, $t_{\alpha/2} = 2.326 = z_{\alpha/2}$ (more accurate than the 2.33 in the z table).
   This manual uses this technique from this point on. [For df = "large" and $\alpha/2 = 0.005$, $t_{\alpha/2} = 2.576 \neq 2.575 = z_{\alpha/2}$ (as found in the z table). This is a discrepancy caused by using different mathematical approximation techniques to construct the tables, and not a true difference. While 2.576 is the more standard value, his manual will continue to use 2.575.]

7. σ unknown, population not normal, n=6: neither normal nor t applies

9. σ known, population not normal, n=200: use z
   $\alpha = 0.10$, $z_{\alpha/2} = z_{0.05} = 1.645$

11. σ unknown, population normal, n=12: use t with df = 11
   $\alpha = 0.01$, $t_{df,\alpha/2} = t_{11,0.005} = 3.106$

13. σ unknown, normal distribution: use t with df = 19
   $\alpha = 0.05$, $t_{df,\alpha/2} = t_{19,0.025} = 2.093$

   a. $E = t_{\alpha/2} \cdot s/\sqrt{n}$
   $= 2.093(569)/\sqrt{20}$
   $= 266$ dollars

   b. $\bar{x} \pm E$
   9004 ± 266
   $8738 < \mu < 9270$ (dollars)

15. From the SPSS display: $8.0518 < \mu < 8.0903$ (grams)
There is 95% confidence that the interval from 8.0518 grams to 8.0903 grams contains the true mean weight of all U.S. dollar coins in circulations.

17. a. $\bar{x} = 3.2$ mg/dL
b. σ unknown, n > 30: use t with df=46 [45]
$\alpha = 0.05$, $t_{df,\alpha/2} = t_{46,0.025} = 2.014$
$\bar{x} \pm t_{\alpha/2} \cdot s/\sqrt{n}$
$3.2 \pm 2.014(18.6)/\sqrt{47}$
3.2 ± 5.5
$-2.3 < \mu < 8.7$ (mg/dl)
Since the confidence interval includes 0, there is a reasonable possibility that the true value is zero – i.e., that the Garlicin treatment has no effect on LDL cholesterol levels.

19. a. $\bar{x} = 98.20$ °F
b. σ unknown, n > 30: use t with df=105 [100]
$\alpha = 0.01$, $t_{df,\alpha/2} = t_{105,0.005} = 2.626$
$\bar{x} \pm t_{\alpha/2} \cdot s/\sqrt{n}$
$98.20 \pm 2.626(0.62)/\sqrt{106}$
98.20 ± 0.16
$98.04 < \mu < 98.36$ (°F)
c. No, the confidence interval does not contain the value 98.6 °F. This suggests that the common belief that 98.6 °F is the normal body temperature may not be correct.

21. a. σ unknown, n > 30: use t with df=336 [300]
$\alpha = 0.05$, $t_{df,\alpha/2} = t_{336,0.025} = 1.968$
$\bar{x} \pm t_{\alpha/2} \cdot s/\sqrt{n}$
$6.0 \pm 1.968(2.3)/\sqrt{337}$
6.0 ± 0.2
$5.8 < \mu < 6.2$ (days)

b. σ unknown, n > 30: use t with df=369 [400]
$\alpha = 0.05$, $t_{df,\alpha/2} = t_{369,0.025} = 1.966$
$\bar{x} \pm t_{\alpha/2} \cdot s/\sqrt{n}$
$6.1 \pm 1.966(2.4)/\sqrt{370}$
1.6 ± 0.2
$5.9 < \mu < 6.3$ (days)

c. The two confidence intervals are very similar and overlap considerably. There is no evidence that the echinacea treatment is effective.

23. a. σ unknown, n ≤ 30: if approximately normal distribution, use t with df=19
$\alpha = 0.05$, $t_{df,\alpha/2} = t_{19,0.025} = 2.093$
$\bar{x} \pm t_{\alpha/2} \cdot s/\sqrt{n}$
$5.0 \pm 2.093(2.4)/\sqrt{20}$
5.0 ± 1.1
$3.9 < \mu < 6.1$ (VAS units)

b. σ unknown, n ≤ 30: if approximately normal distribution, use t with df=19
$\alpha = 0.05$, $t_{df,\alpha/2} = t_{99,0.025} = 2.093$
$\bar{x} \pm t_{\alpha/2} \cdot s/\sqrt{n}$
$4.7 \pm 2.093(2.9)/\sqrt{20}$
4.7 ± 1.4
$3.3 < \mu < 6.1$ (VAS units)

c. The two confidence intervals are very similar and overlap considerably. There is no evidence that the magnet treatment is effective.

25. preliminary values: $n = 6$, $\Sigma x = 9.23$, $\Sigma x^2 = 32.5197$
$\bar{x} = (\Sigma x)/n = (9.23)/6 = 1.538$
$s^2 = [n(\Sigma x^2) - (\Sigma x)^2]/[n(n-1)] = [6(32.5197) - (9.23)^2]/[6(5)] = 3.664$
$s = 1.914$
σ unknown (and assuming the distribution is approximately normal), use t with df=5
$\alpha = 0.05$, $t_{df,\alpha/2} = t_{5,0.025} = 2.571$

$\bar{x} \pm t_{\alpha/2} \cdot s/\sqrt{n}$

$1.538 \pm 2.571(1.914)/\sqrt{6}$

1.538 ± 2.009

$-0.471 < \mu < 3.547$ [which should be adjusted, since negative values are not possible]

$0 < \mu < 3.547$ (micrograms/cubic meter)

Yes. The fact that 5 of the 6 sample values are below $\bar{x}$ raises a question about whether the data meet the requirement that the underlying distribution is normal.

27. preliminary values: $n = 10$, $\Sigma x = 204.0$, $\Sigma x^2 = 5494.72$

$\bar{x} = (\Sigma x)/n = (204.0)/10 = 20.40$

$s^2 = [n(\Sigma x^2) - (\Sigma x)^2]/[n(n-1)] = [10(5494.72) - (204.0)^2]/[10(9)] = 148.124$

$s = 12.171$

a. σ unknown (and assuming the distribution is approximately normal), use t with df=9

$\alpha = 0.05$, $t_{df,\alpha/2} = t_{9,0.05} = 2.262$

$\bar{x} \pm t_{\alpha/2} \cdot s/\sqrt{n}$

$20.40 \pm 2.262(12.171)/\sqrt{10}$

20.40 ± 8.71

$11.7 < \mu < 29.1$ (million dollars)

b. No. Since the data are the top 10 salaries, they are not a random sample.

c. There is a sense in which the data are the population (i.e., the top ten salaries) and are not a sample of any population. Possible populations from which the data could be considered a sample (but not a representative sample appropriate for any statistical inference) would be the salaries of all TV personalities, the salaries of the top 10 salaries of TV personality for different years.

d. No. Since no population can be identified from which these data are a random sample, the confidence interval has no context and makes no sense.

29. preliminary values: $n = 12$, $\Sigma x = 52118$, $\Sigma x^2 = 228{,}072{,}688$

$\bar{x} = (\Sigma x)/n = (52118)/12 = 4343.17$

$s^2 = [n(\Sigma x^2) - (\Sigma x)^2]/[n(n-1)] = [12(228072688) - (52118)^2]/[12(11)] = 155957.06$

$s = 394.91$

σ unknown (and assuming the distribution is approximately normal), use t with df=11

$\alpha = 0.05$, $t_{df,\alpha/2} = t_{12,0.025} = 2.201$

$\bar{x} \pm t_{\alpha/2} \cdot s/\sqrt{n}$

$4343.17 \pm 2.201(394.91)/\sqrt{12}$

4343.17 ± 250.91

$4092.2 < \mu < 4594.1$ (seconds)

31. a. preliminary values: $n = 25$, $\Sigma x = 31.4$, $\Sigma x^2 = 40.74$

$\bar{x} = (\Sigma x)/n = (31.4)/25 = 1.256$

$s^2 = [n(\Sigma x^2) - (\Sigma x)^2]/[n(n-1)] = [25(40.74) - (31.4)^2]/[25(24)] = 32.54/600 = 0.0542$

$s = 0.2329$

σ unknown (and assuming the distribution is approximately normal), use t with df=24

$\alpha = 0.05$, $t_{df,\alpha/2} = t_{24,0.025} = 2.064$

$\bar{x} \pm t_{\alpha/2} \cdot s/\sqrt{n}$

$1.256 \pm 2.064(0.2329)/\sqrt{25}$

1.256 ± 0.096

$1.16 < \mu < 1.35$ (mg)

NOTE: The Minitab output for this exercise is given below.

```
Variable   N      Mean     StDev  SE Mean          95% CI
nicotine  25   1.25600  0.23288  0.04658  (1.15987, 1.35213)
```

b. preliminary values: $n = 25$, $\Sigma x = 22.9$, $\Sigma x^2 = 22.45$

$\bar{x} = (\Sigma x)/n = (22.9)/25 = 0.916$

$s^2 = [n(\Sigma x^2) - (\Sigma x)^2]/[n(n-1)] = [25(22.45) - (22.9)^2]/[25(24)] = 36.84/600 = 0.0614$

$s = 0.2478$

σ unknown (and assuming the distribution is approximately normal), use t with df=24

$\alpha = 0.05$, $t_{df,\alpha/2} = t_{24,0.025} = 2.064$

$\bar{x} \pm t_{\alpha/2} \cdot s/\sqrt{n}$

$0.916 \pm 2.064(0.2478)/\sqrt{25}$

0.916 ± 0.102

$0.81 < \mu < 1.02$ (mg)

NOTE: The Minitab output for this exercise is given below.

```
Variable   N      Mean      StDev    SE Mean              95% CI
nicotine  25  0.916000  0.247790  0.049558  (0.813717, 1.018283)
```

c. There is no overlap in the confidence intervals. Yes; since the CI for the filtered cigarettes is completely below the CI for the unfiltered cigarettes, the filters appear to be effective in reducing the amounts of nicotine.

33. preliminary values: $n = 43$, $\Sigma x = 2738$, $\Sigma x^2 = 307{,}250$

$\bar{x} = (\Sigma x)/n = (2738)/43 = 63.674$

$s^2 = [n(\Sigma x^2) - (\Sigma x)^2]/[n(n-1)] = [43(307250) - (2738)^2]/[43(42)] = 3164.511$

$s = 56.254$

σ unknown and n>30, use t with df=42 [40]

$\alpha = 0.01$, $t_{df,\alpha/2} = t_{42,0.005} = 2.704$

$\bar{x} \pm t_{\alpha/2} \cdot s/\sqrt{n}$

$63.674 \pm 2.704(56.254)/\sqrt{43}$

63.674 ± 23.197

$40.5 < \mu < 86.9$ (years)

Yes, the confidence interval changes considerably from the previous $52.3 < \mu < 57.4$.

Yes, apparently confidence interval limits can be very sensitive to outliers.

When apparent outliers are discovered in data sets they should be carefully examined to determine if an error has been made. If an error has been made that cannot be corrected, the value should be discarded. If the value appears to be valid, it may be informative to construct confidence intervals with and without the outlier.

35. assuming a large population

$\alpha = 0.05$ & df=99 [100], $t_{df,\alpha/2} = t_{99,0.025} = 1.984$

$E = t_{\alpha/2} \cdot s/\sqrt{n}$

$= 1.984(0.0518)/\sqrt{100}$

$= 0.0103$ g

$\bar{x} \pm E$

0.8565 ± 0.0103

$0.8462 < \mu < 0.8668$ (grams)

using the finite population N = 465

$\alpha = 0.05$ & df=99 [100], $t_{df,\alpha/2} = t_{99,0.025} = 1.984$

$E = [t_{\alpha/2} \cdot s/\sqrt{n}] \times \sqrt{(N-n)/(N-1)}$

$= [1.984(0.0518)/\sqrt{100}] \times \sqrt{365/464}$

$= 0.0091$ g

$\bar{x} \pm E$

0.8565 ± 0.0091

$0.8474 < \mu < 0.8656$ (grams)

The second confidence interval is narrower, reflecting the fact that there are more restrictions and less variability (and more certainty) in the finite population situation when n>.05N.

7-5 Estimating a Population Variance

1. We can be 95% confident that the interval from 0.0455grams to 0.0602 grams includes the true value of the standard deviation in the weights for the population of all M&M's.

3. No; the population of last two digits from 00 to 99 follows a uniform distribution and not a normal distribution. One of the requirements for using the methods of this section is that the population values have a distribution that is approximately normal – even if the sample size is large.

5. $\alpha = 0.05$ and df = 8 $\quad \chi_L^2 = \chi_{8,0.975}^2 = 2.180 \quad \chi_R^2 = \chi_{8,0.025}^2 = 17.535$

7. $\alpha = 0.01$ and df = 80 $\quad \chi_L^2 = \chi_{80,0.995}^2 = 51.172 \quad \chi_R^2 = \chi_{80,0.005}^2 = 116.321$

9. $\alpha = 0.05$ and df = 29; $\chi_L^2 = \chi_{df,1-\alpha/2}^2$ and $\chi_R^2 = \chi_{df,\alpha/2}^2$

$$(n-1)s^2/\chi_R^2 < \sigma^2 < (n-1)s^2/\chi_L^2$$
$$(29)(333)^2/45.722 < \sigma^2 < (29)(333)^2/16.047$$
$$70333.3 < \sigma^2 < 200397.6$$
$$265 < \sigma < 448$$

11. $\alpha = 0.01$ and df = 6; $\chi_L^2 = \chi_{df,1-\alpha/2}^2$ and $\chi_R^2 = \chi_{df,\alpha/2}^2$

$$(n-1)s^2/\chi_R^2 < \sigma^2 < (n-1)s^2/\chi_L^2$$
$$(6)(2.019)^2/18.548 < \sigma^2 < (6)(2.019)^2/0.676$$
$$1.3186 < \sigma^2 < 36.1807$$
$$1.148 < \sigma < 6.015 \text{ (cells/microliter)}$$

13. From the upper right section of Table 7-2, n = 19,205.
No. This sample size is too large to be practical for most applications.

15. From the lower left section of Table 7-2, n = 101.
Yes. This sample size is practical for most applications.

17. $\alpha = 0.05$ and df = 189; $\chi_L^2 = \chi_{df,1-\alpha/2}^2$ and $\chi_R^2 = \chi_{df,\alpha/2}^2$

$$(n-1)s^2/\chi_R^2 < \sigma^2 < (n-1)s^2/\chi_L^2$$
$$(189)(645)^2/228.9638 < \sigma^2 < (189)(645)^2/152.8222$$
$$343411 < \sigma^2 < 514511$$
$$586 < \sigma < 717 \text{ (grams)}$$

No. Since the confidence interval includes 696, it is a reasonable possibility for σ.

19. a. $\alpha = 0.05$ and df = 22; $\chi_L^2 = \chi_{df,1-\alpha/2}^2$ and $\chi_R^2 = \chi_{df,\alpha/2}^2$

$$(n-1)s^2/\chi_R^2 < \sigma^2 < (n-1)s^2/\chi_L^2$$
$$(22)(22.9)^2/36.781 < \sigma^2 < (22)(22.9)^2/10.982$$
$$313.67 < \sigma^2 < 1050.54$$
$$17.7 < \sigma < 32.4 \text{ (minutes)}$$

b. $\alpha = 0.05$ and df = 11; $\chi_L^2 = \chi_{df,1-\alpha/2}^2$ and $\chi_R^2 = \chi_{df,\alpha/2}^2$

$$(n-1)s^2/\chi_R^2 < \sigma^2 < (n-1)s^2/\chi_L^2$$
$$(11)(20.8)^2/21.920 < \sigma^2 < (11)(20.8)^2/3.816$$
$$217.11 < \sigma^2 < 1247.13$$
$$14.7 < \sigma < 35.3 \text{ (minutes)}$$

c. The two intervals are similar. No, there does not appear to be a difference in the variation of lengths of PG/PGF-13 movies and R movies.

21. preliminary values: $n = 12$, $\Sigma x = 52118$, $\Sigma x^2 = 228{,}072{,}688$

$\bar{x} = (\Sigma x)/n = (52118)/12 = 4343.2$

$s^2 = [n(\Sigma x^2) - (\Sigma x)^2]/[n(n-1)] = [12(228072688) - (52118)^2]/[12(11)] = 155{,}957.06$

$s = 394.91$

$\alpha = 0.01$ and df = 11; $\chi_L^2 = \chi_{df,1-\alpha/2}^2$ and $\chi_R^2 = \chi_{df,\alpha/2}^2$

$$(n-1)s^2/\chi_R^2 < \sigma^2 < (n-1)s^2/\chi_L^2$$

$$(11)(394.91)^2/26.757 < \sigma^2 < (11)(394.91)^2/2.603$$

$$64115.10 < \sigma^2 < 659057.88$$

$$253.2 < \sigma < 811.8 \text{ (seconds)}$$

23. preliminary values: $n = 6$, $\Sigma x = 9.23$, $\Sigma x^2 = 32.5197$

$\bar{x} = (\Sigma x)/n = (9.23)/6 = 1.538$

$s^2 = [n(\Sigma x^2) - (\Sigma x)^2]/[n(n-1)] = [6(32.5197) - (9.213)^2]/[6(5)] = 3.664$

$s = 1.914$

$\alpha = 0.05$ and df = 5; $\chi_L^2 = \chi_{df,1-\alpha/2}^2$ and $\chi_R^2 = \chi_{df,\alpha/2}^2$

$$(n-1)s^2/\chi_R^2 < \sigma^2 < (n-1)s^2/\chi_L^2$$

$$(5)(3.664)/12.833 < \sigma^2 < (5)(3.664)/0.831$$

$$1.4276 < \sigma^2 < 22.0468$$

$$1.195 < \sigma < 4.695 \text{ (micrograms per cubic meter)}$$

Yes. One of the requirements to use the methods of this section is that the original distribution be approximately normal, and the fact that 5 of the 6 sample values are less than the mean suggests that the original distribution is not normal.

25. preliminary values: $n = 100$, $\Sigma x = 70311$, $\Sigma x^2 = 50{,}278{,}497$

$\bar{x} = (\Sigma x)/n = (70311)/100 = 703.11$

$s^2 = [n(\Sigma x^2) - (\Sigma x)^2]/[n(n-1)] = [100(50278497) - (70311)^2]/[100(99)] = 8506.36$

$s = 92.23$

$\alpha = 0.05$ and df = 99 [100]; $\chi_L^2 = \chi_{df,1-\alpha/2}^2$ and $\chi_R^2 = \chi_{df,\alpha/2}^2$

$$(n-1)s^2/\chi_R^2 < \sigma^2 < (n-1)s^2/\chi_L^2$$

$$(99)(8506.36)/129.561 < \sigma^2 < (99)(8506.36)/74.222$$

$$6499.87 < \sigma^2 < 11346.09$$

$$80.6 < \sigma < 106.5 \text{ (FICO units)}$$

NOTE: The statistical portion of Excel yielded the following results.

```
Confidence Level      Lower Conf. Limit Stan. Dev.   Upper Conf. Limit
0.95                  80.979                 92.23   107.141
```

27. Applying the given formula yields the following χ_L^2 and χ_R^2 values.

$$\chi^2 = (1/2)[\pm z_{\alpha/2} + \sqrt{2(df) - 1}]^2$$
$$= (1/2)[\pm 1.96 + \sqrt{2(189) - 1}]^2$$
$$= (1/2)[\pm 1.96 + 19.416]^2$$
$$= (1/2)[17.456]^2 \text{ and } (1/2)[21.376]^2$$
$$= 152.3645 \text{ and } 228.4771$$

These are close to the 152.8222 and 228.9638 given in exercise #17.

Statistical Literacy and Critical Thinking

1. A point estimate is a single value calculated from sample data that is used to estimate the true value of a population characteristic, called the parameter. In this context the sample proportion that test positive is the best point estimate for the population proportion that would test positive. A confidence interval is a range of values that is likely, with some specific degree of confidence, to include the true value of the population parameter. The major advantage of the confidence interval over the point estimate is its ability to communicate a sense of the accuracy of the estimate.

2. We can be 95% confident that the interval from 2.62% to 4.99% contains the true percentage of all job applicants who would test positive for drug use.

3. The confidence level in Exercise 2 is 95%. In general, the confidence level specifies the proportion of times a given procedure to construct an interval estimate can be expected to produce an interval that will include the true value of the parameter.

4. The respondents are not likely to be representative of the general population for two reasons. The sample is a convenience sample, composed only of those who visit the AOL Web site. The sample is a voluntary response sample, composed only of those who take the time to self-select themselves to be in the survey. Convenience samples are typically not representative racially, socio-economically, etc. Voluntary response samples typically include mainly those with strong opinions on, or a personal interest in, the topic of the survey.

Chapter Quick Quiz

1. We can be 95% confident that the interval from 20.0 to 20.0 contains the true value of the population mean.

2. The interval includes some values greater than 50%, suggesting that the Republican may win; but the interval also includes some values less than 50%, suggesting that the Republican may lose. Statement (2), that the election is too close to call, best describes the results of the survey.

3. The critical value of $t_{\alpha/2}$ for n=20 and $\alpha = 0.05$ is $t_{19,0.025} = 2.093$.

4. The critical value of $z_{\alpha/2}$ for n=20 and $\alpha = 0.10$ is $z_{0.05} = 1.645$.

5. $\alpha = 0.05$, $z_{\alpha/2} = z_{0.025} = 1.96$ and $E = 0.02$; $\hat{p}$ unknown, use $\hat{p} = 0.5$

   $n = [(z_{\alpha/2})^2\hat{p}\hat{q}]/E^2$
   $= [(1.96)^2(0.5)(0.5)]/(0.02)^2$
   $= 2401$

6. $\hat{p} = x/n = 240/600 = 0.40$

7. $\alpha = 0.05$, $z_{\alpha/2} = z_{0.025} = 1.96$ and $\hat{p} = x/n = 240/600 = 0.4000$

   $\hat{p} \pm z_{\alpha/2}\sqrt{\hat{p}\hat{q}/n}$
   $0.4000 \pm 1.96\sqrt{(0.4000)(0.6000)/600}$
   0.4000 ± 0.0392
   $0.361 < p < 0.439$

8. σ unknown, $n > 30$: use t with df=35 and $\alpha = 0.05$, $t_{df,\alpha/2} = t_{35,0.025} = 2.014$

$\bar{x} \pm t_{\alpha/2} \cdot s/\sqrt{n}$

$40.0 \pm 2.030(10.0)/\sqrt{36}$

40.0 ± 3.4

$36.6 < \mu < 43.4$ (years)

9. σ known, $n > 30$: use z and $\alpha = 0.05$, $z_{\alpha/2} = z_{0.025} = 1.96$

$\bar{x} \pm z_{\alpha/2} \cdot \sigma/\sqrt{n}$

$40.0 \pm 1.96(10.0)/\sqrt{36}$

40.0 ± 3.3

$36.7 < \mu < 43.3$ (years)

10. $\alpha = 0.05$, $z_{\alpha/2} = z_{0.025} = 1.96$

$n = [z_{\alpha/2} \cdot \sigma/E]^2$

$= [(1.96)(12)/(0.5)]^2$

$= 2212.76$, rounded up to 2213

Review Exercises

1. $\alpha = 0.05$ and $z_{\alpha/2} = z_{0.025} = 1.96$; $\hat{p} = x/n = 589/745 = 0.7906$

$\hat{p} \pm z_{\alpha/2}\sqrt{\hat{p}\hat{q}/n}$

$0.7906 \pm 1.96\sqrt{(0.7906)(0.2094)/745}$

0.7906 ± 0.0292

$0.761 < p < 0.820$ or $76.1\% < p < 82.0\%$

We can be 95% confident that the interval from 76.1% to 82.0% contains the true percentage of all adults who believe that it is morally wrong to not report all income on their tax returns.

2. $\alpha = 0.01$, $z_{\alpha/2} = z_{0.005} = 2.575$ and $E = 0.02$; $\hat{p}$ unknown, use $\hat{p} = 0.5$

$n = [(z_{\alpha/2})^2\hat{p}\hat{q}]/E^2$

$= [(2.575)^2(0.5)(0.5)]/(0.02)^2$

$= 4144.14$, rounded up to 4145

3. $\alpha = 0.01$, $z_{\alpha/2} = z_{0.005} = 2.575$

$n = [z_{\alpha/2} \cdot \sigma/E]^2$

$= [(2.575)(28785)/(500)]^2$

$= 21975.91$, rounded up to 21,976

No, the required sample size is too large to be practical. It appears that some re-thinking of the requirements is necessary.

4. σ unknown, $n > 30$: use t with df=36 and $\alpha = 0.01$, $t_{df,\alpha/2} = t_{36,0.005} = 2.719$

$\bar{x} \pm t_{\alpha/2} \cdot s/\sqrt{n}$

$2.4991 \pm 2.719(0.0165)/\sqrt{37}$

2.4991 ± 0.0074

$2.4917 < \mu < 2.5065$ (grams)

Since the above interval includes 2.5 grams, it appears on the surface that the manufacturing process is meeting the design specifications – but the interval may not be relevant to determine whether the pennies were manufactured according to specification if the sample came (as so it seems) from worn pennies in circulation.

5. preliminary values: $n = 5$, $\Sigma x = 3344$, $\Sigma x^2 = 2{,}470{,}638$

$\bar{x} = (\Sigma x)/n = (3344)/5 = 668.8$

$s^2 = [n(\Sigma x^2) - (\Sigma x)^2]/[n(n\text{-}1)]$

$= [5(2470638) - (3344)^2]/[5(4)] = 58542.7$

$s = 241.96$

σ unknown and n=5: assuming a normal distribution, use t with df=4

$\alpha = 0.05$, $t_{df,\alpha/2} = t_{4,0.025} = 2.776$

$\bar{x} \pm t_{\alpha/2} \cdot s/\sqrt{n}$

$668.8 \pm 2.776(241.96)/\sqrt{5}$

668.8 ± 300.4

$368.4 < \mu < 969.2$ (hic)

6. preliminary values: $n = 5$, $\Sigma x = 3344$, $\Sigma x^2 = 2{,}470{,}638$

$\bar{x} = (\Sigma x)/n = (3344)/5 = 668.8$

$s^2 = [n(\Sigma x^2) - (\Sigma x)^2]/[n(n\text{-}1)]$

$= [5(2470638) - (3344)^2]/[5(4)] = 58542.7$

$s = 241.96$

$\alpha = 0.05$ and df = 4; $\chi^2_L = \chi^2_{df,1-\alpha/2}$ and $\chi^2_R = \chi^2_{df,\alpha/2}$

$(n\text{-}1)s^2/\chi^2_R < \sigma^2 < (n\text{-}1)s^2/\chi^2_L$

$(4)(58542.7)/11.143 < \sigma^2 < (4)(58542.7)/0.484$

$21015.06 < \sigma^2 < 483823.97$

$145.0 < \sigma < 695.6$ (hic)

7. Let x = the number who believe that cloning should not be allowed.

a. $\hat{p} = x/n = 901/1012 = 0.8903$, rounded to 0.890

b. $\alpha = 0.05$, $z_{\alpha/2} = z_{0.025} = 1.96$

$\hat{p} \pm z_{\alpha/2}\sqrt{\hat{p}\hat{q}/n}$

$0.8903 \pm 1.96\sqrt{(0.8903)(0.1097)/1012}$

0.8903 ± 0.0193

$0.871 < p < 0.910$

c. Yes. Since the entire interval is above 50% = 0.50, there is strong evidence that the majority is opposed to such cloning.

8. a. $\alpha = 0.05$, $z_{\alpha/2} = z_{0.025} = 1.96$ and $E = 0.04$; $\hat{p}$ unknown, use $\hat{p} = 0.5$

$n = [(z_{\alpha/2})^2\hat{p}\hat{q}]/E^2$

$= [(1.96)^2(0.5)(0.5)]/(0.04)^2$

$= 600.25$, rounded up to 601

b. $\alpha = 0.05$, $z_{\alpha/2} = z_{0.025} = 1.96$

$n = [z_{\alpha/2} \cdot \sigma/E]^2$

$= [(1.96)(14227)/(750)]^2$

$= 1382.34$, rounded up to 1383

c. To meet both criteria simultaneously , use the larger sample size of n=1383.

9. preliminary values: $n = 8$, $\Sigma x = 30.72$, $\Sigma x^2 = 160.2186$
   a. $\bar{x} = (\Sigma x)/n = (30.72)/8 = 3.840$ lbs
   b. σ unknown and n=8: assuming a normal distribution, use t with df=7
      $s^2 = [n(\Sigma x^2) - (\Sigma x)^2]/[n(n\text{-}1)]$
      $= [8(160.2186) - (30.72)^2]/[8(7)] = 6.03626$
      $s = 2.4569$
      $\alpha = 0.05$, $t_{df,\alpha/2} = t_{7,0.025} = 2.365$
      $\bar{x} \pm t_{\alpha/2} \cdot s/\sqrt{n}$
      $3.8400 \pm 2.365(2.4569)/\sqrt{8}$
      3.8400 ± 2.0543
      $1.786 < \mu < 5.894$ (lbs)
   c. σ known and a normal distribution: use z
      $\alpha = 0.05$, $z_{\alpha/2} = z_{0.025} = 1.96$
      $\bar{x} \pm z_{\alpha/2} \cdot \sigma/\sqrt{n}$
      $3.8400 \pm 1.96(3.108)/\sqrt{8}$
      3.8400 ± 2.1537
      $1.686 < \mu < 5.994$ (lbs)

10. preliminary values: $n = 8$, $\Sigma x = 30.72$, $\Sigma x^2 = 160.2186$
    $\bar{x} = (\Sigma x)/n = (30.72)/8 = 3.840$
    $s^2 = [n(\Sigma x^2) - (\Sigma x)^2]/[n(n\text{-}1)]$
    $= [8(160.2186) - (30.72)^2]/[8(7)] = 6.03626$
    $s = 2.4569$
    a. $\alpha = 0.05$ and $df = 7$; $\chi_L^2 = \chi_{df,1-\alpha/2}^2$ and $\chi_R^2 = \chi_{df,\alpha/2}^2$
       $(n\text{-}1)s^2/\chi_R^2 < \sigma^2 < (n\text{-}1)s^2/\chi_L^2$
       $(7)(6.03626)/16.013 < \sigma^2 < (7)(6.03626)/1.690$
       $2.6387 < \sigma^2 < 25.0022$
       $1.624 < \sigma < 5.000$ (lbs)
    b. $\alpha = 0.05$ and $df = 7$; $\chi_L^2 = \chi_{df,1-\alpha/2}^2$ and $\chi_R^2 = \chi_{df,\alpha/2}^2$
       $(n\text{-}1)s^2/\chi_R^2 < \sigma^2 < (n\text{-}1)s^2/\chi_L^2$
       $(7)(6.03626)/16.013 < \sigma^2 < (7)(6.03626)/1.690$
       $2.639 < \sigma^2 < 25.002$ (lbs^2)

Cumulative Review Exercises

1. scores in order: 103 105 110 119 119 123 125 125 127 128
   preliminary values: $n = 10$, $\Sigma x = 1184$, $\Sigma x^2 = 140948$
   a. $\bar{x} = (\Sigma x)/n = (1184)/10 = 118.4$ lbs
   b. $\tilde{x} = (x_5 + x_6)/2 = (119 + 123)/2 = 121.0$ lbs
   c. $s^2 = [n(\Sigma x^2) - (\Sigma x)^2]/[n(n\text{-}1)]$
      $= [10(140948) - (1184)^2]/[10(9)] = 84.711$
      $s = 9.2$ lbs

2. Ratio, since differences are meaningful and there is a meaningful zero.

3. From Exercise 1, $\overline{x}$ = 118.4 and s = 9.204.
   $\alpha = 0.05$, $t_{df,\alpha/2} = t_{9,0.025} = 2.262$
   $\overline{x} \pm t_{\alpha/2} \cdot s/\sqrt{n}$
   $118.4 \pm 2.262(9.204)/\sqrt{10}$
   118.4 ± 6.6
   $111.8 < \mu < 125.0$ (lbs)

4. $\alpha = 0.05$, $z_{\alpha/2} = z_{0.05} = 1.96$
   $n = [z_{\alpha/2} \cdot \sigma/E]^2$
   $= [(1.96)(7.5)/(2)]^2 = 54.0225$, rounded up to 55

5. Let D = an applicant tests positive for drugs.
   P(D) = 0.038
   a. $P(\overline{D}) = 1 - P(D)$
   $= 1 - 0.038$
   $= 0.962$

   b. $P(D_1 \text{ and } D_2) = P(D_1) \cdot P(D_2|D_1)$
   $= (0.038)(0.038) = 0.00144$

   c. binomial: n=500 and p=0.038
   normal approximation appropriate since
   $np = 500(0.038) = 19 \geq 5$
   $nq = 500(0.962) = 481 \geq 5$
   $\mu = np = 500(0.038) = 19$
   $\sigma = \sqrt{npq} = \sqrt{500(0.038)(0.962)} = 4.275$
   $P(x \geq 20)$
   $= P(x > 19.5)$
   $= P(z > 0.12)$
   $= 1 - 0.5478$
   $= 0.4522$

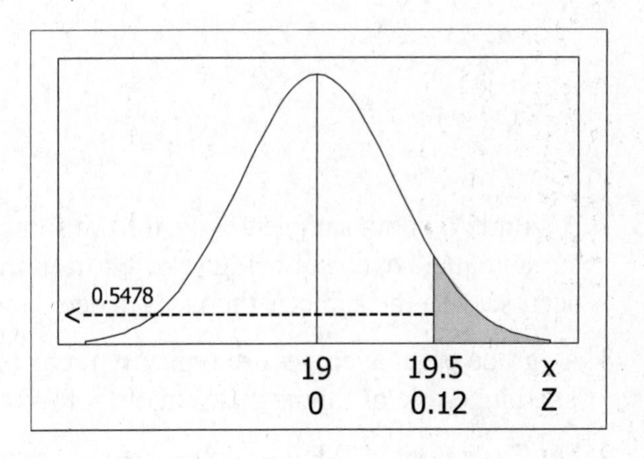

6. a. normal distribution
   $\mu = 21.1$
   $\sigma = 4.8$
   $P(x > 20.0)$
   $= P(z > -0.23)$
   $= 1 - 0.4090$
   $= 0.5910$

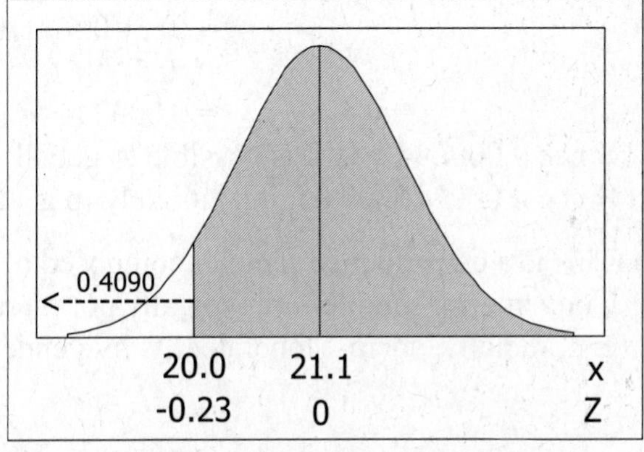

NOTE: Since ACT scores are whole numbers, another valid interpretation of part (a) is $P(x>20) = P_C(x>20.5) = P(z>-0.13) = 1 - 0.4483 = 0.5517$. Presumably the "20.0" was specified in the exercise to discourage this interpretation and to allow for a direct comparison to part (b).

b. normal distribution,
since the original distribution is so

$\mu_{\bar{x}} = \mu = 21.1$

$\sigma_{\bar{x}} = \sigma/\sqrt{n} = 4.8/\sqrt{25} = 0.96$

$P(\bar{x} > 20.0)$
$= P(z > -1.15)$
$= 1 - 0.1251$
$= 0.8749$

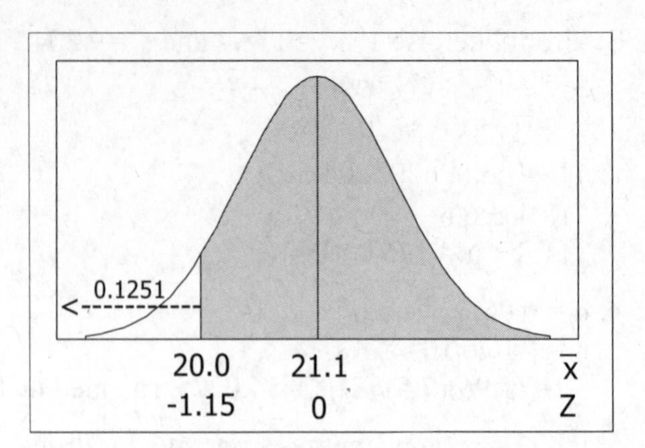

c. normal distribution: $\mu = 21.1$, $\sigma = 4.8$
For P_{90}, A = 0.9000 [0.8997]
and z = 1.28 [from z table]
or z = 1.282 [from last row of t table]

$x = \mu + z\sigma$
$= 21.1 + (1.282)(4.8)$
$= 21.1 + 6.2$
$= 27.3$

7. A simple random sample of size n from some population occurs when every sample of size n has an equal chance of being selected from that population. A voluntary response sample occurs when the subjects themselves decide whether to be included.

8. As grade point averages are typically reported to two decimal places, R = 4.00 – 0.00 = 4.00. The range rule of thumb states that $\sigma \approx R/4 = 4.00/4 = 1.000$.

9. Let C = getting a T-F question correct by random guessing.
$P(C_1 \text{ and } C_2 \text{ and } \ldots \text{and } C_{12}) = P(C_1)\cdot P(C_2)\cdot \ldots \cdot P(C_{12})$
$= (0.5)(0.5)\ldots(0.5)$
$= (0.5)^{12}$
$= 0.000244$

Since 0.000244 > 0, it is possible to get all 12 questions correct by random guessing.
Since 0.000244 < 0.05, it is unlikely to get all 12 questions correct by random guessing.

10. This is a convenience sample, composed of those friends who happen to be available. Convenience samples are typically not representative of the population in a variety of ways – e.g., racially, socio-economically, by gender, etc.

Chapter 8

Hypothesis Testing

8-2 Basics of Hypothesis Testing

1. Given the large sample size and the fact that 20% is so much less than 50%, it is apparent that any confidence interval for the proportion of bosses that are good communicators would fall entirely below 50%. Assuming the magazine has properly interpreted the survey, the results appear to support the claim that "less than 50% of bosses are good communicators" but not necessarily that "less than 50% of the people believe that bosses are good communicators" – those are two different statements that should not be confused.

   Given that the responders constitute a voluntary response sample, and not a random sample, it is likely that they are not representative of the population and consist largely of people with strong feelings on and/or a personal interest in the topic. The results should not be used to support the stated claim.

3. No. Since the claim that the mean is equal to a specific value must be the null hypothesis, the only possible conclusions are to reject that claim or to fail to reject that claim. Hypothesis testing cannot be used to support a claim that a parameter is equal to a particular value.

5. If the claim were not true, and $p \leq 0.5$, then getting 90 heads in a sample of 100 tosses would be an unusual event. There is sufficient evidence to support the claim.

7. If the claim were not true, and $\mu \geq 75$, then getting a mean pulse rate of 74.4 in a sample of students would not be an unusual event. There is not sufficient evidence to support the claim.

9. original claim: $\mu > \$60,000$
   H_o: $\mu = \$60,000$
   H_1: $\mu > \$60,000$

11. original claim: $\sigma = 0.62$ °F
    H_o: $\sigma = 0.62$ °F
    H_1: $\sigma \neq 0.62$ °F

13. original claim: $\sigma < 40$ seconds
    H_o: $\sigma = 40$ seconds
    H_1: $\sigma < 40$ seconds

15. original claim: $p = 0.80$
    H_o: $p = 0.80$
    H_1: $p \neq 0.80$

17. Two-tailed test; place $\alpha/2 = 0.005$ in each tail.
    Use $A = 1-\alpha/2 = 1-0.0050 = 0.9950$ and $z = 2.575$.
    critical values: $\pm z_{\alpha/2} = \pm z_{0.005} = \pm 2.575$

19. Right-tailed test; place $\alpha = 0.02$ in the upper tail.
    Use $A = 1-\alpha = 1-0.000 = 0.9800$ [closest entry = 0.9798] and $z = 2.05$.
    critical value: $+z_{\alpha} = +z_{0.02} = +2.05$

21. Two-tailed test; place $\alpha/2 = 0.025$ in each tail.
Use A = $1-\alpha/2 = 1-0.0250 = 0.9750$ and $z = 1.96$.
critical values: $\pm z_{\alpha/2} = \pm z_{0.025} = \pm 1.96$

23. Left-tailed test; place $\alpha = 0.005$ in the lower tail.
Use A = $\alpha = 0.0050$ and $z = -2.575$.
critical value: $-z_{\alpha} = -z_{0.005} = -2.575$

25. $\hat{p} = x/n = 152/580 = 0.262$
$z_{\hat{p}} = (\hat{p} - p)/\sqrt{pq/n}$
$= (0.262 - 0.250)/\sqrt{(0.25)(0.75)/580} = 0.012/0.0180 = 0.67$

27. $\hat{p} = x/n = 314/1122 = 0.280$
$z_{\hat{p}} = (\hat{p} - p)/\sqrt{pq/n}$
$= (0.280 - 0.250)/\sqrt{(0.25)(0.75)/1122} = 0.030/0.0129 = 2.31$

29. P-value
$= P(z<-1.25)$
$= 0.1056$
Since $0.1056 > 0.05$, fail to reject H_o.

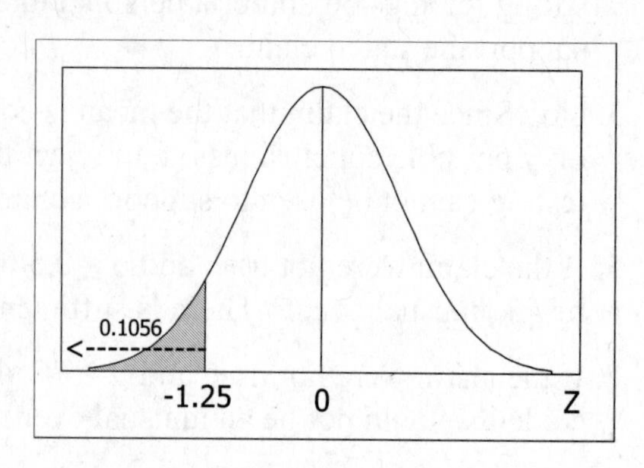

31. P-value
$= 2 \cdot P(z>1.75)$
$= 2 \cdot [1 - 0.9599]$
$= 2 \cdot [0.0401]$
$= 0.0802$
Since $0.0802 > 0.05$, fail to reject H_o.

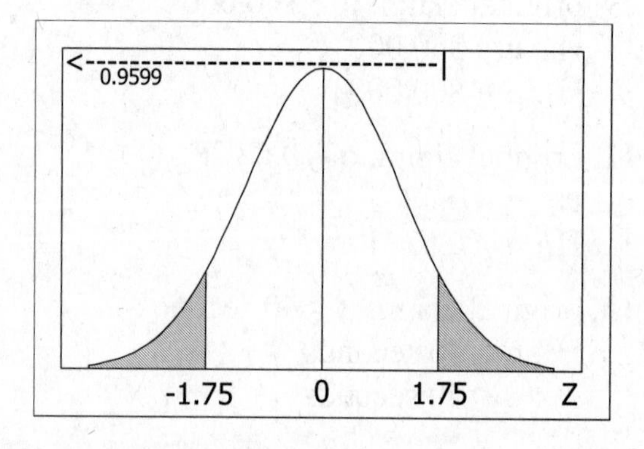

33. P-value
$= 2 \cdot P(z<-2.75)$
$= 2 \cdot (0030)$
$= 0.0060$
Since $0.0060 < 0.05$, reject H_o.

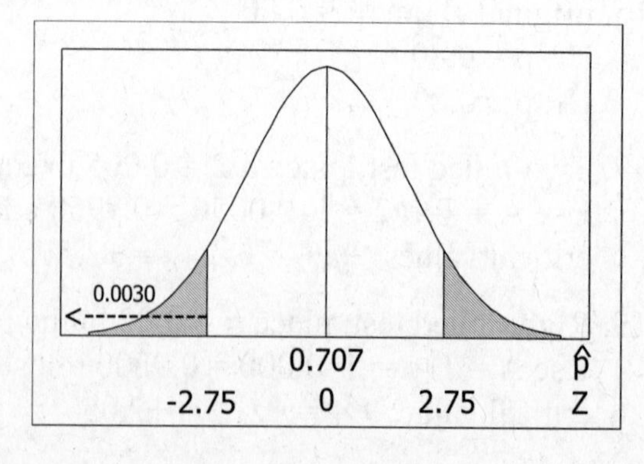

35. P-value
 $= P(z>2.30)$
 $= 1 - 0.9893$
 $= 0.0107$
 Since $0.0107 < 0.05$, reject H_o.

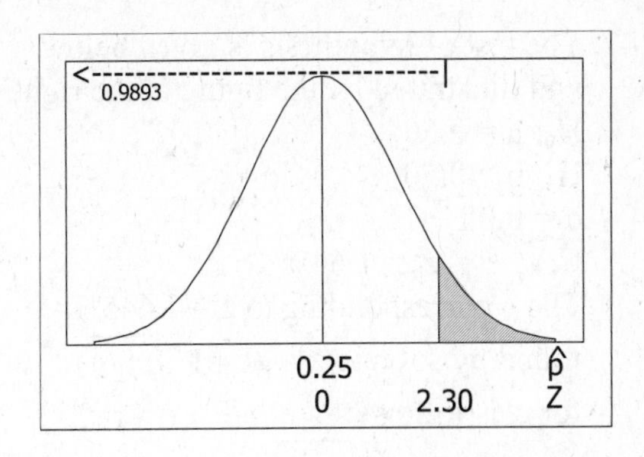

37. There is not sufficient evidence to conclude that the percentage of blue M&M's is greater than 5%.

39. There is not sufficient evidence to reject the claim that the percentage of Americans who know their credit score is equal to 20%.

41. original claim: $p = 0.41$
 H_o: $p = 0.41$
 type I error: rejecting H_o when H_o is actually true
  rejecting the claim that the percentage of non-smokers exposed to secondhand smoke is 41% when that percentage actually is 41%
 type II error: failing to reject H_o when H_1 is actually true
  failing to reject the claim that the percentage of non-smokers exposed to secondhand smoke is 41% what that percentage is actually different from 41%

43. original claim: $p > 0.70$
 H_o: $p = 0.70$
 type I error: rejecting H_o when H_o is actually true
  rejecting the claim that the percentage of college students who use alcohol is 70% when that percentage actually is 70%
 type II error: failing to reject H_o when H_1 is actually true
  failing to reject the claim that the percentage of college students who use alcohol is 70% when that percentage is actually greater than 70%

45. a. Not necessarily. Being 95% confident of a conclusion (i.e., rejecting the null hypothesis at the 0.05 level of significance) does not guarantee being 99% confident of that same conclusion (i.e., rejecting the null hypothesis at the 0.01 level of significance). Rejecting at the 0.01 level requires more evidence (i.e. a larger difference between the null hypothesis and the observed data) than rejecting at the 0.05 level. More simply: Rejection at the 0.05 level of significance means that the observed sample result was among the most extreme 5% of the possible results, but not necessarily among the most extreme 1% of the possible results as required for rejection at the 0.01 level of significance.
 b. Yes. Being 99% confident of a conclusion (i.e., rejecting the null hypothesis at the 0.01 level of significance) guarantees being 95% confident of that same conclusion (i.e., rejecting the null hypothesis at the 0.05 level of significance). Rejecting at the 0.05 level requires less evidence (i.e. a smaller difference between the null hypothesis and the observed data) than rejecting at the 0.01 level. More simply, rejection at the 0.01 level of significance means that the observed sample result was among the most extreme 1% of the possible results, and so it must be included among the most extreme 5% of the possible results as required for rejection at the 0.05 level of significance.

47. The test of hypothesis is given below and illustrated by the figure at the right.
H_o: $p = 0.50$
H_1: $p > 0.50$
$\alpha = 0.05$
C.V. $z = z_\alpha = 1.645$
The c corresponding to $z = 1.645$ is found by solving $z_c = (c-p)/\sqrt{pq/n}$ for c as follows:
$c = p + z_c \cdot \sqrt{pq/n}$
$= 0.50 + (1.645) \cdot \sqrt{(0.50)(0.50)/64}$
$= 0.50 + (1.645) \cdot (0.0625)$
$= 0.6028$

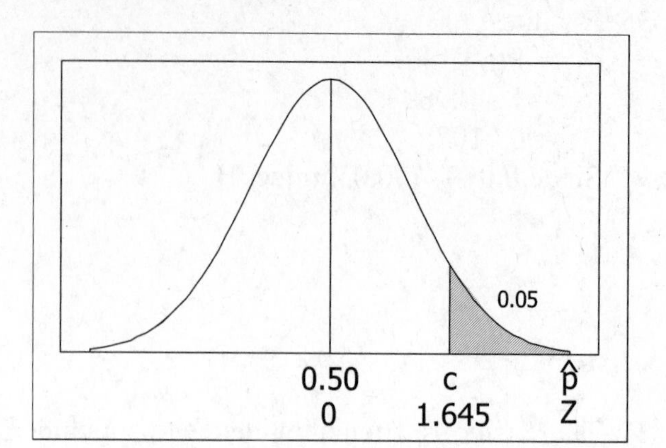

a. The power calculations are given below and illustrated by the figure at the right.
power = P(rejecting $H_o|H_o$ is false)
$= P(\hat{p} > .6028|p=0.65)$
$= P(z > -0.79)$
$= 1 - P(z < -0.79)$
$= 1 - 0.2148 = 0.7852$
The z corresponding to c= 0.6028 is found as follows:
$z_c = (c-p)/\sqrt{pq/n}$
$= (0.6028-0.65)/\sqrt{(0.65)(0.35)/64}$
$= -0.0472/0.0596 = -0.79$

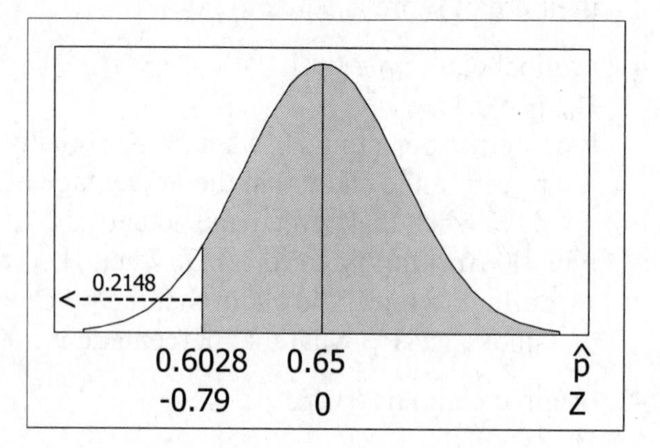

b. The shaded region in the bottom graph represents the probability of getting a sample proportion greater than 0.6028 (i.e., of rejecting the null hypothesis that the proportion is 0.50) whenever the population proportion is actually 0.65. This is precisely the power of the test (i.e., the probability of correctly rejecting a false null hypothesis when a specified alternative is true).

8-3 Testing a Claim About a Proportion

1. There were 1261 + 491 + 384 = 2136 total responses. The sample proportion of yes responses is $\hat{p} = x/n = 491/2136 = 0.230$. The symbol $\hat{p}$ is used to represent a sample proportion.

3. The value of the sample proportion is $\hat{p} = x/n = 123/280 = 0.439$. Since $0.979 > 0.05$, the observed results could easily occur by chance whenever the null hypothesis is true. There is not sufficient evidence to reject the hypothesis that the proportion of correct responses is $p = 0.50$ (i.e., the proportion expected by chance alone) in favor of the claim that $p > 0.50$.

NOTE: To reinforce the concept that all z scores are standardized rescalings obtained by subtracting the mean and dividing by the standard deviation, the manual uses the "usual" z formula written to apply to $\hat{p}$'s

$$z_{\hat{p}} = (\hat{p} - \mu_{\hat{p}})/\sigma_{\hat{p}}.$$

When the normal approximation to the binomial applies, the $\hat{p}$'s are normally distributed with

$$\mu_{\hat{p}} = p \text{ and } \sigma_{\hat{p}} = \sqrt{pq/n}.$$

And so the formula for the z statistic may also be written as $z_{\hat{p}} = (\hat{p} - p)/\sqrt{pq/n}$.
In addition, the manual continues to use the more accurate $z_{0.01} = 2.326$ taken from the "large" row of the t table rather than the 2.33 obtained by reading the z table backwards.

5. a. $\hat{p} = x/n = 530/1000 = 0.530$

   $z_{\hat{p}} = (\hat{p} - \mu_{\hat{p}})/\sigma_{\hat{p}}$
   $= (\hat{p} - p)/\sqrt{pq/n}$
   $= (0.530 - 0.50)/\sqrt{(0.50)(0.50)/1000}$
   $= 0.030/0.0158$
   $= 1.90$

   b. $z = \pm z_{\alpha/2} = \pm z_{0.005} = \pm 2.575$

   c. P-value $= 2 \cdot P(z>1.90)$
   $= 2 \cdot (1 - 0.9713)$
   $= 2 \cdot (0.0287)$
   $= 0.0574$

   d. Do not reject H_o; there is not sufficient evidence to reject the claim that the percentage of all college applications that are submitted online is 50%.

   e. No. A hypothesis test will either "reject" or "fail to reject" a claim that a population parameter is equal to a specified value.

7. original claim: $p = 0.75$

   $\hat{p} = x/n = 1640/2246 = 0.730$

   Ho: $p = 0.75$
   H_1: $p \neq 0.75$
   $\alpha = 0.05$
   C.V. $z = \pm z_{\alpha/2} = 1.96$
   calculations:
   $z_{\hat{p}} = (\hat{p} - \mu_{\hat{p}})/\sigma_{\hat{p}} = -2.17$ [TI-83/84 Plus]
   P-value = 0.0301 [TI-83/84 Plus]

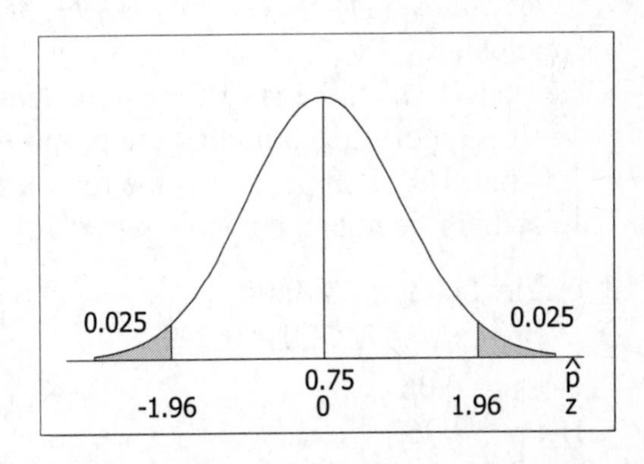

   conclusion:
   Reject H_o; there is sufficient evidence to reject the claim that $p = 0.75$ and conclude that $p \neq 0.75$ (in fact, that $p < 0.75$). There is sufficient evidence to reject the claim that the proportion of adults who use cell phones while driving is 75%.

NOTE: The "in fact, that $p < 0.75$" in the conclusion for Exercise 7 is given in parentheses. Some instructors prefer not to suggest one-sided conclusions from two sided tests. This manual follows the common sense approach consistent with the text's interpretation of two-sided confidence intervals that are entirely above or below a specified value.

9. original claim: $p = 0.75$

   $\hat{p} = x/n = 589/745 = 0.791$

   Ho: $p = 0.75$

   H_1: $p \neq 0.75$

   $\alpha = 0.01$

   C.V. $z = \pm z_{\alpha/2} = \pm z_{0.005} = \pm 2.575$

   calculations:

   $z_{\hat{p}} = (\hat{p} - \mu_{\hat{p}})/\sigma_{\hat{p}}$

   $= (0.791 - 0.75)/\sqrt{(0.75)(0.25)/745}$

   $= 0.041/0.01586 = 2.56$

   P-value $= 2 \cdot P(z>2.56) = 2 \cdot (1 - 0.9948) = 2 \cdot (0.0052) = 0.0104$

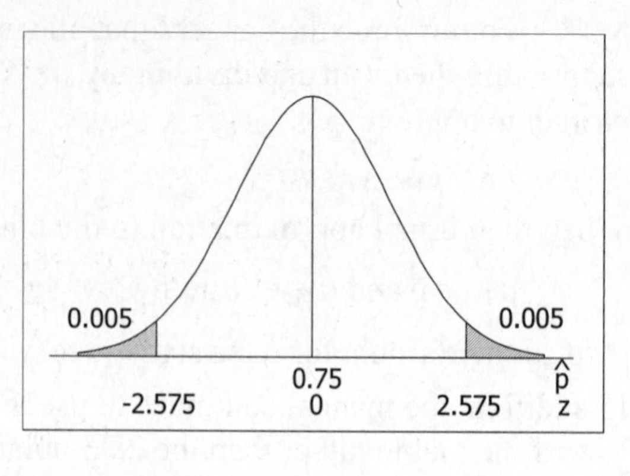

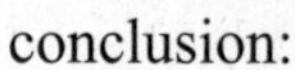

   conclusion:

   Do not reject H_o; there is not sufficient evidence to reject the claim that $p = 0.75$. There is not sufficient evidence to reject the claim that the proportion of adults who say it is morally wrong to not report all income is 75%.

11. original claim: $p > 1/3$

   $\hat{p} = x/n = 327/839 = 0.390$

   Ho: $p = 1/3$

   H_1: $p > 1/3$

   $\alpha = 0.01$

   C.V. $z = z_{\alpha} = z_{0.01} = 2.326$

   calculations:

   $z_{\hat{p}} = (\hat{p} - \mu_{\hat{p}})/\sigma_{\hat{p}}$

   $= (0.390 - 0.333)/\sqrt{(1/3)(2/3)/839}$

   $= 0.056/0.1627 = 3.47$

   P-value $= P(z>3.47) = 1 - 0.9997 = 0.0003$

   conclusion:

   Reject H_o; there is sufficient evidence to conclude that $p > 1/3$. There is sufficient evidence to support the claim that the proportion of tennis challenges that are successful is greater than 1/3. It appears that the referees are erring on more than 1/3 of the challenged calls – which is not an enviable record, even if the proportion of challenged calls is quite small.

13. original claim: $p > 0.06$

   $\hat{p} = x/n = 72/724 = 0.099$

   Ho: $p = 0.06$

   H_1: $p > 0.06$

   $\alpha = 0.05$

   C.V. $z = z_{\alpha} = z_{0.05} = 1.645$

   calculations:

   $z_{\hat{p}} = (\hat{p} - \mu_{\hat{p}})/\sigma_{\hat{p}}$

   $= (0.099 - 0.06)/\sqrt{(0.06)(0.94)/724}$

   $= 0.039/0.008826 = 4.47$

   P-value $= P(z>4.47) = 1 - 0.9999 = 0.0001$

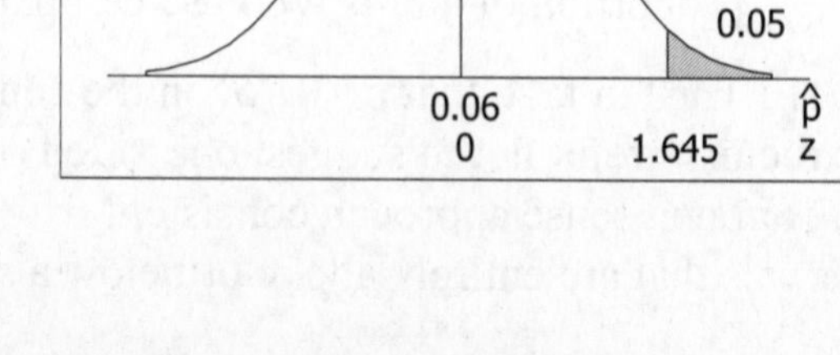

   conclusion:

   Reject H_o; there is not sufficient evidence to conclude that $p > 0.06$. There is sufficient evidence to support the claim that the proportion of Tamiflu recipients who experience nausea is greater than the 6% rate experienced by those who took the placebo. Yes; nausea does appear to be a legitimate concern for recipients of Tamiflu.

15. original claim: $p \neq 0.000340$

$\hat{p} = x/n = 135/420{,}095 = 0.000321$

Ho: $p = 0.000340$

H_1: $p \neq 0.000340$

$\alpha = 0.005$

C.V. $z = \pm z_{\alpha/2} = \pm z_{0.0025} = \pm 2.81$

calculations:

$z_{\hat{p}} = (\hat{p} - \mu_{\hat{p}})/\sigma_{\hat{p}}$

$= (0.000321 - 0.000340)/\sqrt{\dfrac{(0.000340)(0.999660)}{420{,}095}}$

$= -0.000019/0.0000284 = -0.66$

P-value $= 2 \cdot P(z<-0.66) = 2 \cdot (0.2546) = 0.5092$

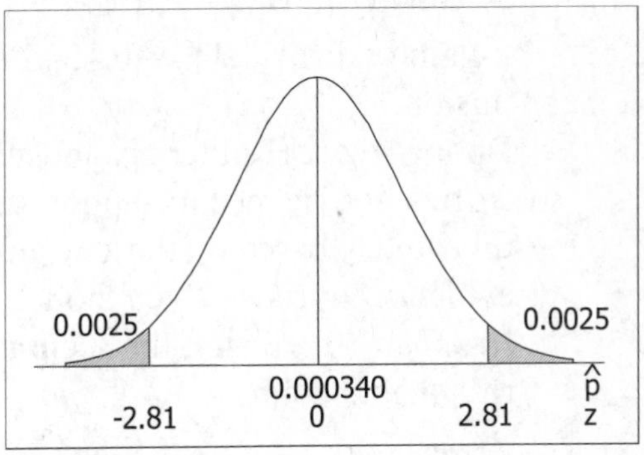

conclusion:

Do not reject H_o; there is not sufficient evidence to conclude that $p \neq 0.000340$. There is not sufficient evidence to support the claim that the proportion cell phone users who develop such cancers is different from the rate for people who do not use cell phones. No; based on this study, cell phone users have no extra reasons to be concerned about cancer of the brain or nervous system.

17. original claim: $p < 0.20$

$\hat{p} = x/n = 1299/(1299 + 5686) = 1299/6985 = 0.186$

Ho: $p = 0.20$

H_1: $p < 0.20$

$\alpha = 0.01$

C.V. $z = -z_{\alpha} = -z_{0.01} = -2.326$

calculations:

$z_{\hat{p}} = (\hat{p} - \mu_{\hat{p}})/\sigma_{\hat{p}}$

$= (0.186 - 0.20)/\sqrt{(0.20)(0.80)/6985}$

$= -0.014/0.004786 = -2.93$

P-value $= P(z<-2.93) = 0.0017$

conclusion:

Reject H_o; there is sufficient evidence to conclude that $p < 0.20$. There is sufficient evidence to support the claim that less than 20% of Michigan gas pumps are inaccurate. No; from the perspective of the consumer, the rate does not appear to be low enough. While the point estimate of 0.186 indicates the rate is lower than 20%, it should probably be about 1/10 of that.

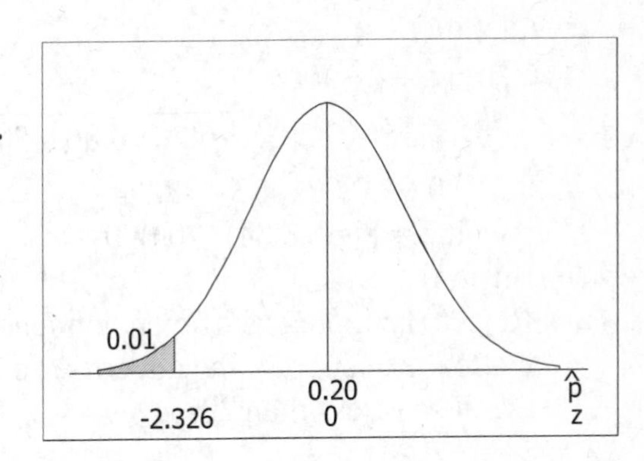

19. original claim: $p < 0.80$

$\hat{p} = x/n = 74/98 = 0.755$

Ho: $p = 0.80$

H_1: $p < 0.80$

$\alpha = 0.05$

C.V. $z = -z_{\alpha} = -z_{0.05} = -1.645$

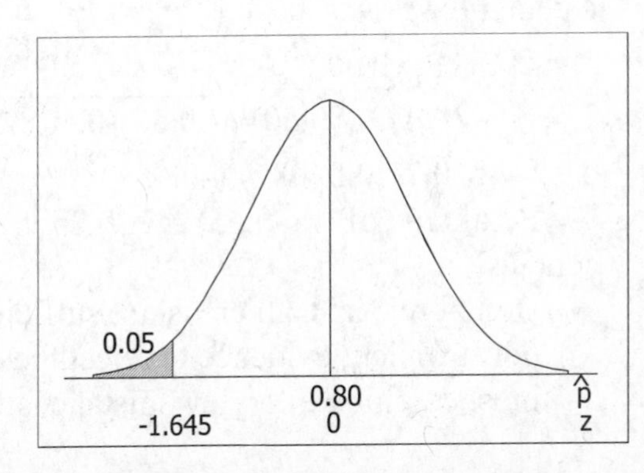

calculations:

$z_{\hat{p}} = (\hat{p} - \mu_{\hat{p}})/\sigma_{\hat{p}}$

$= (0.755 - 0.80)/\sqrt{(0.80)(0.20)/98}$

$= -0.45/0.04041 = -1.11$

P-value $= P(z<-1.11) = 0.1335$

conclusion:

Do not reject H_o; there is not sufficient evidence to conclude that $p < 0.80$. There is not sufficient evidence to support the claim polygraph tests are correct less than 80% of the time. Yes; based on these results, polygraph test results should probably be prohibited as evidence in trials. Even though the point estimate of 75.5% accuracy does not support the less than 80% claim, the accuracy rate is still far too small to make conclusions beyond a reasonable doubt.

21. NOTE: The value for x is not given. In truth, any $726 \leq x \leq 744$ rounds to the given $\hat{p} = x/5000 = 15\%$. For want of more precise information, use $\hat{p} = 0.15$.

original claim: $p < 0.20$

$\hat{p} = x/n = x/5000 = 0.15$

Ho: $p = 0.20$

H_1: $p < 0.20$

$\alpha = 0.01$

C.V. $z = -z_\alpha = -z_{0.01} = -2.326$

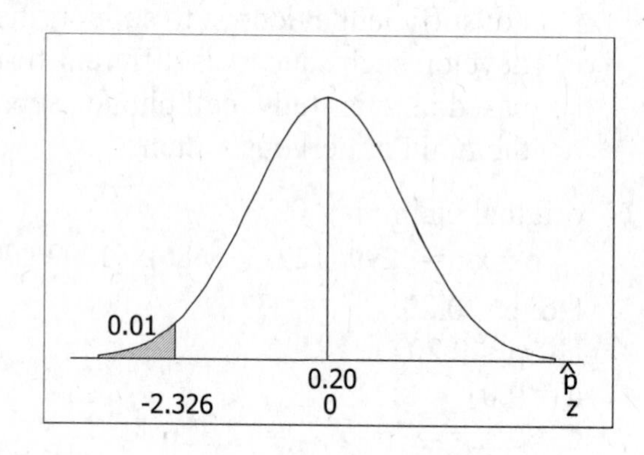

calculations:

$z_{\hat{p}} = (\hat{p} - \mu_{\hat{p}})/\sigma_{\hat{p}}$

$= (0.15 - 0.20)/\sqrt{(0.20)(0.80)/5000}$

$= -0.05/0.005657 = -8.84$

P-value $= P(z<-8.84) = 0.0001$

conclusion:

Reject H_o; there is sufficient evidence to conclude that $p < 0.20$. There is sufficient evidence to support the advertiser's claim that the proportion of households tuned to *60 Minutes* is less than 20%.

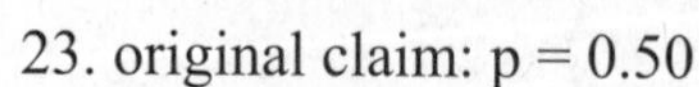

23. original claim: $p = 0.50$

use $x = (.473)(150) = 70.95$, rounded to 71

$\hat{p} = x/n = 71/150 = 0.473$

Ho: $p = 0.50$

H_1: $p \neq 0.50$

$\alpha = 0.05$ [assumed]

C.V. $z = \pm z_{\alpha/2} = \pm z_{0.025} = \pm 1.96$

calculations:

$z_{\hat{p}} = (\hat{p} - \mu_{\hat{p}})/\sigma_{\hat{p}}$

$= (0.473 - 0.50)/\sqrt{(0.50)(0.50)/150}$

$= -0.27/0.04082 = -0.65$

P-value $= 2 \cdot P(z<-0.65) = 2 \cdot (0.2578) = 0.5156$

conclusion:

Do not reject H_o; there is not sufficient evidence to reject the claim that $p = 0.50$. There is not sufficient evidence to reject the claim that the proportion of executives who say the most common interview mistake is failure to know the company is 50%. The important lesson is that a person going for a job interview should prepare by learning about the company at which he is applying.

25. NOTE: The value for x is not given. In truth, any $2183 \leq x \leq 2213$ rounds to the given $\hat{p} = x/3011 = 73\%$. For want of more precise information, use $\hat{p} = 0.73$.

original claim: $p = 0.75$

$\hat{p} = x/n = x/3011 = 0.73$

Ho: $p = 0.75$

H_1: $p \neq 0.75$

$\alpha = 0.05$ [assumed]

C.V. $z = \pm z_{\alpha/2} = \pm z_{0.025} = \pm 1.96$

calculations:

$z_{\hat{p}} = (\hat{p} - \mu_{\hat{p}})/\sigma_{\hat{p}}$

$= (0.73 - 0.75)/\sqrt{(0.75)(0.25)/3011}$

$= -0.02/0.007891 = -2.53$

P-value $= 2 \cdot P(z<-2.53) = 2 \cdot (0.0057) = 0.0114$

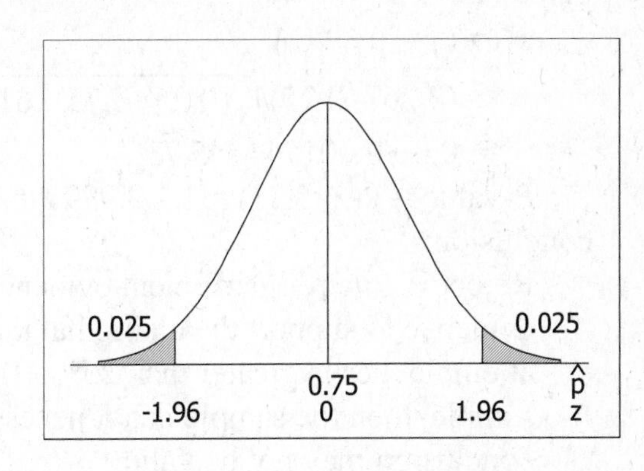

conclusion:

Reject H_o; there is sufficient evidence to reject the claim that $p = 0.75$ and conclude that $p \neq 0.75$ (in fact, that $p < 0.75$). There is sufficient evidence to reject the claim that the proportion of adults who use the Internet is ¾. While the difference between 0.73 and 0.75 may be of little practical significance, in the interest of accuracy the reporter should not write that ¾ of all adults use the Internet.

27. NOTE: The problem as stated is not possible. No whole number x yields $x/59 = 43\%$, as $25/59 = 42.4\%$ and $26/59 = 44.1\%$. Use the closest possible value, $x = 25$.

original claim: $p = 0.50$

$\hat{p} = x/n = 25/59 = 0.424$

Ho: $p = 0.50$

H_1: $p \neq 0.50$

$\alpha = 0.05$

C.V. $z = \pm z_{\alpha/2} = \pm z_{0.025} = \pm 1.96$

calculations:

$z_{\hat{p}} = (\hat{p} - \mu_{\hat{p}})/\sigma_{\hat{p}}$

$= (0.424 - 0.50)/\sqrt{(0.50)(0.50)/59}$

$= -0.076/0.06509 = -1.17$

P-value $= 2 \cdot P(z<-1.17) = 2 \cdot (0.1210) = 0.2420$

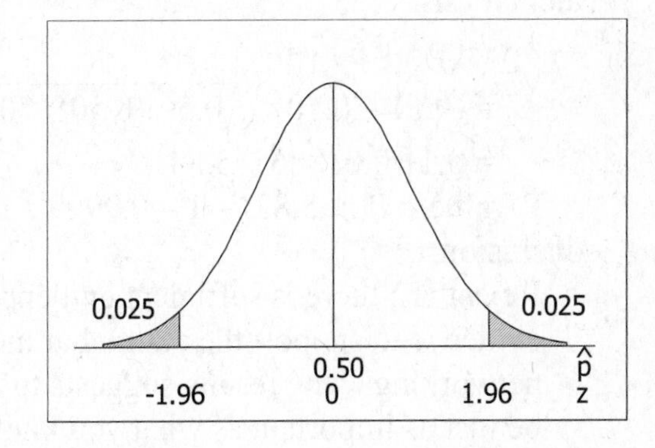

conclusion:

Do not reject H_o; there is not sufficient evidence to reject the claim that $p = 0.50$. There is not sufficient evidence to reject the claim that women with 12 years of education or less have no ability to predict the gender of their babies. Conclude that the guesses of these women do not differ significantly from the results of random guesses.

29. NOTE: The value for x is not given. In truth, any $15{,}720 \leq x \leq 16{,}336$ rounds to the given $\hat{p} = x/61{,}647 = 26\%$. For want of more precise information, use $\hat{p} = 0.26$.

original claim: $p > 0.25$

$\hat{p} = x/n = x/61{,}647 = 0.26$

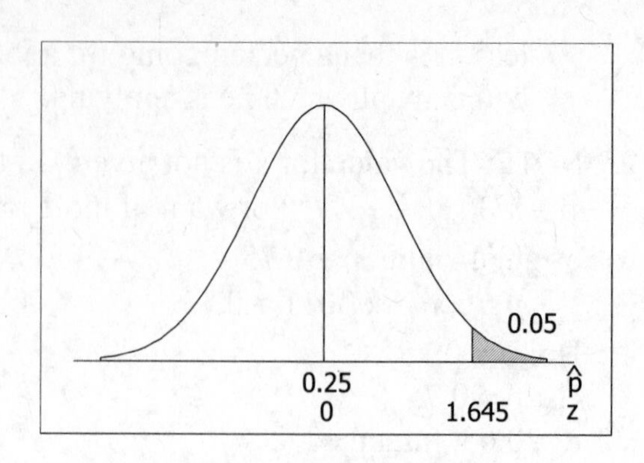

Ho: $p = 0.25$
H_1: $p > 0.25$
$\alpha = 0.05$
C.V. $z = z_\alpha = z_{0.05} = 1.645$
calculations:
$z_{\hat{p}} = (\hat{p} - \mu_{\hat{p}})/\sigma_{\hat{p}}$
$= (0.26 - 0.25)/\sqrt{(0.25)(0.75)/61{,}647}$
$= 0.01/0.001744 = 5.73$
P-value = $P(z>5.73) = 1 - 0.9999 = 0.0001$
conclusion:
Reject H_o; there is sufficient evidence to conclude that $p > 0.25$. There is sufficient evidence to support the claim that the proportion of employees who say that bosses scream at employees is greater than 25%. If the survey was done with a voluntary response sample, then the sample is not necessarily representative of the population and the above conclusion may not be valid.

31. NOTE: The value for x is not given. In truth, any $426 \leq x \leq 432$ rounds to the given $\hat{p} = x/703 = 61\%$. For want of more precise information, use $\hat{p} = 0.61$.
original claim: $p > 0.50$
$\hat{p} = x/n = x/703 = 0.61$
Ho: $p = 0.50$
H_1: $p > 0.50$
$\alpha = 0.05$
C.V. $z = z_\alpha = z_{0.05} = 1.645$
calculations:
$z_{\hat{p}} = (\hat{p} - \mu_{\hat{p}})/\sigma_{\hat{p}}$
$= (0.61 - 0.50)/\sqrt{(0.50)(0.50)/703}$
$= 0.11/0.01886 = 5.83$
P-value = $P(z>5.83) = 1 - 0.9999 = 0.0001$
conclusion:

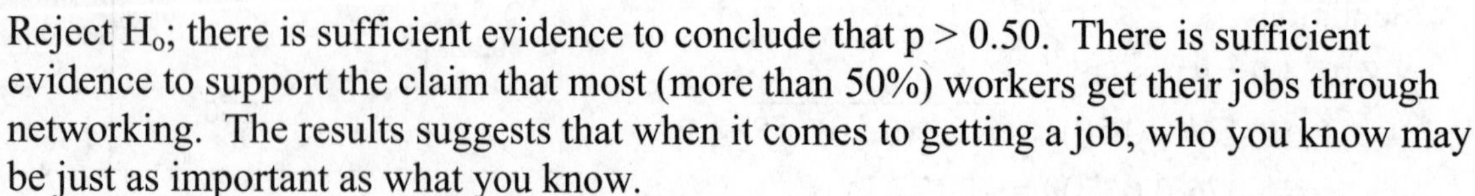

Reject H_o; there is sufficient evidence to conclude that $p > 0.50$. There is sufficient evidence to support the claim that most (more than 50%) workers get their jobs through networking. The results suggests that when it comes to getting a job, who you know may be just as important as what you know.

33. There are 100 total M&M's, 13 of which are red.
original claim: $p = 0.20$

$\hat{p} = x/n = 13/100 = 0.130$
Ho: $p = 0.20$
H_1: $p \neq 0.20$
$\alpha = 0.05$ [assumed]
C.V. $z = \pm z_{\alpha/2} = \pm z_{0.025} = \pm 1.96$
calculations:
$z_{\hat{p}} = (\hat{p} - \mu_{\hat{p}})/\sigma_{\hat{p}}$
$= (0.130 - 0.20)/\sqrt{(0.20)(0.80)/100}$
$= -0.070/0.04000 = -1.75$
P-value = $2 \cdot P(z<-1.75) = 2 \cdot (0.0401) = 0.0802$

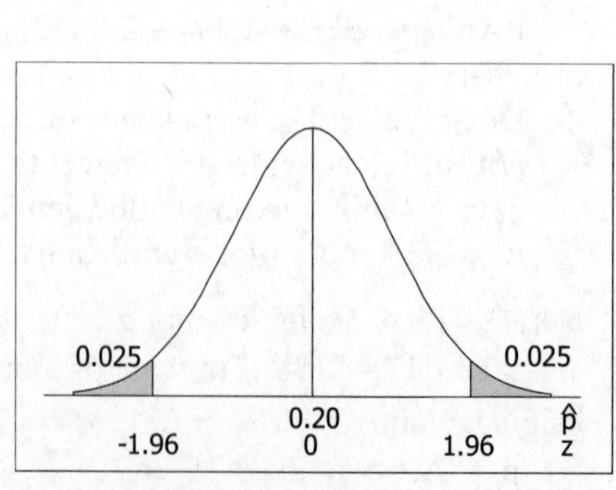

conclusion:

Do not reject H_o; there is not sufficient evidence to reject the claim that $p = 0.20$. There is not sufficient evidence to reject the claim that the proportion of M&M's that are red is 20%.

35. There are 54 total bears, 35 of which are males.

original claim: $p = 0.50$

$\hat{p} = x/n = 35/54 = 0.648$

Ho: $p = 0.50$

H_1: $p \neq 0.50$

$\alpha = 0.05$

C.V. $z = \pm z_{\alpha/2} = \pm z_{0.025} = \pm 1.96$

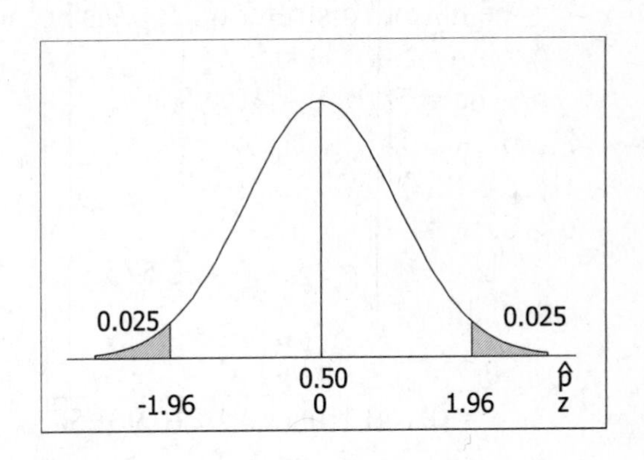

calculations:

$z_{\hat{p}} = (\hat{p} - \mu_{\hat{p}})/\sigma_{\hat{p}}$

$= (0.648 - 0.50)/\sqrt{(0.50)(0.50)/54}$

$= 0.148/0.06804 = 2.18$

P-value $= 2 \cdot P(z>2.18) = 2 \cdot (1 - 0.9854) = 2 \cdot (0.0146) = 0.0292$

conclusion:

Reject H_o; there is sufficient evidence to reject the claim that $p = 0.50$ and conclude that $p \neq 0.50$ (in fact, that $p > 0.50$). There is sufficient evidence to reject the claim that the bears were selected from a population in which the percentage of males is equal to 50%.

37. There are 35 total movies, 12 of which are rated R.

original claim: $p = 0.55$

$\hat{p} = x/n = 12/35 = 0.343$

Ho: $p = 0.55$

H_1: $p \neq 0.55$

$\alpha = 0.01$

C.V. $z = \pm z_{\alpha/2} = \pm z_{0.005} = \pm 2.575$

```
MINITAB RESULT
Binomial with n = 35 and p = 0.55
 x    P( X <= x )
12     0.0109367
```

calculations:

P-value = P(a result as extreme as or more extreme than the observed result| H_o is true)

$= 2 \cdot P(12 \text{ or fewer occurrences} | p = 0.55)$

$= 2 \cdot P(x \leq 12 | p = 0.55)$

$= 2 \cdot (0.0109367)$

$= 0.0218$

conclusion:

Do not reject H_o; there is not sufficient evidence to reject the claim that $p = 0.55$. There is not sufficient evidence to reject the claim that the movies were selected from a population in which the percentage of R-rated movies is 55%.

39. original claim: $p = 0.10$

    $\hat{p} = x/n = 0/50 = 0$

    Yes, the methods of this section can be used. In general, the appropriateness of a test depends on the design of the experiment and not the particular results. For this problem, the normal distribution applies because

    $np = 50(0.1) = 5 \geq 5$

    $nq = 50(0.9) = 45 \geq 5$

    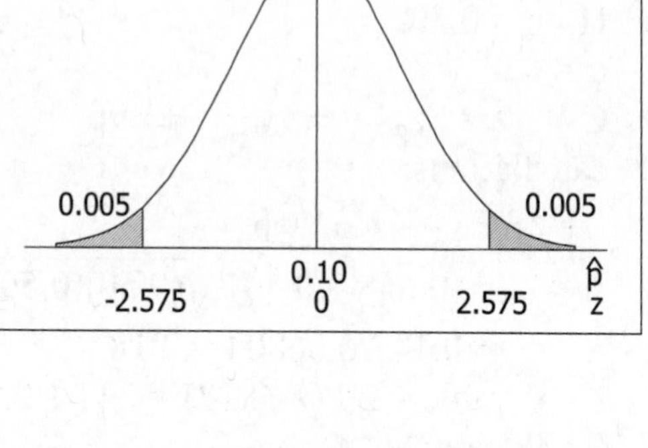

    H_o: $p = 0.10$

    H_1: $p \neq 0.10$

    $\alpha = 0.01$

    C.V. $z = \pm z_{\alpha/2} = \pm z_{0.005} = \pm 2.575$

    calculations:

    $z_{\hat{p}} = (\hat{p} - \mu_{\hat{p}})/\sigma_{\hat{p}}$

    $= (0 - 0.10)/\sqrt{(0.10)(0.90)/50}$

    $= -0.10/0.0424 = -2.36$

    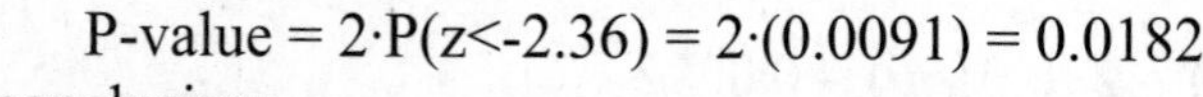

    P-value $= 2 \cdot P(z<-2.36) = 2 \cdot (0.0091) = 0.0182$

    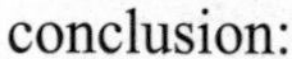

    conclusion:

    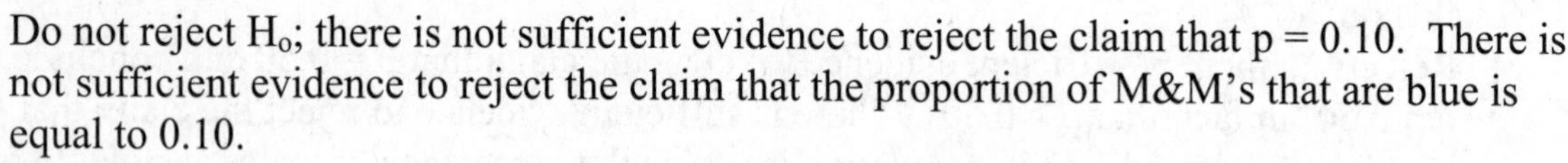

    Do not reject H_o; there is not sufficient evidence to reject the claim that $p = 0.10$. There is not sufficient evidence to reject the claim that the proportion of M&M's that are blue is equal to 0.10.

8-4 Testing a Claim About a Mean: σ Known

1. In order to use the methods of this section to test the claim that $\mu = 1.5$,
   a. the sample must be a simple random sample.
   b. the population standard deviation σ must be known.
   c. the population distribution must be approximately normal (since $n \leq 30$).

3. A one-tailed test at the 0.01 level of significance rejects the null hypothesis if the sample statistic falls into the extreme 1% of the sampling distribution in the appropriate tail. The corresponding (two-sided) confidence interval test that places 1% each tail would be a 98% confidence interval.

5. original claim: $\mu = 5$ cm

    H_o: $\mu = 5$ cm

    H_1: $\mu \neq 5$ cm

    $\alpha = 0.05$ [assumed]

    C.V. $z = \pm z_{\alpha/2} = \pm z_{0.025} = \pm 1.96$

    calculations:

    $z_{\bar{x}} = (\bar{x} - \mu_{\bar{x}})/\sigma_{\bar{x}}$

    $= 1.34$ [TI-83/84]

    P-value $= 0.1797$ [TI-83/84]

    conclusion:

    Do not reject H_o; there is not sufficient evidence to reject the claim that $\mu = 5$. There is not sufficient evidence to reject the claim that women have a mean wrist breadth equal to 5 cm.

7. original claim: $\mu > 210$ sec
   H_o: $\mu = 210$ sec
   H_1: $\mu > 210$ sec
   $\alpha = 0.05$
   C.V. $z = z_\alpha = z_{0.05} = 1.645$
   calculations:
   $z_{\bar{x}} = (\bar{x} - \mu_{\bar{x}})/\sigma_{\bar{x}}$
   $= (252.5 - 210)/(54.5/\sqrt{40})$
   $= 42.5/8.6172 = 4.93$
   P-value = P(z>4.93) = 1 – 0.9999 = 0.0001

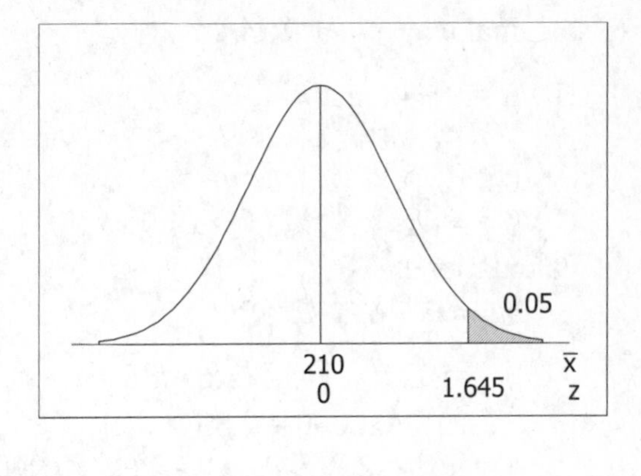

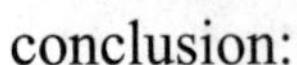
   conclusion:

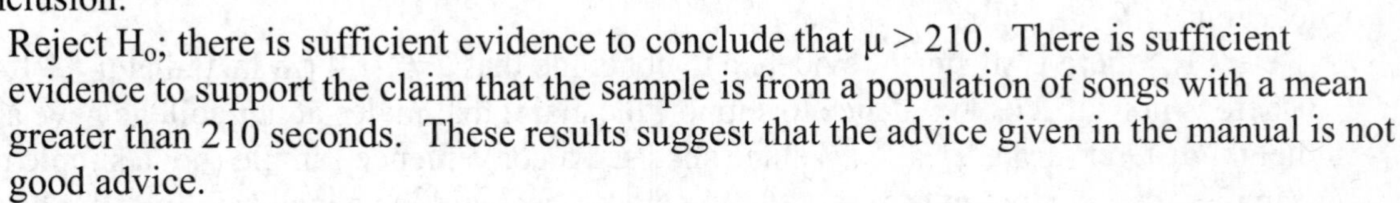
   Reject H_o; there is sufficient evidence to conclude that $\mu > 210$. There is sufficient evidence to support the claim that the sample is from a population of songs with a mean greater than 210 seconds. These results suggest that the advice given in the manual is not good advice.

9. original claim: $\mu = 0.8535$ grams
   H_o: $\mu = 0.8535$ grams
   H_1: $\mu \neq 0.8535$ grams
   $\alpha = 0.05$
   C.V. $z = \pm z_{\alpha/2} = \pm z_{0.025} = \pm 1.96$
   calculations:
   $z_{\bar{x}} = (\bar{x} - \mu_{\bar{x}})/\sigma_{\bar{x}}$
   $= (0.8635 - 0.8535)/(0.0565/\sqrt{19})$
   $= 0.0100/0.01296$
   $= 0.77$
   P-value = 2·P(z>0.77) = 2·(1 – 0.7794) = 2·(0.2206) = 0.4412

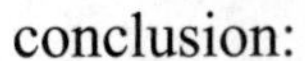
   conclusion:
   Do not reject H_o; there is not sufficient evidence to reject the claim that $\mu = 0.8535$. There is not sufficient evidence to reject the claim that the mean weight of green M&M's is 0.8535 grams. Yes; green M&M's appear to have weights consistent with the label.

11. original claim: $\mu > 0$ lbs
   H_o: $\mu = 0$ lbs
   H_1: $\mu > 0$ lbs
   $\alpha = 0.01$
   C.V. $z = z_\alpha = z_{0.01} = 2.326$
   calculations:
   $z_{\bar{x}} = (\bar{x} - \mu_{\bar{x}})/\sigma_{\bar{x}}$
   $= (3.0 - 0)/(4.9/\sqrt{40})$
   $= 3.0/0.7748 = 3.87$
   P-value = P(z>3.87) = 1 – 0.9999 = 0.0001

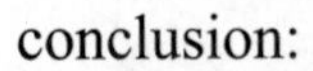
   conclusion:
   Reject H_o; there is sufficient evidence to conclude that $\mu > 0$. There is sufficient evidence to support the claim that the mean weight loss is greater than 0. The diet is effective in that the weight loss is statistically significant – but a mere 3.0 lbs weight loss after following the regimen for an entire year suggests the diet may have no practical significance.

13. original claim: $\mu \neq 91.4$ cm

H_o: $\mu = 91.4$ cm
H_1: $\mu \neq 91.4$ cm
$\alpha = 0.05$
C.V. $z = \pm z_{\alpha/2} = \pm z_{0.025} = \pm 1.96$
calculations:
$z_{\bar{x}} = (\bar{x} - \mu_{\bar{x}})/\sigma_{\bar{x}}$
$= (92.8 - 91.4)/(3.6/\sqrt{36})$
$= 1.4/0.6000 = 2.33$
P-value = $2 \cdot P(z>2.33) = 2 \cdot (1 - 0.9901) = 2 \cdot (0.0099) = 0.0198$
conclusion:
Reject H_o; there is sufficient evidence to conclude that $\mu \neq 91.4$ (in fact, that $\mu > 91.4$). There is not sufficient evidence to support the claim that males at her college have a sitting height different from 91.4. Yes; since she used a convenience sample (not a simple random sample), the requirements of this section are not met and the conclusion may not be valid.

15. original claim: $\mu <$ \$500,000
H_o: $\mu =$ \$500,000
H_1: $\mu <$ \$500,000
$\alpha = 0.05$
C.V. $z = -z_{\alpha} = -z_{0.05} = -1.645$
calculations:
$z_{\bar{x}} = (\bar{x} - \mu_{\bar{x}})/\sigma_{\bar{x}}$
$= (415953 - 500000)/(463364/\sqrt{40})$
$= -84047/73264.2813 = -1.15$
P-value = $P(z<-1.15) = 0.1251$
conclusion:
Do not reject H_o; there is not sufficient evidence to conclude that $\mu <500{,}000$. There is not sufficient evidence to support the claim that the mean salary of an NCAA football coach is less than \$500,000.

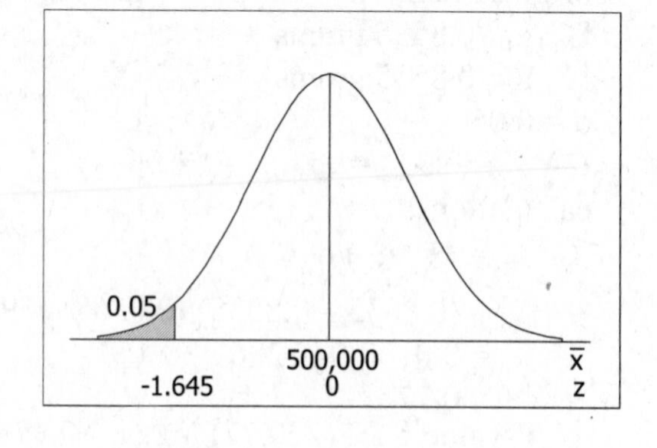

17. original claim: $\mu \neq 235.8$ cm
H_o: $\mu = 235.8$ cm
H_1: $\mu \neq 235.8$ cm
$\alpha = 0.05$
C.V. $z = \pm z_{\alpha/2} = \pm z_{0.025} = \pm 1.96$
calculations:
$z_{\bar{x}} = (\bar{x} - \mu_{\bar{x}})/\sigma_{\bar{x}}$
$= (235.4 - 235.8)/(4.5/\sqrt{40})$
$= -0.4/0.7115 = -0.56$
P-value = $2 \cdot P(z<-0.56) = 2 \cdot (0.2877) = 0.5754$
conclusion:
Do not reject H_o; there is not sufficient evidence to conclude that $\mu \neq 235.8$. There is not sufficient evidence to support the claim that the new baseballs have a bounce height different from the 235.8 cm of the old ones. No; the new baseballs are not significantly different.

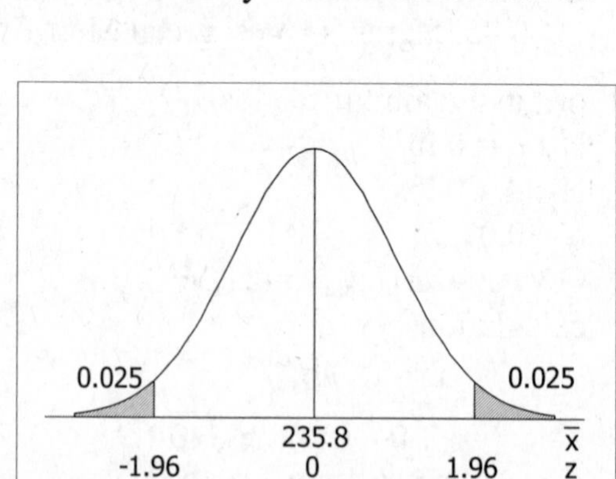

19. summary statistics: $n = 12$ $\Sigma x = 9131$ $\Sigma x^2 = 6{,}985{,}297$ $\bar{x} = (\Sigma x)/n = 9131/12 = 760.9$
 original claim: $\mu = 678$ FICO units
 H_o: $\mu = 678$ FICO units
 H_1: $\mu \neq 678$ FICO units
 $\alpha = 0.05$
 C.V. $z = \pm z_{\alpha/2} = \pm z_{0.025} = \pm 1.96$
 calculations:

$$z_{\bar{x}} = (\bar{x} - \mu_{\bar{x}})/\sigma_{\bar{x}}$$
$$= (760.9 - 678)/(58.3/\sqrt{12})$$
$$= 82.9/16.8298$$
$$= 4.93$$

 P-value = $2 \cdot P(z>4.93) = 2 \cdot (1 - 0.9999) = 2 \cdot (0.0001) = 0.0002$
 conclusion:

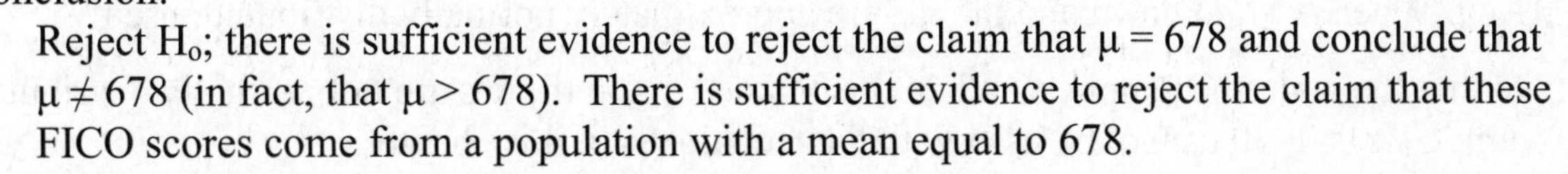

 Reject H_o; there is sufficient evidence to reject the claim that $\mu = 678$ and conclude that $\mu \neq 678$ (in fact, that $\mu > 678$). There is sufficient evidence to reject the claim that these FICO scores come from a population with a mean equal to 678.

21. The mean length of the $n = 50$ screws is $\bar{x} = 0.74682$ inches.
 original claim: $\mu = 0.75$ inches
 H_o: $\mu = 0.75$ inches
 H_1: $\mu \neq 0.75$ inches
 $\alpha = 0.05$
 C.V. $z = \pm z_{\alpha/2} = \pm z_{0.025} = \pm 1.96$
 calculations:

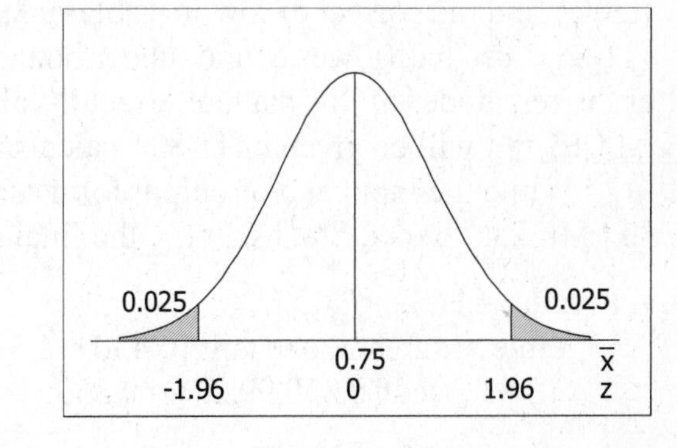

$$z_{\bar{x}} = (\bar{x} - \mu_{\bar{x}})/\sigma_{\bar{x}}$$
$$= (0.74682 - 0.75)/(0.012/\sqrt{50})$$
$$= -0.00318/0.001697 = -1.87$$

 P-value = $2 \cdot P(z<-1.87) = 2 \cdot (0.0307) = 0.0614$
 conclusion:
 Do not reject H_o; there is not sufficient evidence to reject the claim that $\mu = 0.75$. There is not sufficient evidence to reject the claim that the screws have a mean length of ¾ in. as indicated on the label. Yes; the lengths appear to be consistent with the label.

23. a. In general, the power of a test is the probability of correctly rejecting the hull hypothesis when some specified alternative is actually correct. In this context, a power of 0.2296 indicates that when the true population mean is actually 170 lbs, the test has a 22.96% probability of correctly rejecting the null hypothesis that the mean is 166.3 lbs in order to conclude that the mean is greater than 166.3 lbs.
 b. In general, β = P(type II error)
 = P(not rejecting $H_o|H_o$ is false)
 = 1 – P(rejecting $H_o|H_o$ is false)
 In this context, $\beta = 1 - P(\text{rejecting } H_o{:}\mu = 166.3 \mid \mu = 170))$
 $= 1 - 0.2296$
 $= 0.7704$

8-5 Testing a Claim About a Mean: σ Not Known

1. Yes; since $n \leq 30$, the sample should be from a population that is approximately normally distributed. We consider the normality requirement to be satisfied for such data if there are no outliers and the histogram of the sample data is approximately bell-shaped. More formally, a normal quantile plot could be used to determine whether the sample data are approximately normally distributed.

3. A t test is a hypothesis test that uses the Student t distribution, typically to perform a test about μ when the true value of σ is not know. The letter t is used because that was the notation chosen by William Gosset (1786-1937), the developer of the distribution. He wrote under the pseudonym Student, and t is most prominent letter in "student" (considering that "s" was already being used to denote the standard deviation).

5. Use t. When σ is unknown and the x's are approximately normally distributed, use t.

7. Neither the z nor the t applies. When σ is unknown and the x's are not normally distributed, sample sizes $n \leq 30$ cannot be used with the techniques in this chapter.

NOTE: Exercises 9–12 may be worked as follows.
table: find the correct df row in Table A-3, and see what values surround the given t.
TI-83+: use tcdf(lower bound, upper bound, df) to get the probability between 2 values
For the remainder of this manual, exact P-values for the t distribution USING THE UNROUNDED VALUES of t will be given in TI-83+ calculator format as described above and used in exercises 9–12. While the input and/or output formats may be different for other technologies [e.g., Minitab, Excel, Statdisk, etc], the final answers should be the same.

9. P-value = $P(t_{24} > 0.430)$
table for area in one tail: [$0.430 < 1.318$] P- value > 0.10
TI-83+: tcdf(0.430,99,24) = 0.3355

11. P-value = $2 \cdot P(t_8 < -1.905)$
table for area in two tails: [$-2.306 < -1.905 < -1.860$] $0.05 <$ P- value < 0.10
TI-83+: 2·tcdf(-99,-1.905,8) = 0.0932

13. original claim: $\mu > 210$ sec
H_o: $\mu = 210$ sec
H_1: $\mu > 210$ sec
$\alpha = 0.05$ and df = 39
C.V. $t = t_\alpha = t_{0.05} = 1.685$
calculations:
$t_{\bar{x}} = (\bar{x} - \mu)/s_{\bar{x}}$
$= 4.93$ [Minitab]
P-value = 0.000 [Minitab]

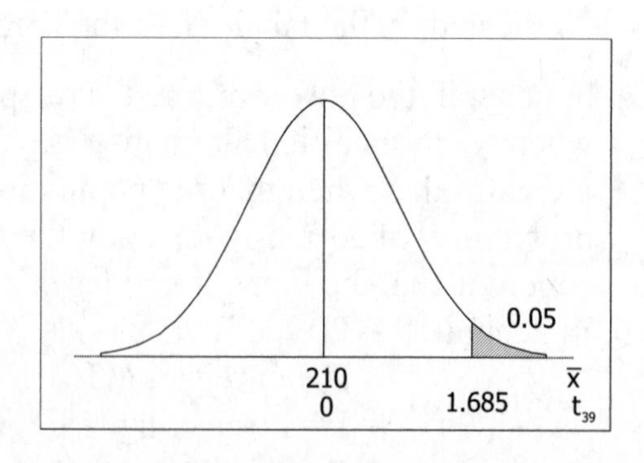

conclusion:

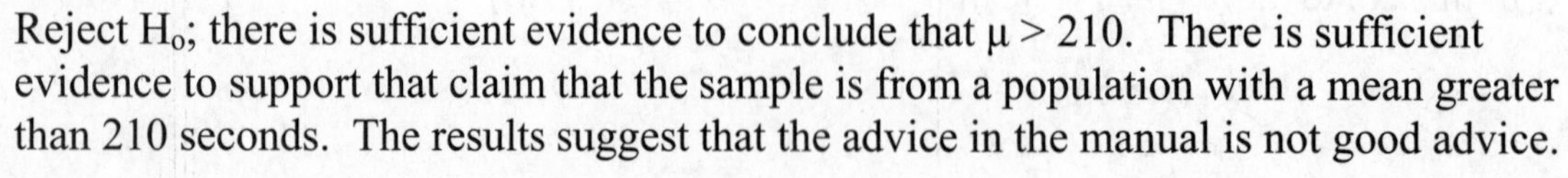
Reject H_o; there is sufficient evidence to conclude that $\mu > 210$. There is sufficient evidence to support that claim that the sample is from a population with a mean greater than 210 seconds. The results suggest that the advice in the manual is not good advice.

15. original claim: $\mu < 21.1$ mg
   H_o: $\mu = 21.1$ mg
   H_1: $\mu < 21.1$ mg
   $\alpha = 0.05$ and df = 24
   C.V. $t = -t_\alpha = -t_{0.05} = -1.711$
   calculations:
   $t_{\bar{x}} = (\bar{x} - \mu)/s_{\bar{x}}$
   $= (13.2 - 21.1)/(3.7/\sqrt{25})$
   $= -7.9/0.7400$
   $= -10.676$

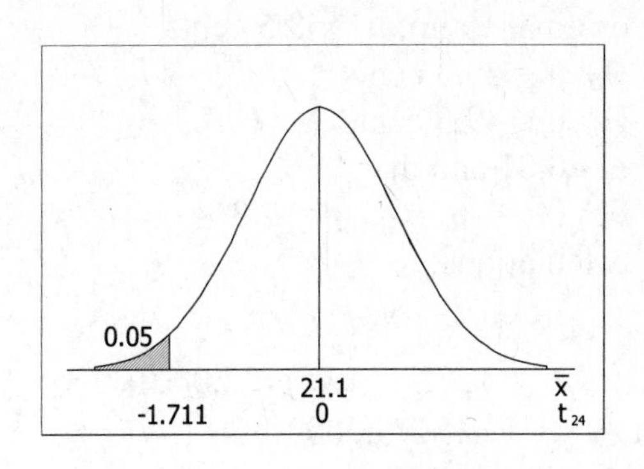

   P-value = $P(t_{24} < -10.676)$ = tcdf(-99,-10.676,24) = 6.752E-11 = 0.00000000007
   conclusion:
   Reject H_o; there is sufficient evidence to conclude that $\mu < 21.1$. There is sufficient evidence to support the claim that filtered 100 mm cigarettes have a mean tar amount less than 21.1 mg. The results suggest that filters are effective in reducing the amount of tar.

17. original claim: $\mu = 2.5$ grams
   H_o: $\mu = 2.5$ grams
   H_1: $\mu \neq 2.5$ grams
   $\alpha = 0.05$ and df = 36
   C.V. $t = \pm t_{\alpha/2} = \pm t_{0.025} = \pm 2.028$
   calculations:
   $t_{\bar{x}} = (\bar{x} - \mu)/s_{\bar{x}}$
   $= (2.49910 - 2.5)/(0.01648/\sqrt{37})$
   $= -0.00090/0.002709$
   $= -0.332$

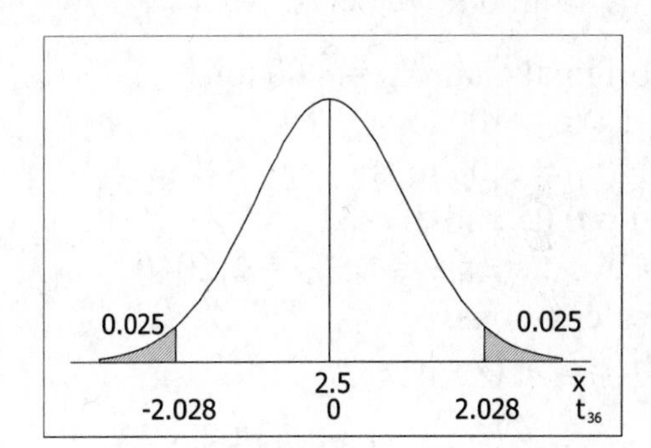

   P-value = $2 \cdot P(t_{36} < -1.756) = 2 \cdot$tcdf(-99,-0.332,36) = 0.7417

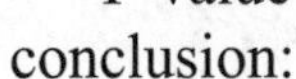

   conclusion:
   Do not reject H_o; there is not sufficient evidence to reject the claim that $\mu = 2.5$. There is not sufficient evidence to reject the claim that the sample is from a population with mean weight 2.5 grams. Yes; the pennies appear to conform to the specifications.

19. original claim: $\mu > 4.5$ years
   H_o: $\mu = 4.5$ years
   H_1: $\mu > 4.5$ years
   $\alpha = 0.05$ and df = 80
   C.V. $t = t_\alpha = t_{0.05} = 1.664$
   calculations:
   $t_{\bar{x}} = (\bar{x} - \mu)/s_{\bar{x}}$
   $= (4.8 - 4.5)/(2.2/\sqrt{81})$
   $= 0.3/0.2444$
   $= 1.227$
   P-value = $P(t_{80} > 1.227)$ = tcdf(1.227 ,99,80) = 0.1117
   conclusion:
   Do not reject H_o; there is not sufficient evidence to conclude that $\mu > 4.5$. There is not sufficient evidence to support the claim that the mean time for all college students to earn their bachelor's degrees is greater than 4.5 years.

21. original claim: $\mu = 49.5$ cents
    H_o: $\mu = 49.5$ cents
    H_1: $\mu < 49.5$ cents
    $\alpha = 0.01$ and df = 99
    C.V. $t = -t_\alpha = -t_{0.01} = -2.364$
    calculations:
    $t_{\bar{x}} = (\bar{x} - \mu)/s_{\bar{x}}$
    $= (23.8 - 49.5)/(32.0/\sqrt{100})$
    $= -25.7/3.2000$
    $= -8.031$
    P-value = $2 \cdot P(t_{99} < -8.031) = 2 \cdot \text{tcdf}(-99,-8.031,99) = 2.06\text{E-}12 = 0.000000000002$
    conclusion:
    Reject H_o; there is sufficient evidence to conclude that $\mu < 49.5$. There is sufficient evidence to conclude that the cents portion of all checks has a mean that is less than 49.5 cents. The results suggest that the cents portions of checks are not uniformly distributed from 0 to 99 cents.

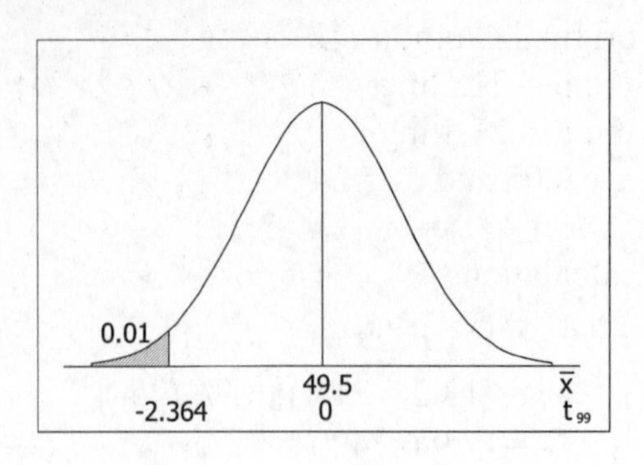

23. original claim: $\mu = 8.00$ tons
    H_o: $\mu = 8.00$ tons
    H_1: $\mu \neq 8.00$ tons
    $\alpha = 0.05$ and df = 31
    C.V. $t = \pm t_{\alpha/2} = \pm t_{0.025} = \pm 2.040$
    calculations:
    $t_{\bar{x}} = (\bar{x} - \mu)/s_{\bar{x}}$
    $= (7.78 - 8.00)/(1.08/\sqrt{32})$
    $= -0.22/0.1909 = -1.152$
    P-value = $2 \cdot P(t_{17} < -1.152) = 2 \cdot \text{tcdf}(-99,-1.152,31) = 0.2580$
    conclusion:
    Do not reject H_o; there is not sufficient evidence to reject the claim that $\mu = 8.00$. There is not sufficient evidence to reject the claim that all cars have a mean greenhouse gas emission of 8.00 tons.

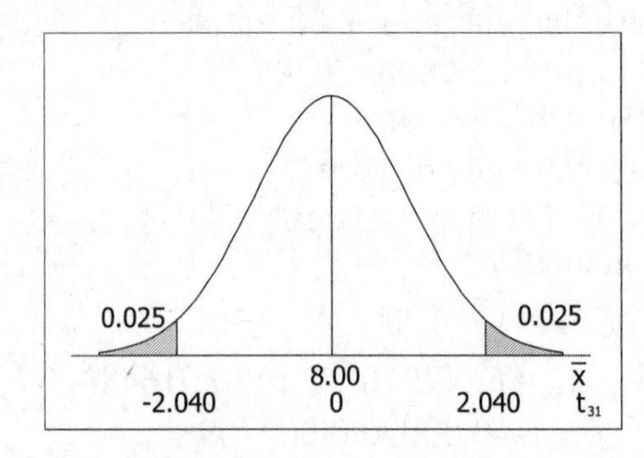

25. preliminary values: n = 6, $\Sigma x = 4222$, $\Sigma x^2 = 3{,}342{,}798$, $\bar{x} = 703.67$, $s^2 = 74383.47$, $s = 272.73$
    original claim: $\mu < 1000$ hic
    H_o: $\mu = 1000$ hic
    H_1: $\mu < 1000$ hic
    $\alpha = 0.01$ and df = 5
    C.V. $t = -t_\alpha = -t_{0.01} = -3.365$
    calculations:
    $t_{\bar{x}} = (\bar{x} - \mu)/s_{\bar{x}}$
    $= (703.67 - 1000)/(272.73/\sqrt{6})$
    $= -296.33/111.3429$
    $= -2.661$
    P-value = $P(t_5 < -2.661) = \text{tcdf}(-99,-2.661,5) = 0.0224$
    conclusion:
    Do not reject H_o; there is not sufficient evidence to conclude that $\mu < 1000$. There is not sufficient evidence to support the claim that the population mean is less than 1000 hic. No; since one of the sample values is 1210, there is proof that not all of the child booster seats meet the specified requirement.

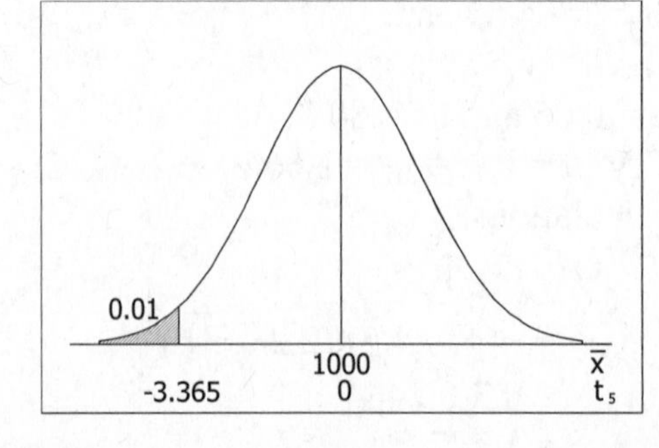

27. preliminary: $n = 5$, $\Sigma x = 32061$, $\Sigma x^2 = 220{,}431{,}831$, $\overline{x} = 6412.2$, $s^2 = 3{,}712{,}581.7$, $s = 1926.80$
    original claim: $\mu = \$5000$
    H_o: $\mu = \$5000$
    H_1: $\mu \neq \$5000$
    $\alpha = 0.05$ and df = 4
    C.V. $t = \pm t_{\alpha/2} = \pm t_{0.025} = \pm 2.776$
    calculations:
    $t_{\overline{x}} = (\overline{x} - \mu)/s_{\overline{x}}$
    $= (6412.2 - 5000)/(1926.80/\sqrt{5})$
    $= 1412.2/861.6927$
    $= 1.639$
    P-value = $2 \cdot P(t_4 > 1.639) = 2 \cdot \text{tcdf}(1.639,99,4) = 0.1766$
    conclusion:
    Do not reject H_o; there is not sufficient evidence to reject the claim that $\mu = 5000$. There is not sufficient evidence to reject the claim that the mean damage cost is \$5000.

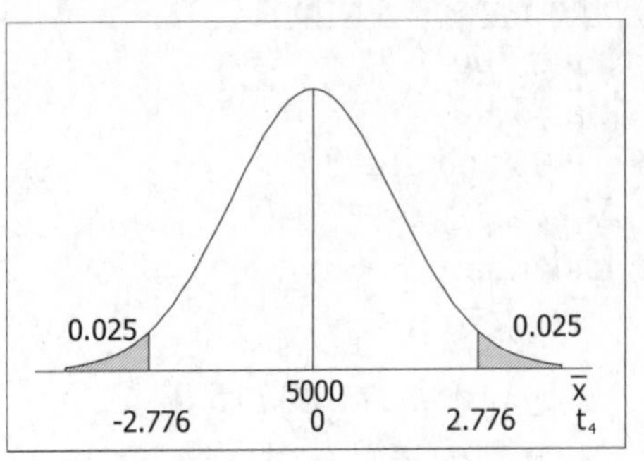

29. summary statistics from statistical software: $n = 50$, $\overline{x} = 0.74682$, $s = 0.012322$
    original claim: $\mu = 0.75$ inches
    H_o: $\mu = 0.75$ inches
    H_1: $\mu \neq 0.75$ inches
    $\alpha = 0.05$ and df = 49
    C.V. $t = \pm t_{\alpha/2} = \pm t_{0.025} = \pm 2.009$
    calculations:
    $t_{\overline{x}} = (\overline{x} - \mu)/s_{\overline{x}}$
    $= (0.74682 - 0.75)/(0.012322/\sqrt{50})$
    $= -0.00318/0.001743 = -1.825$
    P-value = $2 \cdot P(t_{49} < -1.825) = 2 \cdot \text{tcdf}(-99, -1.825,49) = 0.0741$
    conclusion:
    Do not reject H_o; there is not sufficient evidence to reject the claim that $\mu = 0.75$. There is not sufficient evidence to reject the claim that the screws have a mean length of ¾ in. as indicated on the label. Yes; the lengths appear to be consistent with the label.

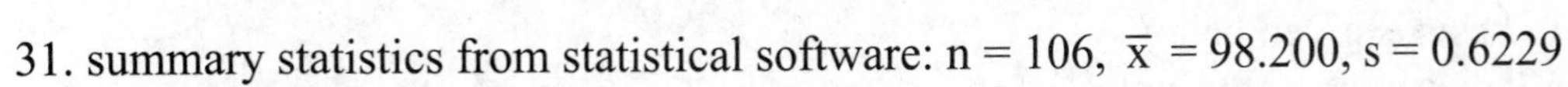

31. summary statistics from statistical software: $n = 106$, $\overline{x} = 98.200$, $s = 0.6229$
    original claim: $\mu = 98.6$ °F
    H_o: $\mu = 98.6$ °F
    H_1: $\mu \neq 98.6$ °F
    $\alpha = 0.05$ [assumed] and df = 105
    C.V. $t = \pm t_{\alpha/2} = \pm t_{0.025} = \pm 1.984$
    calculations:
    $t_{\overline{x}} = (\overline{x} - \mu)/s_{\overline{x}}$
    $= (98.2000 - 98.6)/(0.6229/\sqrt{106})$
    $= -0.4000/0.06050 = -6.611$
    P-value = $2 \cdot P(t_{105} < -6.611) = 2 \cdot \text{tcdf}(-99,-6.611,105) = 1.627\text{E-}9 = 0.000000002$
    conclusion:
    Reject H_o; there is sufficient evidence to reject the claim that $\mu = 98.6$ and conclude that $\mu \neq 98.6$ (in fact, that $\mu < 98.6$). There is sufficient evidence to reject the claim that the mean body temperature of the population is 98.6 °F. Yes; it appears there is sufficient evidence to conclude that the common belief is wrong.

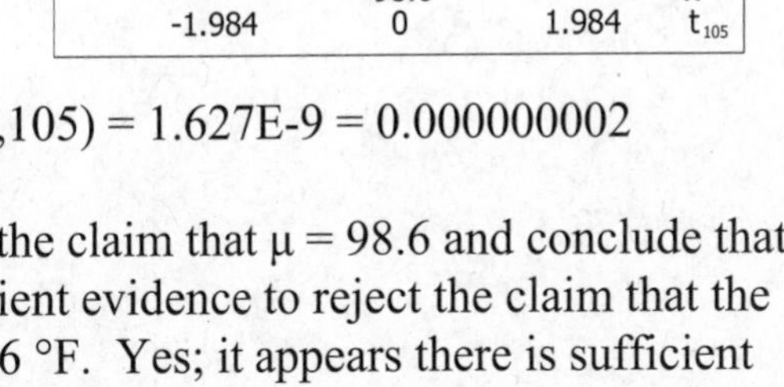

33. The two methods may be compared as follows.

method of this section

H_o: $\mu = 100$
H_1: $\mu \neq 100$
$\alpha = 0.05$ and df = 31
C.V. $t = \pm t_{\alpha/2} = \pm 2.040$
calculations:
$t_{\bar{x}} = (\bar{x} - \mu)/s_{\bar{x}}$
$= (105.3 - 100)/(15.0/\sqrt{32})$
$= 5.3/2.6517 = 1.999$
P-value = $2 \cdot P(t_{31} > 1.999)$
$= 2 \cdot \text{tcdf}(1.999,99,31)$
$= 0.0545$
conclusion:
Do not reject H_o; there is not enough evidence to reject the claim that $\mu = 100$.

alternative method

H_o: $\mu = 100$
H_1: $\mu \neq 100$
$\alpha = 0.05$
C.V. $z = \pm z_{\alpha/2} = \pm 1.960$
calculations:
$z_{\bar{x}} = (\bar{x} - \mu)/\sigma_{\bar{x}}$
$= (105.3 - 100)/(15.0/\sqrt{32})$
$= 5.3/2.6517 = 1.999$
P-Value) = $2 \cdot P(z > 1.999)$
$= 2 \cdot \text{normalcdf}(1.999,99)$
$= 0.0456$
conclusion:
Reject H_o; there is enough evidence to reject the claim that $\mu = 100$ and conclude that $\mu \neq 100$ (in fact, that $\mu > 100$).

The two methods lead to different conclusions – because of the unwarranted artificial precision created in the alternative method by assuming that the s value could be used for σ.

35. $A = z_\alpha \cdot [(8 \cdot df + 3)/(8 \cdot df + 1)]$
$= 1.645 \cdot [(8 \cdot 74 + 3)/(8 \cdot 74 + 1)]$
$= 1.645 \cdot [595/593]$
$= 1.6505$

$t = \sqrt{df \cdot (e^{A \cdot A/df} - 1)}$
$= \sqrt{74 \cdot (e^{(1.6505) \cdot (1.6505)/74} - 1)}$
$= \sqrt{74 \cdot (e^{.0368} - 1)}$
$= \sqrt{74 \cdot (.03750)}$
$= 1.6659$

This agrees exactly with the given $t_{74,.05} = 1.666$.

37. NOTE: Throughout this exercise the manual uses the s = 26.32 given in Example 1. Going back to the original Data Set 1 in Appendix B to find a more precise value will give slightly different answers (beginning at the third decimal place) in parts (a) and (b).
The test of hypothesis is given below and illustrated by the figure at the right.
H_o: $\mu = 166.3$ lbs
H_1: $\mu > 166.3$ lbs
$\alpha = 0.05$ and df = 39
C.V. $t = t_\alpha = t_{0.05} = 1.685$
The c corresponding to t = 1.685 is found by solving $t_c = (c - \mu)/(s/\sqrt{n})$ for c as follows.
$c = \mu + t_c \cdot (s/\sqrt{n})$
$= 166.3 + 1.685(26.32/\sqrt{40})$
$= 166.3 + 7.01$
$= 173.31$

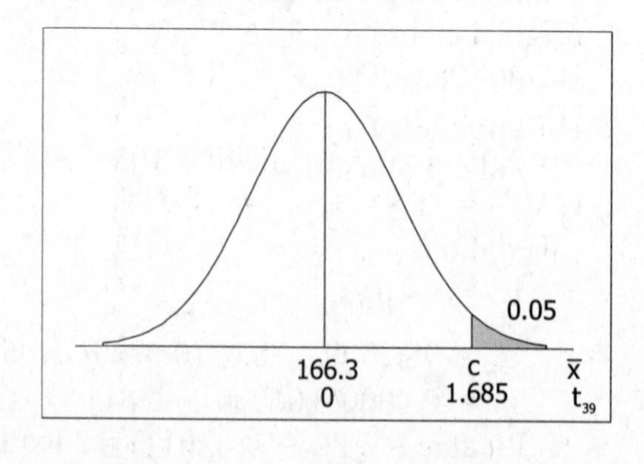

a. The power calculations are given below and illustrated by the figure at the right.
The t corresponding to c= 173.31 is

$t_c = (c-\mu)/(s/\sqrt{n})$
$= (173.31 - 180)/(26.32/\sqrt{40})$
$= -6.688/4.1616 = -1.607$

power = P(rejecting $H_o|H_o$ is false)
$= P(\bar{x}>173.31|\mu=180)$
$= P(t_{39}>-1.607)$
= tcdf(-1.607,99,39)
= 0.9419

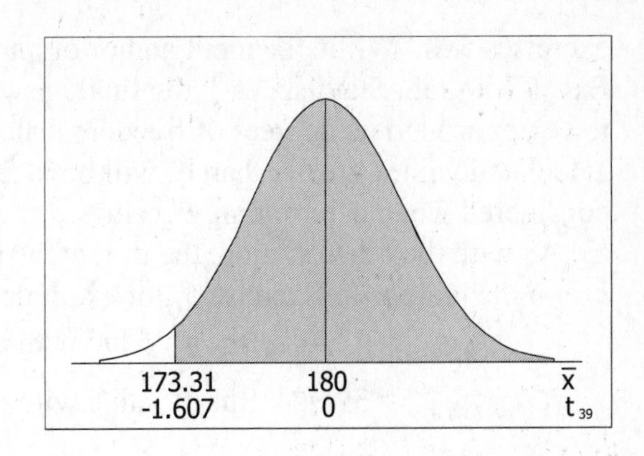

b. β = P(type II error) = P(not rejecting $H_o|H_o$ is false)
= 1 – P(rejecting $H_o|H_o$ is false)
= 1 – 0.9419
= 0.0481

The test is very effective in recognizing that the mean is greater than 166.3 lbs when the mean is actually equal to 180 lbs, for it will do so 94.19% of the time and fail to do so only 4.81% of the time.

8-6 Testing a Claim About a Standard Deviation or Variance

1. The normality requirement for an hypothesis test about a standard deviation is stricter than the normality requirement for an hypothesis test about a mean. Deviations from normality that were tolerated when testing a claim about a mean may be serious enough to invalidate the results when testing a claim about a standard deviation.

3. No. Unlike tests and confidence intervals involving the mean, which do not require normality when n>30, test and confidence intervals involving standard deviations require approximate normality for all sample sizes.

5. a. test statistic: $\chi^2 = (n-1)s^2/\sigma^2 = (24)(645)^2/(696)^2 = 20.612$
b. critical values for α = 0.05 and df = 24: $\chi^2 = \chi^2_{1-\alpha/2} = \chi^2_{0.975} = 12.401$
$\chi^2 = \chi^2_{\alpha/2} = \chi^2_{0.025} = 39.364$

c. P-value limits: 15.659 < 20.612
P-value > 0.20
P-value exact: $2 \cdot \chi^2$cdf(0,20.612,24) = 0.6770
d. conclusion: Do not reject H_o; there is not sufficient to conclude that σ ≠ 696

NOTES ON THE CHI-SQUARE DISTRIBUTION:
(1) Following the pattern used with the z and t distributions, this manual uses the closest entry from Table A-4 for the χ^2 as if it were the precise value necessary and does not use interpolation. This sacrifices little accuracy – and even interpolation does not yield precise values. When more accuracy is needed, refer either to more complete tables or to computer-produced values.
(2) The P-value portion of exercises 5–8 may be worked as follows.
table: find the correct df row in Table A-4, and see what values surround the calculated χ^2
TI-83/84+: use χ^2cdf(lower bound, upper bound, df) to get the probability between 2 values
For the remainder of this manual, exact P-values for the χ^2 distribution USING THE UNROUNDED VALUES of χ^2 will be given in TI-83/84+ calculator format as described above and used in

exercises 5–8. While the input and/or output formats may be different for other technologies [e.g., Excel, Minitab, Statdisk, etc], the final answers should be the same. Since the χ^2 distribution tends to center around its degrees of freedom, calculated values less than df will be in the lower tail and calculated values greater than df will be in the upper tail – and those are the tails that should be considered when determining P-values.
(3) As with the t distribution, the df may be used as a subscript to identify which χ^2 distribution to use in the tables. In Exercise 5, for example, one could indicate

$\chi^2_{24,0.975} = 12.401$ [the χ^2 value with df = 24 and 0.975 in the upper tail]

$\chi^2_{24,0.025} = 39.364$ [the χ^2 value with df = 24 and 0.025 in the upper tail]

7. a. test statistic: $\chi^2 = (n-1)s^2/\sigma^2 = (14)(4.8)^2/(3.5)^2 = 26.331$
   b. critical value for $\alpha = 0.01$ and df = 14: $\chi^2 = \chi^2_\alpha = \chi^2_{0.01} = 29.141$
   c. P-value limits: $26.119 < 26.331 < 29.141$
      $0.01 < \text{P-value} < 0.025$
      P-value exact: χ^2cdf(26.331,999,14) = 0.0235
   d. conclusion: Do not reject H_o; there is not sufficient evidence to conclude that $\sigma > 3.5$.

NOTE: The formula $\chi^2 = (n-1)s^2/\sigma^2$ used in the calculations contains df = n-1. When the exact df needed in the problem does not appear in the table and the closest χ^2 value is used to determine the critical region, some instructors recommend using the same df in the calculations that were used to determine the C.V. This manual typically uses the closest entry to determine the C.V. and the n from the problem in the calculations – even though this introduces a slight discrepancy.

9. There are n=37 weights in the data set.
   original claim: $\sigma > 0.0230$ grams
   H_o: $\sigma = 0.0230$ grams
   H_1: $\sigma > 0.0230$ grams
   $\alpha = 0.05$ and df = 36
   C.V. $\chi^2 = \chi^2_\alpha = \chi^2_{0.05} = 55.758$
   calculations:
   $\chi^2 = (n-1)s^2/\sigma^2$
   $= (36)(0.03910)^2/(0.0230)^2$
   $= 104.040$
   P-value = χ^2cdf(104.040,999,36) = 1.578E-8 = 0.00000002
   conclusion:
   Reject H_o; there is sufficient evidence to conclude that $\sigma > 0.0230$. There is sufficient evidence to support the claim that the weights of pre-1983 pennies have a standard deviation greater than 0.0230 grams. Based on these results, it appears that the weights pre-1983 pennies vary more than the weights of post-1983 pennies – which is partly, if not mostly, the result of being in circulation longer and having a wider variety in the amounts of wear.

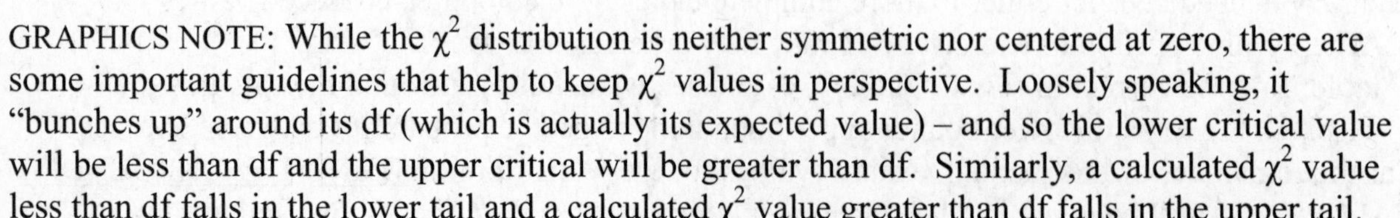

GRAPHICS NOTE: While the χ^2 distribution is neither symmetric nor centered at zero, there are some important guidelines that help to keep χ^2 values in perspective. Loosely speaking, it "bunches up" around its df (which is actually its expected value) – and so the lower critical value will be less than df and the upper critical will be greater than df. Similarly, a calculated χ^2 value less than df falls in the lower tail and a calculated χ^2 value greater than df falls in the upper tail.

To illustrate χ^2 tests of hypotheses, this manual uses a "generic" figure resembling a χ^2 distribution with df=4. Actually χ^2 distributions with df=1 and df=2 have no upper limit and approach the y axis asymptotically, while χ^2 distributions with df>30 are essentially symmetric and normal-looking. Because the χ^2 distribution is positively skewed, the expected value df is noted slightly to the right of the figure's peak.

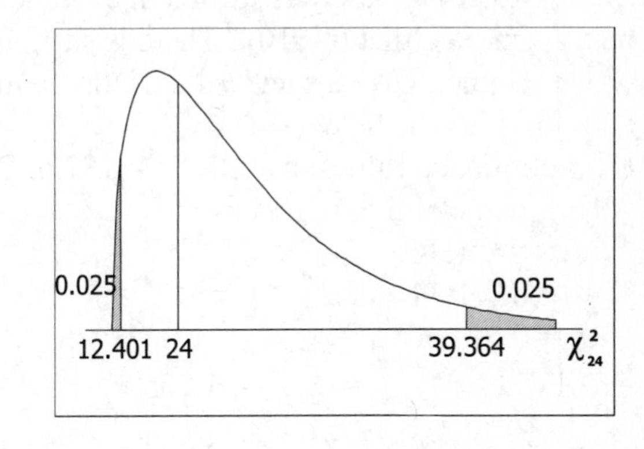

11. original claim: $\sigma \neq 3.2$ mg
    H_o: $\sigma = 3.2$ mg
    H_1: $\sigma \neq 3.2$ mg
    $\alpha = 0.05$ and df = 24
    C.V. $\chi^2 = \chi^2_{1-\alpha/2} = \chi^2_{0.975} = 12.401$
    $\chi^2 = \chi^2_{\alpha/2} = \chi^2_{0.025} = 39.364$
    calculations:
    $\chi^2 = (n-1)s^2/\sigma^2$
    $= (24)(3.7)^2/(3.2)^2 = 32.086$
    P-value = $2 \cdot \chi^2$cdf(32.086,999,24) = 0.2498
    conclusion:
    Do not reject H_o; there is not sufficient evidence to conclude that $\sigma \neq 3.2$. There is not sufficient evidence to support the claim that the tar content of such cigarettes has a standard deviation different from 3.2 mg.

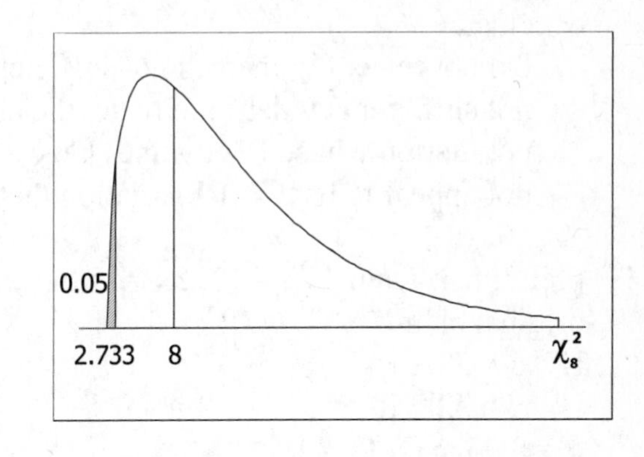

13. There are n=9 heights.
    original claim: $\sigma < 2.5$ inches
    H_o: $\sigma = 2.5$ inches
    H_1: $\sigma < 2.5$ inches
    $\alpha = 0.05$ and df = 8
    C.V. $\chi^2 = \chi^2_{1-\alpha} = \chi^2_{0.95} = 2.733$
    calculations:
    $\chi^2 = (n-1)s^2/\sigma^2$
    $= (8)(1.5)^2/(2.5)^2$
    $= 2.880$
    P-value = χ^2cdf(0,2.880,8) = 0.0583
    conclusion:
    Do not reject H_o; there is not sufficient evidence to conclude that $\sigma < 2.5$. There is not sufficient evidence to support the claim that the heights of supermodels have a standard deviation that is less than 2.5 inches, the value for the general female population. The conclusion indicates that while supermodels may be more homogeneous than the general population, that cannot be concluded from this sample of size n=9.

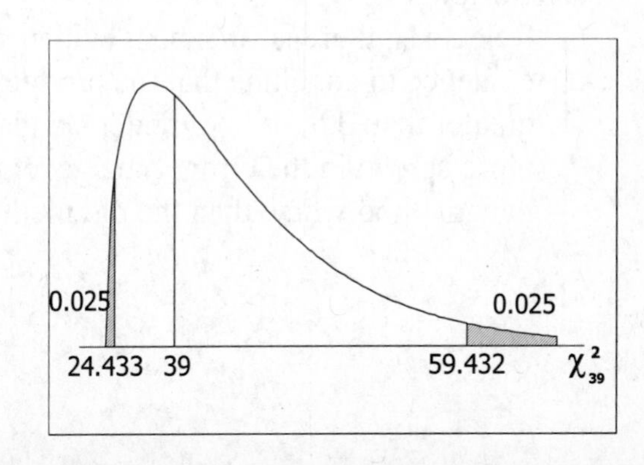

15. original claim: $\sigma = 10$ beats/minute
    H_o: $\sigma = 10$ beats/minute
    H_1: $\sigma \neq 10$ beats/minute
    $\alpha = 0.05$ and df = 39
    C.V. $\chi^2 = \chi^2_{1-\alpha/2} = \chi^2_{0.975} = 24.433$
    $\chi^2 = \chi^2_{\alpha/2} = \chi^2_{0.025} = 59.432$

calculations:

$\chi^2 = (n-1)s^2/\sigma^2$

$= (39)(12.5)^2/(10)^2 = 60.9375$

P-value = $2 \cdot \chi^2$cdf(60.9375,999,39) = 0.0277

conclusion:

Reject H_o; there is sufficient evidence to reject the claim that $\sigma = 10$ conclude that $\sigma \neq 10$ (in fact, that $\sigma > 10$). There is sufficient evidence to reject the claim that the pulse rates of women have a standard deviation equal to 10 beats/minute.

17. preliminary values: n = 10, $\Sigma x = 187.6$, $\Sigma x^2 = 3532.04$, $\bar{x} = 18.76$, $s^2 = 1.4071$, s = 1.1862

original claim: $\sigma = 1.34$

H_o: $\sigma = 1.34$

H_1: $\sigma \neq 1.34$

$\alpha = 0.01$ and df = 9

C.V. $\chi^2 = \chi^2_{1-\alpha/2} = \chi^2_{0.995} = 1.735$

$\chi^2 = \chi^2_{\alpha/2} = \chi^2_{0.005} = 23.589$

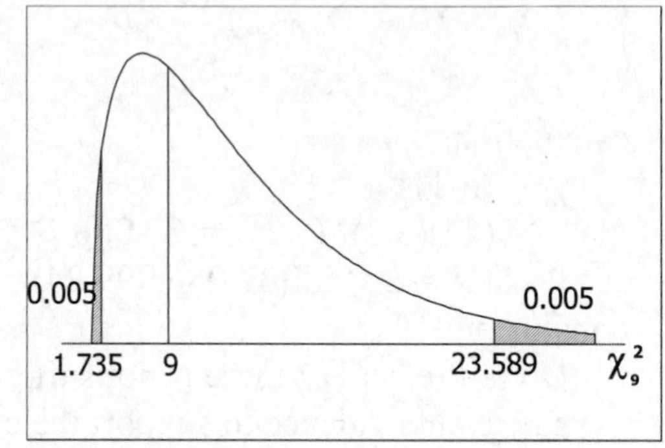

calculations:

$\chi^2 = (n-1)s^2/\sigma^2$

$= (9)(1.1862)^2/(1.34)^2$

$= 7.053$

P-value = $2 \cdot \chi^2$cdf(0,7.053,9) = 0.7368

conclusion:

Do not reject H_o; there is not sufficient evidence to reject the claim that $\sigma = 1.34$. There is not sufficient evidence to reject the claim that recent Miss America winners are from a population whose BMI values have a standard deviation of 1.34. No; recent winners do not appear to have MBI variation that is different from that of the 1920's and 1930's.

19. preliminary values: n = 12, $\Sigma x = -196$, $\Sigma x^2 = 33452$, $\bar{x} = -16.3$, $s^2 = 2750.06$, s = 52.441

original claim: $\sigma > 32.2$ ft

H_o: $\sigma = 32.2$ ft

H_1: $\sigma > 32.2$ ft

$\alpha = 0.05$ and df = 11

C.V. $\chi^2 = \chi^2_{\alpha} = \chi^2_{0.05} = 19.675$

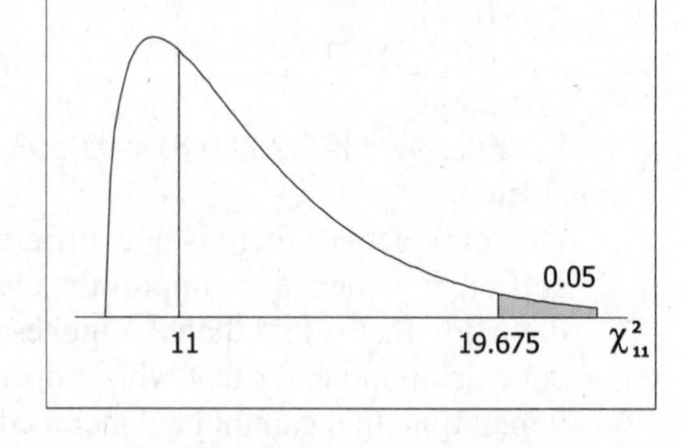

calculations:

$\chi^2 = (n-1)s^2/\sigma^2$

$= (11)(52.441)^2/(32.2)^2$

$= 29.176$

P-value = χ^2cdf(29.176,999,11) = 0.0021

conclusion:

Reject H_o; there is sufficient evidence to conclude that $\sigma > 32.2$. There is sufficient evidence to conclude that the production method has errors with a standard deviation greater than 32.2 ft. A greater standard deviation means that the new altimeters will have more spread in the errors (i.e., generally larger errors) than in the past – which make the new method worse than the old method. Yes; the company should take action.

21. lower $\chi^2 = \frac{1}{2}(-z_{\alpha/2} + \sqrt{2 \cdot df - 1})^2$
$= \frac{1}{2}(-2.575 + \sqrt{2 \cdot (99) - 1})^2$
$= \frac{1}{2}(-2.575 + \sqrt{197})^2$
$= \frac{1}{2}(131.3469) = 65.673$
upper $\chi^2 = \frac{1}{2}(z_{\alpha/2} + \sqrt{2 \cdot df - 1})^2$
$= \frac{1}{2}(2.575 + \sqrt{2 \cdot (99) - 1})^2$
$= \frac{1}{2}(2.575 + \sqrt{197})^2$
$= \frac{1}{2}(275.9143) = 137.957$
Since the χ^2_{99} is shifted to the left (centered at df=99) compared to the χ^2_{100} (centered at df=100) used to give approximate values in Exercise 16, both values are slightly smaller as expected.

Statistical Literacy and Critical Thinking

1. For testing a claim that $\mu > 0$, that must be the alternative hypothesis – and the null hypothesis must be that $\mu = 0$. A P-value of 0.0091 indicates that the probability of obtaining these results when the null hypothesis is true is only 0.0091. Such a P-value calls for the rejection of the null hypothesis and support of the original claim that the mean rainfall amount is greater than 0 inches. The memory aid indicates that a small (e.g., less than 0.05) P-value calls for the rejection of the null hypothesis and that a large (i.e., greater than 0,05) P-value indicates there is not sufficient evidence to reject the null hypothesis.
2. Yes; since P-value = $0.0041 < 0.05$, there is statistical significance. No; since 0.019 inches is such a trivial amount (and is it even possible to measure heights of males that accurately?), there is not practical significance.
3. A voluntary response sample is one in which the respondents themselves decide whether or not to participate. Since such samples tend to include mostly those with a special interest in the topic and/or strong feelings about the topic, they are not necessarily representative of the general population and should not be used to make inferences.
4. A procedure is robust against departures from normality if it works well (i.e., is "correct" $1-\alpha$ of the time) even when the sample data are from a population that does not follow a normal distribution. Yes; the t test of a population mean is robust against departures from normality. No; the $\chi 2$ test of a population standard deviation is not robust against departures from normality.

Chapter Quick Quiz

1. Since the claim that the proportion of males is greater than 0.5 does not contain the equality, it must be the alternative hypothesis. The null hypothesis is that the proportion of males is equal to 0.5. In symbolic form, Ho: $p = 0.5$ and H_1: $p > 0.5$.

2. The t distribution is appropriate for the indicated test. The others are not appropriate for the following reasons.
   normal – used to perform a test about μ when σ is known..
   chi-square – used to perform a test about σ.
   binomial – used to perform an exact test about p.
   uniform – not an appropriate sampling distribution for any sample statistic.

3. The chi-square distribution is appropriate for the indicated test. The others are not appropriate for the following reasons.
   normal – used to perform a test about μ when σ is known..
   t – used to perform a test about μ when σ is unknown.
   binomial – used to perform an exact test about p.
   uniform – not an appropriate sampling distribution for any sample statistic.

4. True; the null hypothesis is either rejected or not rejected, but it cannot be supported.

5. P-value = $2 \cdot P(z>1.50) = 2 \cdot (1 - 0.9332) = 2 \cdot (0668) = 0.1336$

6. $\hat{p} = x/n = 30/100 = 0.30$
   $z_{\hat{p}} = (\hat{p} - \mu_{\hat{p}})/\sigma_{\hat{p}} = (\hat{p} - p)/\sqrt{pq/n} = (0.30 - 0.40)/\sqrt{(0.40)(0.60)/100} = -0.10/0.04899 = -2.04$

7. The appropriate distribution is the t distribution with n-1 = 19 degrees of freedom.
   The critical values for the two-tailed test with $\alpha = 0.05$ are $\pm t_{19,0.025} = \pm 2.093$.

8. P-value = $2 \cdot P(z>1.20) = 2 \cdot (1 - 0.8849) = 2 \cdot (1151) = 0.2302$

9. Do not reject H_o; there is not sufficient evidence to conclude that $p < 0.25$. There is not sufficient evidence to support the claim that $p < 0.25$.

10. False. Since hypothesis tests deal with a sample and not the entire population, there is always the chance of an error. When the null hypothesis is true, for example, we expect to incorrectly conclude that it is false α of the time.

Review Exercises

1. original claim: $p < 0.25$
   $\hat{p} = x/n = 261/1088 = 0.240$
   Ho: $p = 0.25$
   H_1: $p < 0.25$
   $\alpha = 0.05$
   C.V. $z = -z_\alpha = -z_{0.05} = -1.645$
   calculations:
   $z_{\hat{p}} = (\hat{p} - \mu_{\hat{p}})/\sigma_{\hat{p}}$
   $= (0.240 - 0.25)/\sqrt{(0.25)(0.75)/1088}$
   $= -0.01011/0.01313$
   $= -0.77$
   P-value = $P(z<-0.77) = 0.2206$

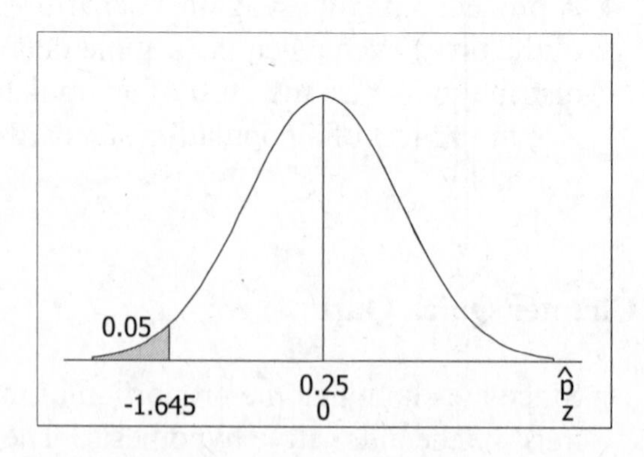

   conclusion:
   Do not reject H_o; there is not sufficient evidence to conclude that $p < 0.25$. There is not sufficient evidence to support the claim that less than ¼ of such adults smoke.

2. original claim: $p > 0.50$

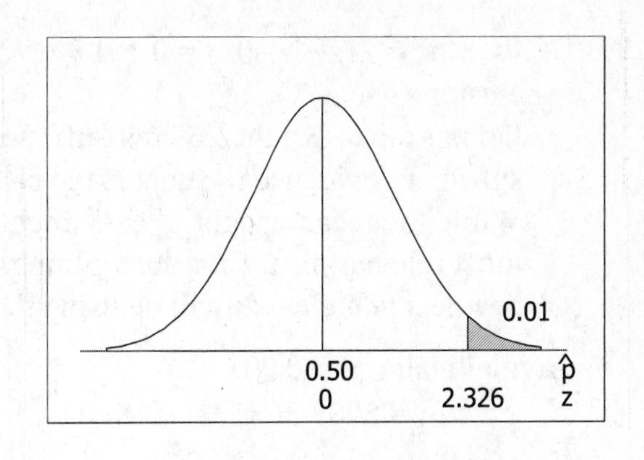

$\hat{p} = x/n = 802/1486 = 0.540$

Ho: $p = 0.50$

H_1: $p > 0.50$

$\alpha = 0.01$

C.V. $z = z_{\alpha} = z_{0.01} = 2.326$

calculations:

$z_{\hat{p}} = (\hat{p} - \mu_{\hat{p}})/\sigma_{\hat{p}}$

$= (0.540 - 0.50)/\sqrt{(0.50)(0.50)/1486}$

$= 0.03970/0.01297$

$= 3.06$

P-value = $P(z>3.06) = 1 - 0.9989 = 0.0011$

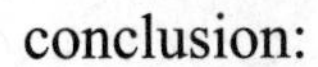

conclusion:

Reject H_o; there is sufficient evidence to conclude that $p > 0.50$. There is sufficient to support the claim that most college students seeking bachelor's degrees earn those degrees within five years.

NOTE: Although it was not specifically stated so, it must be assumed that the "1486 college students who are seeking bachelor's degrees" were followed for five years before the study was completed.

3. original claim: $\mu < 3700$ lbs

H_o: $\mu = 3700$ lbs

H_1: $\mu < 3700$ lbs

$\alpha = 0.01$ and df = 31

C.V. $t = -t_{\alpha} = -t_{0.01} = -2.453$

calculations:

$t_{\bar{x}} = (\bar{x} - \mu)/s_{\bar{x}}$

$= (3605.3 - 3700)/(501.7/\sqrt{32})$

$= -94.7/88.6889$

$= -1.068$

P-value = $P(t_9 < -1.068)$ = tcdf(-99,-1.068,31) = 0.1469

conclusion:

Do not reject H_o; there is not sufficient evidence to conclude that $\mu < 3700$. There is not sufficient evidence to support the claim that the mean weight of cars is less than 3700 lbs. While the mean weight of cars might be a factor in determining long run wear and tear, the most relevant factor for determining the required strength for highways is the weight of the heaviest vehicle that will be using the highway.

4. original claim: $\mu < 3700$ lbs

H_o: $\mu = 3700$ lbs

H_1: $\mu < 3700$ lbs

$\alpha = 0.01$

C.V. $z = -z_{\alpha} = -z_{0.01} = -2.326$

calculations:

$z_{\bar{x}} = (\bar{x} - \mu_{\bar{x}})/\sigma_{\bar{x}}$

$= (3605.3 - 3700)/(520/\sqrt{32})$

$= -94.7/91.9239$

$= -1.03$

P-value = $P(z < -1.03) = 0.1515$

conclusion:

Do not reject H_o; there is not sufficient evidence to conclude that $\mu < 3700$. There is not sufficient evidence to support the claim that the mean weight of cars is less than 3700 lbs. While the mean weight of cars might be a factor in determining long run wear and tear, the most relevant factor for determining the required strength for highways is the weight of the heaviest vehicle that will be using the highway.

5. original claim: $p < 0.20$

$\hat{p} = x/n = 5787/30617 = 0.1890$

H_o: $p = 0.20$

H_1: $p < 0.20$

$\alpha = 0.01$

C.V. $z = -z_\alpha = -z_{0.01} = -2.326$

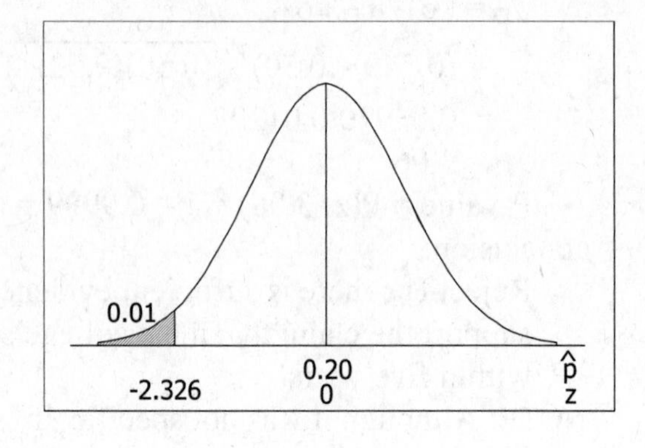

calculations:

$z_{\hat{p}} = (\hat{p} - \mu_{\hat{p}})/\sigma_{\hat{p}}$

$= (0.1890 - 0.20)/\sqrt{(0.20)(0.80)/30617}$

$= -0.01099/0.002286$

$= -4.81$

P-value = $P(z<-4.81) = 0.0001$

conclusion:

Reject H_o; there is sufficient evidence to conclude that $p < 0.20$. There is sufficient evidence to support the claim that fewer than 20% of adults consumed herbs within the past 12 months.

6. original claim: $\mu < 281.8$ lbs

H_o: $\mu = 281.8$ lbs

H_1: $\mu < 281.8$ lbs

$\alpha = 0.01$ and df = 174

C.V. $t = -t_\alpha = -t_{0.01} = -2.345$

calculations:

$t_{\bar{x}} = (\bar{x} - \mu)/s_{\bar{x}}$

$= (267.1 - 281.8)/(22.1/\sqrt{175})$

$= -14.7/1.6706$

$= -8.799$

P-value = $P(t_9 < -8.799)$ = tcdf(-99,-8.799,174) ≈ 0

conclusion:

Reject H_o; there is sufficient evidence to conclude that $\mu < 281.8$. There is sufficient evidence to support the claim that the thinner cans have a mean axial load that is less than 281.8 lbs. Yes; even though the thinner cans are not as strong as the thicker cans currently in use, they apparently can easily withstand the necessary pressure of 158 to 165 lbs.

7. original claim: $\mu = 74$
   H_o: $\mu = 74$
   H_1: $\mu \neq 74$
   $\alpha = 0.05$
   C.V. $z = \pm z_{\alpha/2} = \pm z_{0.025} = \pm 1.96$
   calculations:
   $z_{\bar{x}} = (\bar{x} - \mu_{\bar{x}})/\sigma_{\bar{x}}$
   $= (74.4 - 74)/(12.5/\sqrt{100})$
   $= 0.4/1.25$
   $= 0.32$

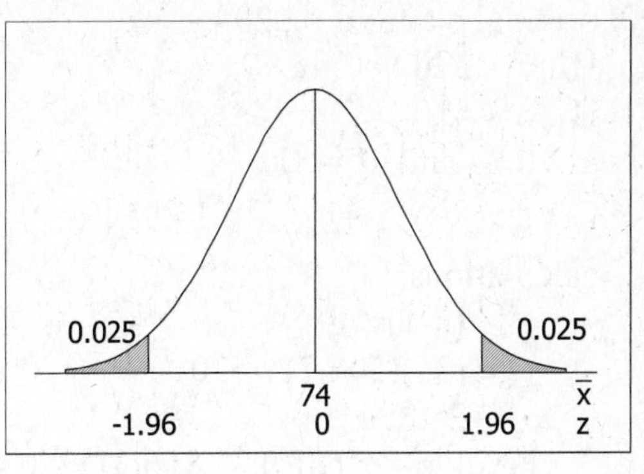
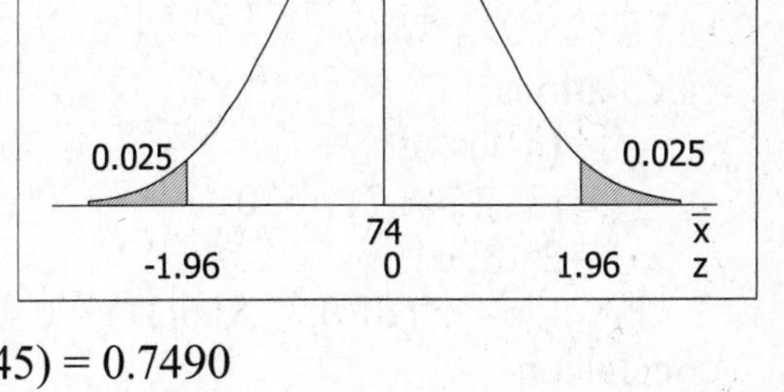

   P-value = $2 \cdot P(z>0.32) = 2 \cdot (1 - 0.6255) = 2 \cdot (0.3745) = 0.7490$
   conclusion:
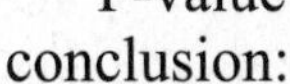
   Do not reject H_o; there is not sufficient evidence to reject the claim that $\mu = 74$. There is not sufficient evidence to reject the claim that the sample comes from a population with a mean equal to 74. Yes; based on these results, the calculator's random number generator appears to be working correctly.

8. original claim: $\mu = 74$
   H_o: $\mu = 74$
   H_1: $\mu \neq 74$
   $\alpha = 0.05$ and df = 99
   C.V. $t = \pm t_{\alpha/2} = \pm t_{0.025} = \pm 1.984$
   calculations:
   $t_{\bar{x}} = (\bar{x} - \mu)/s_{\bar{x}}$
   $= (74.4 - 74)/(11.7/\sqrt{100})$
   $= 0.4/1.17$
   $= 0.342$
   P-value = $2 \cdot P(t_{99} > 0.342) = 2 \cdot \text{tcdf}(0.342,99,99) = 0.7332$
   conclusion:
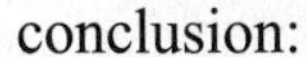
   Do not reject H_o; there is not sufficient evidence to reject the claim that $\mu = 74$. There is not sufficient evidence to reject the claim that the sample comes from a population with a mean equal to 74. Yes; based on these results, the calculator's random number generator appears to be working correctly.
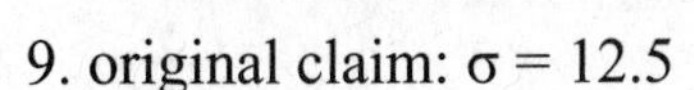

9. original claim: $\sigma = 12.5$
   H_o: $\sigma = 12.5$
   H_1: $\sigma \neq 12.5$
   $\alpha = 0.05$ and df = 99
   C.V. $\chi^2 = \chi^2_{1-\alpha/2} = \chi^2_{0.975} = 74.222$
   $\chi^2 = \chi^2_{\alpha/2} = \chi^2_{0.025} = 129.561$
   calculations:
   $\chi^2 = (n-1)s^2/\sigma^2$
   $= (99)(11.7)^2/(12.5)^2$
   $= 86.734$
   P-value = $2 \cdot \chi^2\text{cdf}(0,86.734,99) = 0.3883$
   conclusion:
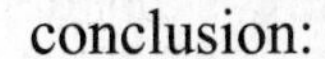
   Do not reject H_o; there is not sufficient evidence to reject the claim that $\sigma = 12.5$. There is not sufficient evidence to reject the claim that the sample comes from a population with a standard deviation equal to 12.5.

10. original claim: $\sigma < 520$
 H_o: $\sigma = 520$
 H_1: $\sigma < 520$
 $\alpha = 0.01$ and df = 31
 C.V. $\chi^2 = \chi^2_{1-\alpha} = \chi^2_{0.99} = 14.954$
 calculations:
  $\chi^2 = (n-1)s^2/\sigma^2$
  $= (31)(501.7)^2/(520)^2$
  $= 28.856$
 P-value = χ^2cdf(0,28.856,31) = 0.4233
 conclusion:
  Do not reject H_o; there is not sufficient evidence to conclude that $\sigma < 520$. There is not sufficient evidence to support the claim that the standard deviation of the weights of cars is less than 520 lbs.

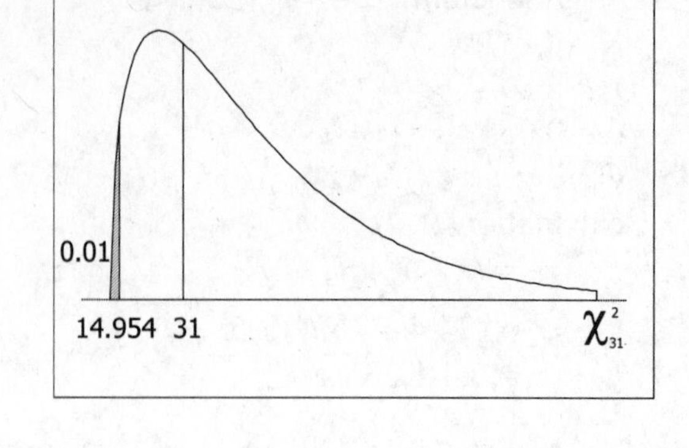

Cumulative Review Exercises

1. scores in numerical order: 10.54 10.75 10.82 10.93 10.94 10.97 11.06 11.07 11.08
 summary statistics: n = 9, $\Sigma x = 98.16$, $\Sigma x^2 = 1070.8508$
 a. $\bar{x} = (\Sigma x)/n = (98.16)/9 = 10.907$ seconds
 b. $\tilde{x} = 10.94$ seconds
 c. $s^2 = [n(\Sigma x^2) - (\Sigma x)^2]/[n(n-1)]$
  $= [9(1070.8508) - (98.16)^2]/[9(8)]$
  $= 2.2716/72$
  $= 0.03155$
  s = 0.178 seconds
 d. $s^2 = 0.03155$, rounded to 0.032 seconds^2
 e. R = 11.08 – 10.54 = 0.54 seconds

2. a. Ratio, since differences are meaningful and there is a meaningful zero.
 b. Continuous, since time can be any value on a continuum.
 c. No; the times were not selected at random from some population, but they were determined by being the winning times for 9 consecutive Olympics.
 d. The sample statistics in Exercise 1 do not consider the chronological order of the data.
 e. A time series plot will reveal tendencies over time, while still giving a general idea (by looking from the vertical axis) of the central tendency and variation of the data.

3. σ unknown, use t with df=8
 $\alpha = 0.05$, $t_{df,\alpha/2} = t_{8,0.025} = 2.306$
  $\bar{x} \pm t_{\alpha/2} \cdot s/\sqrt{n}$
  $10.907 \pm 2.306(0.178)/\sqrt{9}$
  10.907 ± 0.136
  $10.770 < \mu < 11.043$ (seconds)
 No; this result cannot be used to estimate winning times in the future because there is a pattern of decreasing times and no fixed population mean.

4. original claim: $\mu < 11$ seconds
   H_o: $\mu = 11$ seconds
   H_1: $\mu < 11$ seconds
   $\alpha = 0.05$ and df = 8
   C.V. $t = -t_\alpha = -t_{0.05} = -1.860$
   calculations:
   $t_{\bar{x}} = (\bar{x} - \mu)/s_{\bar{x}}$
   $= (10.907 - 11)/(0.178/\sqrt{9})$
   $= -0.0933/0.05921$
   $= -1.576$

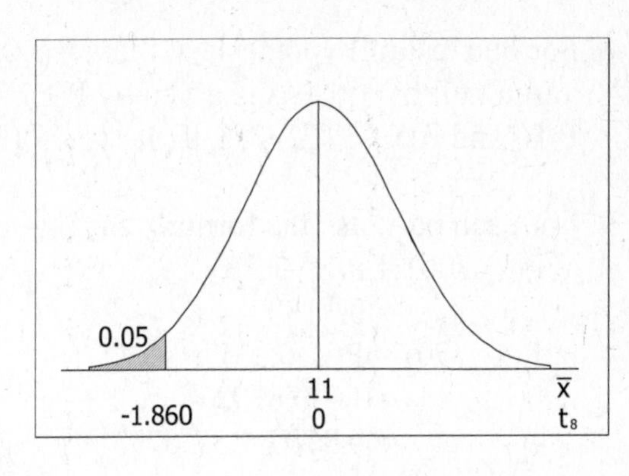

   P-value = $P(t_9 < -1.576)$ = tcdf(-99,-1.576,8) = 0.0768
   conclusion:
   Do not reject H_o; there is not sufficient evidence to conclude that $\mu < 11$. There is not sufficient evidence to support the claim that the mean winning time is less than 11 seconds. Since there is a decreasing pattern of winning times and no fixed population mean, this result does not apply to future times – but even though the mean for past winning times is not significantly less than 11 seconds, the decreasing pattern makes it likely that future winning times will be less than 11 seconds.

5. Some parts of this exercise are better answered from the frequency distribution representation at the right, where the x column gives the (approximate) midpoint value of each bar.
   a. Yes; the histogram is approximately bell-shaped.
   b. $\Sigma f = 100$
   c. Using the first two midpoints, 0.72 – 0.70 = 0.02 grams
   d. Using the formula for summarized grouped data,
   $\bar{x} \approx (\Sigma f{\cdot}x)/(\Sigma f)$
   $= 85.70/100$
   $= 0.857$ grams
   e. No. All we can tell, for example, is that there are 30 values in the class centered at 0.86 grams, but we cannot determine their individual weights.

x	f	f·x
0.70	1	0.70
0.72	0	0.00
0.74	1	0.74
0.76	2	1.52
0.78	5	3.90
0.80	8	6.40
0.82	8	6.56
0.84	13	10.92
0.86	30	25.80
0.88	15	13.20
0.90	3	2.70
0.92	4	3.68
0.94	6	5.64
0.96	1	0.96
0.98	2	1.96
1.00	0	0.00
1.02	1	1.02
	100	85.70

6. Using a vertical scale that does not start at zero exaggerates differences between the classes and gives a distorted impression of the data. The visual impression of the first class (x=1), for example, is that it contains about twice as many data points as the second class (x=2) – and in reality those values are 16 and 13.

7. The requested frequency distribution is given by the first two columns of the table at the right. Using the formula for grouped data,
   $\bar{x} = (\Sigma f{\cdot}x)/(\Sigma f)$
   $= 347/100$
   $= 3.47$

x	f	f·x
1	16	16
2	13	26
3	22	66
4	21	84
5	13	65
6	15	90
	100	347

8. For one test of hypothesis at the $\alpha = 0.05$ level, the probability of making an error by rejecting a true null hypothesis is given by $P(E) = 0.05$. For two such independent tests, $P(E_1 \text{ and } E_2) = P(E_1) \cdot P(E_2|E_1) = (0.05)(0.05) = 0.0025$.

9. For each part, use the formula $z = (x - \mu)/\sigma$
   a. $P(x<700) = P(z<-1.18)$
   $= 0.1190$
   b. $P(x>750) = P(z>0.33)$
   $= 1 - 0.6293$
   $= 0.3707 = 37.07\%$
   c. The $\bar{x}$'s are normally distributed with $\mu_{\bar{x}} = \mu = 739$ and $\sigma_{\bar{x}} = \sigma/\sqrt{n} = 33/\sqrt{50} = 4.6669$.
   $P(\bar{x}<730) = P(z<-1.93)$
   $= 0.0268$
   d. From Table A-2, the z score with 0.9000 [closest entry is 0.8997] below it is 1.28.
   $x = \mu + z\sigma$
   $= 739 + (1.28)(33)$
   $= 781.24$ mm

10. The sample heights appear to come from a normal distribution. The histogram shows a distribution that is approximately bell-shaped, and the normal quantile plot indicates that the data points approximate a straight line.

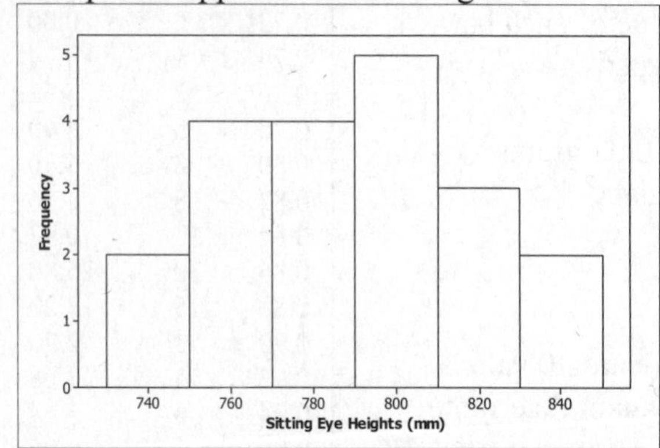

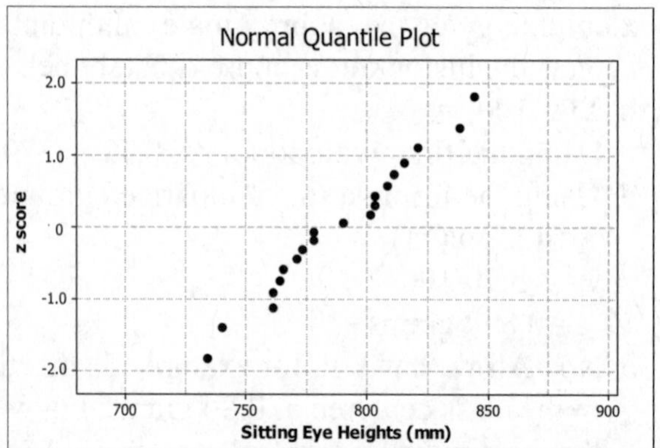

NOTE: Other representations of the data may give different impressions, and so it is always a good idea not to base a conclusion about normality on a single figure. The reasonable representation of the data given at the right, for example, suggests that the data may be bimodal – perhaps representing two different subpopulations. While frequency distributions and histograms are sensitive to how the classes are defined, the normal quantile plot is not subject to such arbitrary decisions.

height (mm)	frequency
725 – 749	2
750 – 774	6
775 – 799	3
800 – 824	7
825 – 846	2
	20

Chapter 9

Inferences from Two Samples

9-2 Inferences About Two Proportions

1. There are two requirements for using the methods of this section, and each of them is violated.
   (1) The samples should be two sample random samples that are independent. These samples are convenience samples, not simple random samples. These samples are likely not independent. Since she surveyed her friends, she may well have males and females that are dating each other (or least that associate with each other) – and people tend to associate with those that have similar behaviors.
   (2) The number of successes for each sample should be at least 5, and the number of failures for each sample should be at least 5. This is not true for the males, for which x=4.
   NOTE: This is the same requirement from previous chapters for using the normal distribution to approximate the binomial that required $np \geq 5$ and $nq \geq 5$.
   Using $\hat{p}$ =x/n to estimate p and $\hat{q}$ = 1- x/n = (n-x)/n to estimate q,

   $n\hat{p} \geq 5$ $\qquad$ $n\hat{q} \geq 5$
   $n[x/n] \geq 5$ $\qquad$ $n[(n-x)/n] \geq 5$
   $x \geq 5$ $\qquad$ $(n-x) \geq 5$

   These inequalities state that the number of successes must be greater than 5, and the number of failures must be greater than 5.

3. In this context,
   $\hat{p}_1 = 15/1583 = 0.00948$
   $\hat{p}_2 = 8/157 = 0.05096$
   $\overline{p} = (15+8)/(1583+157) = 23/1740 = 0.01322$
   p_1 denotes the rue proportion of all Zocor users who experience headaches
   p_2 denotes the true proportion of all placebo users who experience headaches

5. $x = (0.158)(8834) = 1395.772 \approx 1396$
   NOTE: Any $1392 \leq x \leq 1400$ rounds to x/8834 = 15.8%.

7. $\hat{p}_1 = x_1/n_1 = 13/36 = 0.361$ $\qquad$ $\hat{p}_1 - \hat{p}_2 = 0.361 - 0.519 = -0.157$
   $\hat{p}_2 = x_1/n_1 = 14/27 = 0.519$
   a. $\overline{p} = (x_1+x_2)/(n_1+n_2) = (13+14)/(36+27) = 27/63 = 0.429$
   b. $z_{\hat{p}_1-\hat{p}_2} = (\hat{p}_1-\hat{p}_2 - \mu_{\hat{p}_1-\hat{p}_2})/\sigma_{\hat{p}_1-\hat{p}_2}$
   $= (-0.517 - 0)/\sqrt{(0.429)(0.571)/36 + (0.429)(0.571)/27} = -0.517/0.1260 = -1.25$
   c. for $\alpha = 0.05$, the critical values are $z = \pm z_{\alpha/2} = \pm z_{0.025} = \pm 1.96$
   d. P-value $= 2 \cdot P(z<-1.25) = 2 \cdot (0.1056) = 0.2112$

NOTE: While the intermediate steps for calculating $\sigma_{\hat{p}_1-\hat{p}_2}$ and other values show decimals with a fixed number of decimal places, this manual always uses the exact values – i.e., the original fractions recalled from the calculator memory.

9. $\hat{p}_1 = x_1/n_1 = 13/36 = 0.361$ $\quad\quad\quad$ $\hat{p}_1 - \hat{p}_2 = 0.361 - 0.519 = -0.157$

   $\hat{p}_2 = x_1/n_1 = 14/27 = 0.519$

   a. $E = z_{\alpha/2}\sigma_{\hat{p}_1-\hat{p}_2}$

   $= 1.96\sqrt{(0.361)(0.639)/36 + (0.519)(0.481)/27}$

   $= 1.96(0.1251)$

   $= 0.2452$

   b. $(\hat{p}_1 - \hat{p}_2) \pm E$

   -0.1574 ± 0.2452

   $-0.4026 < p_1 - p_2 < 0.0878$

11. Let the Lipitor users group 1.

    original claim: $p_1 - p_2 = 0$

    H_o: $p_1 - p_2 = 0$

    H_1: $p_1 - p_2 \neq 0$

    $\alpha = 0.05$

    C.V. $z = \pm z_{\alpha/2} = z_{0.025} = \pm 1.96$

    calculations:

    $z_{\hat{p}_1-\hat{p}_2} = (\hat{p}_1 - \hat{p}_2 - \mu_{\hat{p}_1-\hat{p}_2})/\sigma_{\hat{p}_1-\hat{p}_2}$

    $= -0.73$ [TI-83/84+]

    P-value = 0.4638 [TI-83/84+]

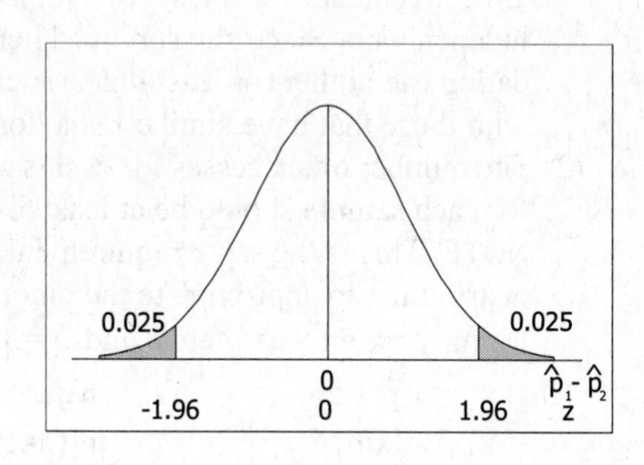

    conclusion:

    Do not reject H_o; there is not sufficient evidence to reject the claim that $p_1 - p_2 = 0$. There is not sufficient evidence to reject the claim that the rate of infection was the same for those treated with Lipitor and those given a placebo.

13. Let the 1993 students be group 1.

    $\hat{p}_1 = x_1/n_1 = 171/560 = 0.305$ $\quad\quad\quad$ $\hat{p}_1 - \hat{p}_2 = 0.305 - 0.365 = -0.060$

    $\hat{p}_2 = x_2/n_2 = 263/720 = 0.365$

    $\bar{p} = (171+263)/(560+720)$

    $= 434/1280 = 0.339$

    original claim: $p_1 - p_2 < 0$

    H_o: $p_1 - p_2 = 0$

    H_1: $p_1 - p_2 < 0$

    $\alpha = 0.05$

    C.V. $z = -z_\alpha = -z_{0.05} = -1.645$

    calculations:

    $z_{\hat{p}_1-\hat{p}_2} = (\hat{p}_1 - \hat{p}_2 - \mu_{\hat{p}_1-\hat{p}_2})/\sigma_{\hat{p}_1-\hat{p}_2}$

    $= (-0.060 - 0)/\sqrt{(0.339)(0.661)/560 + (0.339)(0.661)/720} = -0.060/0.02668 = -2.25$

    P-value = $P(z<-2.25) = 0.0122$

    conclusion:

    Reject H_o; there is sufficient evidence to conclude that $p_1 - p_2 < 0$. There is sufficient evidence to support the claim that the proportion of college students using illegal drugs in 1993 was less than it is now.

15. Let those not wearing seat belts be group 1.
$\hat{p}_1 = x_1/n_1 = 31/2823 = 0.01098$ $\hat{p}_1 - \hat{p}_2 = 0.01098 - 0.00206 = 0.00892$
$\hat{p}_2 = x_2/n_2 = 16/7765 = 0.00206$ $\alpha = 0.10$
$(\hat{p}_1 - \hat{p}_2) \pm z_{\alpha/2}\sigma_{\hat{p}_1-\hat{p}_2}$
$0.00892 \pm 1.645\sqrt{(0.01098)(0.98902)/2823 + (0.00206)(0.99794)/7765}$
0.00892 ± 0.00334
$0.00558 < p_1 - p_2 < 0.01226$
Since the confidence interval does not include the value 0, it suggests that the two population proportions are not equal and that seat belts are effective because the proportion of non-users who are killed is greater than the proportion of users who are killed.

17. Let the women be group 1.
$\hat{p}_1 = x_1/n_1 = 347/386 = 0.899$ $\hat{p}_1 - \hat{p}_2 = 0.899 - 0.850 = 0.049$
$\hat{p}_2 = x_2/n_2 = 305/359 = 0.850$
$\bar{p} = (437+305)/(386+359)$
$= 652/745 = 0.875$
original claim: $p_1 - p_2 \neq 0$
H_o: $p_1 - p_2 = 0$
H_1: $p_1 - p_2 \neq 0$
$\alpha = 0.05$
C.V. $z = \pm z_{\alpha/2} = \pm z_{0.025} = \pm 1.96$
calculations:

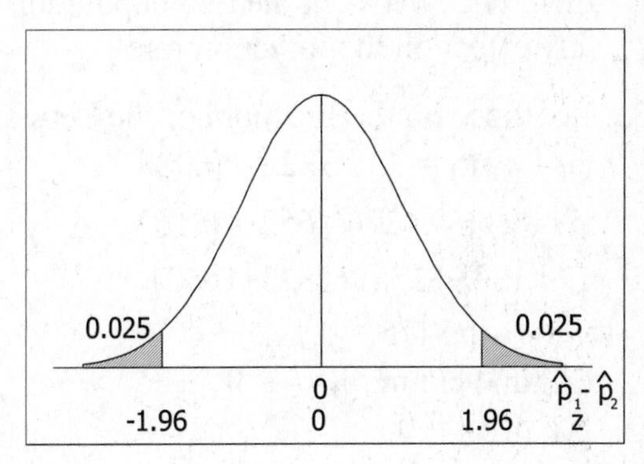

$z_{\hat{p}_1-\hat{p}_2} = (\hat{p}_1 - \hat{p}_2 - \mu_{\hat{p}_1-\hat{p}_2})/\sigma_{\hat{p}_1-\hat{p}_2}$
$= (0.049 - 0)/\sqrt{(0.875)(0.125)/386 + (0.875)(0.125)/359} = 0.049/0.02424 = 2.04$
P-value $= 2 \cdot P(z>2.04) = 2 \cdot (1 - 0.9793) = 2 \cdot (0207) = 0.0414$
conclusion:
Reject H_o; there is sufficient evidence to conclude that $p_1 - p_2 \neq 0$ (in fact, that $p_1 - p_2 > 0$). There is sufficient evidence to support the claim that the percentage of women who agree is different from the percentage of men who agree. Yes; there does appear to be a difference in the way that women and men feel about the issue.

19. Let the games with the roof closed be group 1.
$\hat{p}_1 = x_1/n_1 = 36/53 = 0.679$ $\hat{p}_1 - \hat{p}_2 = 0.679 - 0.577 = 0.102$
$\hat{p}_2 = x_2/n_2 = 15/26 = 0.577$
$\bar{p} = (36+15)/(53+26)$
$= 51/79 = 0.646$
original claim: $p_1 - p_2 > 0$
H_o: $p_1 - p_2 = 0$
H_1: $p_1 - p_2 > 0$
$\alpha = 0.05$
C.V. $z = z_\alpha = z_{0.05} = 1.645$
calculations:

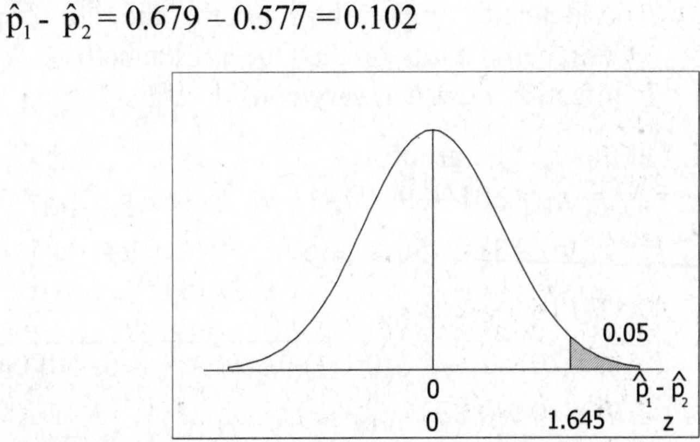

$z_{\hat{p}_1-\hat{p}_2} = (\hat{p}_1 - \hat{p}_2 - \mu_{\hat{p}_1-\hat{p}_2})/\sigma_{\hat{p}_1-\hat{p}_2}$
$= (0.102 - 0)/\sqrt{(0.646)(0.354)/53 + (0.646)(0.354)/26} = 0.102/0.1145 = 0.89$
P-value $= P(z>0.89) = 1 - 0.8133 = 0.1867$

conclusion:

Do not reject H_o; there is not sufficient evidence to conclude that $p_1-p_2 > 0$. There is not sufficient evidence to support the claim that the proportion of wins is higher with the roof closed than with it open. No; the closed roof does not appear to be a significant advantage.

21. Let those treated with echinacea be group 1.

$\hat{p}_1 = x_1/n_1 = 40/45 = 0.889$ $\quad$ $\hat{p}_1 - \hat{p}_2 = 0.889 - 0.854 = 0.035$

$\hat{p}_2 = x_2/n_2 = 88/103 = 0.854$ $\quad$ $\alpha = 0.05$

$(\hat{p}_1 - \hat{p}_2) \pm z_{\alpha/2}\sigma_{\hat{p}_1-\hat{p}_2}$

$0.0345 \pm 1.96\sqrt{(0.889)(0.111)/45 + (0.854)(0.146)/103}$

0.0345 ± 0.1143

$-0.0798 < p_1-p_2 < 0.1489$

Since the confidence interval includes the value 0, it suggests that there is no significant difference between the two population proportions. No; Echinacea does not appear to have any effect on the infection rate.

23. Let those on the Freedom of the Seas ship be group 1.

$\hat{p}_1 = x_1/n_1 = 338/3823 = 0.088$ $\quad$ $\hat{p}_1 - \hat{p}_2 = 0.088 - 0.167 = -0.079$

$\hat{p}_2 = x_2/n_2 = 276/1652 = 0.167$

$\bar{p} = (338+276)/(3823+1652)$

$= 614/5475 = 0.112$

original claim: $p_1-p_2 < 0$

H_o: $p_1-p_2 = 0$

H_1: $p_1-p_2 < 0$

$\alpha = 0.01$

C.V. $z = -z_\alpha = -z_{0.01} = -2.326$

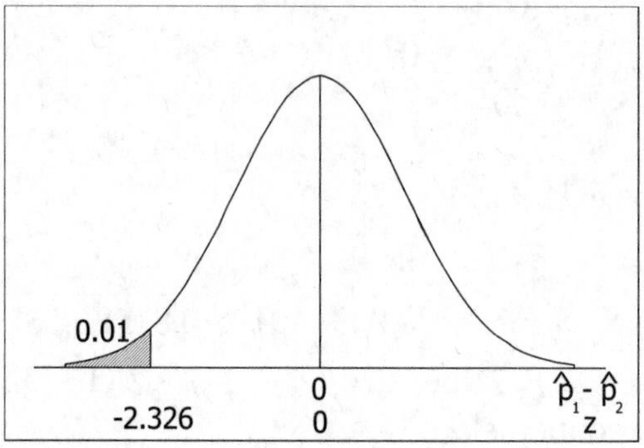

calculations:

$z_{\hat{p}_1-\hat{p}_2} = (\hat{p}_1-\hat{p}_2 - \mu_{\hat{p}_1-\hat{p}_2})/\sigma_{\hat{p}_1-\hat{p}_2}$

$= (-0.079 - 0)/\sqrt{(0.112)(0.888)/3823 + (0.112)(0.888)/1652} = -0.079/0.009291 = -8.47$

P-value = $P(z<-8.47) = 0.0001$

conclusion:

Reject H_o; there is sufficient evidence to conclude that $p_1-p_2 < 0$. There is sufficient evidence to support the claim that the rate of sickness on the Freedom of the Seas ship is less than the rate on the Queen Elizabeth II. Yes; based on these results, it appears that infection rates can vary considerably.

25. Let the men be group 1.

$\hat{p}_1 = x_1/n_1 = 201/489 = 0.411$ $\quad$ $\hat{p}_1 - \hat{p}_2 = 0.411 - 0.360 = 0.051$

$\hat{p}_2 = x_2/n_2 = 126/350 = 0.360$ $\quad$ $\alpha = 0.01$

$(\hat{p}_1 - \hat{p}_2) \pm z_{\alpha/2}\sigma_{\hat{p}_1-\hat{p}_2}$

$0.0510 \pm 2.575\sqrt{(0.411)(0.589)/489 + (0.360)(0.640)/350}$

0.0510 ± 0.0874

$-0.0364 < p_1-p_2 < 0.1385$

Since the confidence interval includes zero, it suggests there is no significant difference between the two population proportions. There does not appear to be a significant difference between rates at which challenges of men and women tennis players are upheld.

27. Let the females be group 1.
$\hat{p}_1 = x_1/n_1 = 1397/2739 = 0.510$ $\hat{p}_1 - \hat{p}_2 = 0.510 - 0.322 = 0.188$
$\hat{p}_2 = x_2/n_2 = 436/1352 = 0.322$
$\bar{p} = (1397+436)/(2739+1352)$
$= 1833/4091 = 0.448$
original claim: $p_1-p_2 \neq 0$
H_o: $p_1-p_2 = 0$
H_1: $p_1-p_2 \neq 0$
$\alpha = 0.01$
C.V. $z = \pm z_{\alpha/2} = \pm z_{0.005} = \pm 2.575$
calculations:

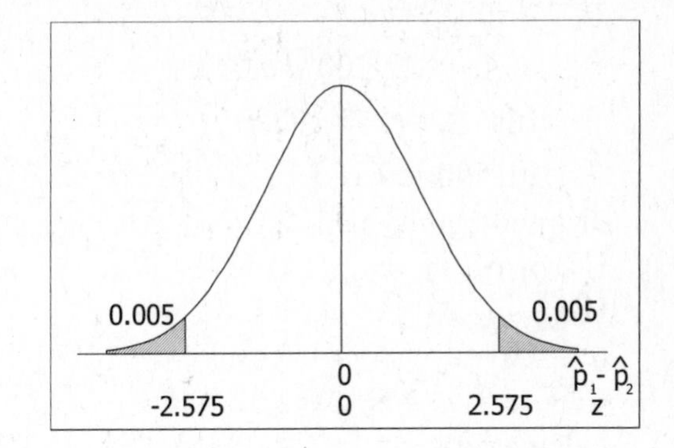

$z_{\hat{p}_1-\hat{p}_2} = (\hat{p}_1-\hat{p}_2 - \mu_{\hat{p}_1-\hat{p}_2})/\sigma_{\hat{p}_1-\hat{p}_2}$
$= (0.188 - 0)/\sqrt{(0.488)(0.512)/2739 + (0.488)(0.512)/1352} = 0.188/0.01653 = 11.35$
P-value $= 2\cdot P(z>11.35) = 2\cdot(1 - 0.9999) = 2\cdot(0.0001) = 0.0002$
conclusion:
Reject H_o; there is sufficient evidence to conclude that $p_1-p_2 \neq 0$ (in fact, that $p_1-p_2 > 0$). There is sufficient evidence to support the claim that female and male survivors have different rates of thyroid disease.

NOTE: The information given for exercises #29, #30, #33, #34, #35 and #36 is not sufficient to determine the exact values of x_1 and x_2. As a consequence, fractions should not be used and the usual level of accuracy given in the answers is not justified. Different methods for dealing with this lack of precision may result in answers slightly different from those given in the text or in this manual. The final conclusions of the tests of hypotheses or the inferences made from the confidence intervals are not affected by these minor differences.

29. Let the males be group 1.
$\hat{p}_1 = x_1/n_1 = x_1/731 = 0.69$ [as any $501 \leq x_1 \leq 508$ gives $\hat{p}_1 = 69\%$, use 0.69 in all calculations]
$\hat{p}_2 = x_2/n_2 = x_2/770 = 0.70$ [as any $536 \leq x_2 \leq 542$ gives $\hat{p}_2 = 70\%$, use 0.70 in all calculations]
$\hat{p}_1 - \hat{p}_2 = 0.69 - 0.70 = -0.01$ $\alpha = 0.10$
$(\hat{p}_1 - \hat{p}_2) \pm z_{\alpha/2}\sigma_{\hat{p}_1-\hat{p}_2}$
$-0.01 \pm 1.645\sqrt{(0.69)(0.31)/731 + (0.70)(0.30)/770}$
-0.01 ± 0.0391
$-0.0491 < p_1-p_2 < 0.0291$
Since the confidence interval includes zero, there is no significant difference between the proportion of "yes" answers from males and females

31. Let the 1976 returns be group 1.
NOTE: $x_1 = (.276)(250) = 69$ and $x_2 = (.073)(300) = 22$.
$\hat{p}_1 = x_1/n_1 = 69/250 = 0.276$ $\quad\quad$ $\hat{p}_1 - \hat{p}_2 = 0.276 - 0.073 = 0.203$
$\hat{p}_2 = x_2/n_2 = 22/300 = 0.073$
$\bar{p} = (69+22)/(250+300)$
$= 91/550 = 0.165$
original claim: $p_1 - p_2 > 0$
H_o: $p_1 - p_2 = 0$
H_1: $p_1 - p_2 > 0$
$\alpha = 0.01$
C.V. $z = z_\alpha = z_{0.01} = 2.326$
calculations:
$z_{\hat{p}_1-\hat{p}_2} = (\hat{p}_1 - \hat{p}_2 - \mu_{\hat{p}_1-\hat{p}_2})/\sigma_{\hat{p}_1-\hat{p}_2}$
$= (0.203 - 0)/\sqrt{(0.165)(0.835)/250 + (0.165)(0.835)/300} = 0.203/0.03182 = 6.37$
P-value = $P(z>6.37) = 1 - 0.9999 = 0.0001$
conclusion:
Reject H_o; there is sufficient evidence to conclude that $p_1 - p_2 > 0$. There is sufficient evidence to support the claim that the percentage of returns designating funds for campaigns was greater in 1976 than it is now.

33. Let the Viagra users be group 1.
$\hat{p}_1 = x_1/n_1 = x_1/734 = 0.16$ [as any $114 \le x_1 \le 121$ gives $\hat{p}_1 = 16\%$, use 0.16 in all calculations]
$\hat{p}_2 = x_2/n_2 = x_2/725 = 0.04$ [as any $26 \le x_2 \le 32$ gives $\hat{p}_2 = 4\%$, use 0.04 in all calculations]
$\hat{p}_1 - \hat{p}_2 = 0.16 - 0.04 = 0.12$
$\bar{p} = [(0.16)(734) + (0.04)(725)]/(734 + 725)$
$= 146.44/1459 = 0.100$
original claim: $p_1 - p_2 > 0$
H_o: $p_1 - p_2 = 0$
H_1: $p_1 - p_2 > 0$
$\alpha = 0.01$
C.V. $z = z_\alpha = z_{0.01} = 2.326$
calculations:
$z_{\hat{p}_1-\hat{p}_2} = (\hat{p}_1 - \hat{p}_2 - \mu_{\hat{p}_1-\hat{p}_2})/\sigma_{\hat{p}_1-\hat{p}_2}$
$= (0.12 - 0)/\sqrt{\frac{(0.100)(0.900)}{734} + \frac{(0.100)(0.900)}{725}}$
$= 0.12/0.01573 = 7.63$
P-value = $P(z>7.63) = 1 - 0.9999 = 0.0001$

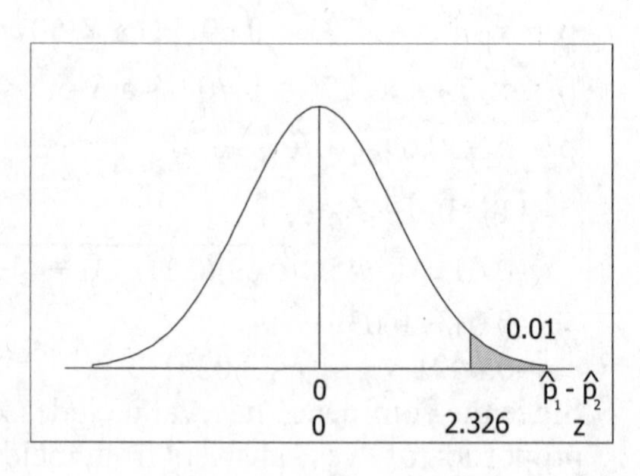

conclusion:
Reject H_o; there is sufficient evidence to conclude that $p_1 - p_2 > 0$. There is sufficient evidence to support the claim that the proportion of persons experiencing headaches is greater for those treated with Viagra. Yes; headaches do appear to be a concern for those who take Viagra.

35. Let the women be group 1 and use $n_1 = n_2 = 61647/2 = 30823.5$
$\hat{p}_1 = x_1/n_1 = 0.27$ [as many x_1 values give $\hat{p}_1 = 27\%$, use 0.27 in all calculations]
$\hat{p}_2 = x_2/n_2 = 0.25$ [as many x_2 values give $\hat{p}_2 = 25\%$, use 0.25 in all calculations]

$\hat{p}_1 - \hat{p}_2 = 0.27 - 0.25 = 0.02$ $\alpha = 0.05$

$(\hat{p}_1 - \hat{p}_2) \pm z_{\alpha/2}\sigma_{\hat{p}_1-\hat{p}_2}$

$0.02 \pm 1.96\sqrt{(0.27)(0.73)/30823.5 + (0.25)(0.75)/30823.5}$

0.02 ± 0.0069

$0.0131 < p_1\text{-}p_2 < 0.0269$

Since the confidence interval does not include zero, there is a significant difference between the two proportions. Since the confidence interval includes only positive values, the proportion of women who say that female bosses are harshly critical is greater than the proportion of men who say so. Despite the large sample size, the fact that the respondents constitute a voluntary response sample means that the results should not be used to make inferences about the general population.

37. For all parts of this exercise

$\hat{p}_1 = x_1/n_1 = 112/200 = 0.560$ $\hat{p}_1 - \hat{p}_2 = 0.560 - 0.440 = 0.120$

$\hat{p}_2 = x_2/n_2 = 88/200 = 0.440$

$\bar{p} = (112+88)/(200+200)$

$= 200/400 = 0.500$

Use with $\alpha = 0.05$ for all parts.

a. $(\hat{p}_1 - \hat{p}_2) \pm z_{\alpha/2}\sigma_{\hat{p}_1-\hat{p}_2}$

$0.1200 \pm 1.96\sqrt{(0.560)(0.440)/200 + (0.440)(0.560)/200}$

0.1200 ± 0.0973

$0.0227 < p_1\text{-}p_2 < 0.2173$

Since the interval does not include 0, conclude that p_1 and p_2 are different.
Since the interval lies entirely above 0, conclude that $p_1 - p_2 > 0$.

b. for group 1

$\hat{p} \pm z_{\alpha/2}\sqrt{\hat{p}\hat{q}/n}$

$0.560 \pm 1.96\sqrt{(0.56)(0.44)/200}$

0.560 ± 0.069

$0.491 < p < 0.629$

for group 2

$\hat{p} \pm z_{\alpha/2}\sqrt{\hat{p}\hat{q}/n}$

$0.440 \pm 1.96\sqrt{(0.44)(0.56)/200}$

0.440 ± 0.069

$0.371 < p < 0.509$

Since the intervals overlap, the implication is that p_1 and p_2 could have the same value.

c. original claim: $p_1\text{-}p_2 = 0$

H_o: $p_1\text{-}p_2 = 0$

H_1: $p_1\text{-}p_2 \neq 0$

$\alpha = 0.05$

C.V. $z = \pm z_{\alpha/2} = \pm z_{0.025} = \pm 1.96$

calculations:

$z_{\hat{p}_1-\hat{p}_2} = (\hat{p}_1 - \hat{p}_2 - \mu_{\hat{p}_1-\hat{p}_2})/\sigma_{\hat{p}_1-\hat{p}_2}$

$= (0.12 - 0)/\sqrt{\frac{(0.5625)(0.4375)}{40} + \frac{(0.5625)(0.4375)}{40}}$

$= 0.12/0.05 = 2.40$

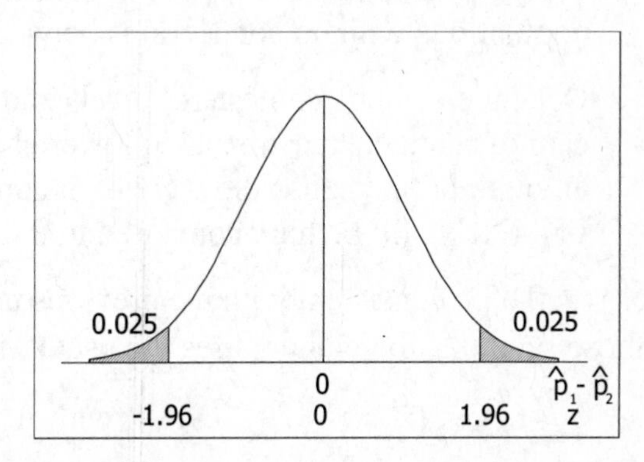

P-value $= 2\cdot P(z>2.40) = 2\cdot(1 - 0.9918) = 2\cdot(0.0082) = 0.0164$

conclusion:

Reject H_o; there is sufficient evidence to reject the claim that $p_1\text{-}p_2 = 0$ and conclude that $p_1\text{-}p_2 \neq 0$ (in fact, that $p_1\text{-}p_2 > 0$).

d. Based on the preceding results, conclude that p_1 and p_2 are unequal and that $p_1 > p_2$. The overlapping interval method of part (b) appears to be the least effective method for comparing two populations.

39. Let the females be group 1.

$\hat{p}_1 = x_1/n_1 = 1397/2739 = 0.510$ $\hat{p}_1 - \hat{p}_2 = 0.510 - 0.322 = 0.188$

$\hat{p}_2 = x_2/n_2 = 436/1352 = 0.322$

original claim: $p_1 - p_2 = 0.15$

H_o: $p_1 - p_2 = 0.15$

H_1: $p_1 - p_2 \neq 0.15$

$\alpha = 0.01$

C.V. $z = \pm z_{\alpha/2} = \pm z_{0.005} = \pm 2.575$

calculations:

$z_{\hat{p}_1-\hat{p}_2} = (\hat{p}_1 - \hat{p}_2 - \mu_{\hat{p}_1-\hat{p}_2})/\sigma_{\hat{p}_1-\hat{p}_2}$

$= (0.188 - 0.15)/\sqrt{\frac{(0.510)(0.490)}{2739} + \frac{(0.322)(0.678)}{1352}}$

$= 0.038/0.01590 = 2.36$

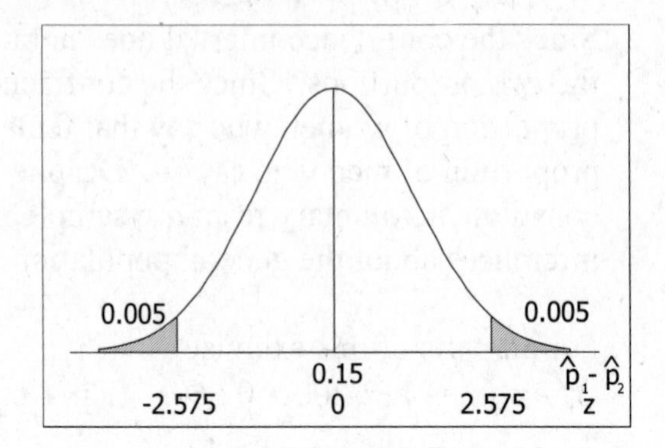

P-value $= 2 \cdot P(z>2.36) = 2 \cdot (1 - 0.9909) = 2 \cdot (0.0091) = 0.0182$

conclusion:

Do not reject H_o; there is not sufficient evidence to reject the claim that $p_1 - p_2 = 0.15$. There is not sufficient evidence to reject the claim that the rate of thyroid disease among female survivors is 15 percentage points higher than that for male survivors.

9-3 Inferences About Two Means: Independent Samples

1. Reversing the designation of which sample is considered group1 and which sample is considered group 2 changes the sign of the point estimate and the signs of the endpoints of the interval estimate. The confidence interval using the new designation is $1.6 < \mu_1 - \mu_2 < 12.2$.

3. A one-tailed test of hypothesis at the 0.01 level of significance corresponds to a two-sided confidence interval at the 2(0.01) = 0.02 level of significance – i.e., to an interval with a confidence level of 98%.

5. Independent, since the men and women in the sample were selected randomly from their own populations with no connections between them.

7. Dependent, since cholesterol levels and determined by many factors that the Lipitor treatment cannot change. Treatments to lower cholesterol typically reduce everyone's levels by a certain amount, but persons who were high compared to the others before the treatment, for example, will likely still be high compared to the others after the treatment.

NOTE: To be consistent with the previous notation, reinforcing patterns and concepts presented in those sections, the manual uses the usual t formula written to apply to $\bar{x}_1 - \bar{x}_2$'s

$$t_{\bar{x}_1-\bar{x}_2} = (\bar{x}_1 - \bar{x}_2 - \mu_{\bar{x}_1-\bar{x}_2})/s_{\bar{x}_1-\bar{x}_2}, \text{ with } \mu_{\bar{x}_1-\bar{x}_2} = \mu_1 - \mu_2 \text{ and } s_{\bar{x}_1-\bar{x}_2} = \sqrt{s_1^2/n_1 + s_2^2/n_2}\ .$$

9. Let the children treated with low humidity be group 1.

$\overline{x}_1 - \overline{x}_2 = 0.98 - 1.09 = -0.11$

original claim: $\mu_1 - \mu_2 = 0$

H_o: $\mu_1 - \mu_2 = 0$

H_1: $\mu_1 - \mu_2 \neq 0$

$\alpha = 0.05$ and df = 45

C.V. $t = \pm t_{\alpha/2} = t_{0.025} = \pm 2.014$

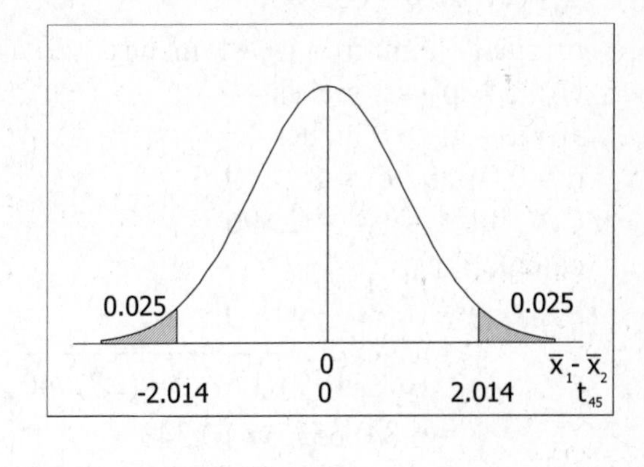

calculations:

$t_{\overline{x}_1-\overline{x}_2} = (\overline{x}_1 - \overline{x}_2 - \mu_{\overline{x}_1-\overline{x}_2})/s_{\overline{x}_1-\overline{x}_2}$

$= (-0.11 - 0)/\sqrt{(1.22)^2/46 + (1.11)^2/46}$

$= -0.11/0.2432 = -0.452$

P-value = 2·tcdf(-99,-0.452,45) = 0.6532

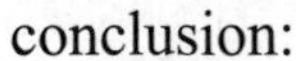

conclusion:

Do not reject H_o; there is not sufficient evidence to reject the claim that $\mu_1 - \mu_2 = 0$. There is not sufficient evidence to reject the claim that the two groups are from populations with the same mean. The results suggest that increasing the humidity does not have a significant effect on the treatment of croup.

11. Let the unfiltered cigarettes be group 1.

$\overline{x}_1 - \overline{x}_2 = 21.1 - 13.2 = 7.9$

$\alpha = 0.10$ and df = 24

$(\overline{x}_1 - \overline{x}_2) \pm t_{\alpha/2} s_{\overline{x}_1-\overline{x}_2}$

$7.9 \pm 1.711\sqrt{(3.2)^2/25 + (3.7)^2/25}$

7.9 ± 1.674

$6.2 < \mu_1 - \mu_2 < 9.6$ (mg)

Yes; since the confidence interval includes only positive values, the results suggest that the filtered cigarettes have less tar than the unfiltered ones.

13. Let the checks be group 1.

$\overline{x}_1 - \overline{x}_2 = 23.8 - 47.6 = -23.8$

original claim: $\mu_1 - \mu_2 < 0$ cents

H_o: $\mu_1 - \mu_2 = 0$ cents

H_1: $\mu_1 - \mu_2 < 0$ cents

$\alpha = 0.05$ and df = 99

C.V. $t = -t_\alpha = -t_{0.05} = -1.660$

calculations:

$t_{\overline{x}_1-\overline{x}_2} = (\overline{x}_1 - \overline{x}_2 - \mu_{\overline{x}_1-\overline{x}_2})/s_{\overline{x}_1-\overline{x}_2}$

$= (-23.8 - 0)/\sqrt{(32.0)^2/100 + (33.5)^2/100}$

$= -23.8/4.6328 = -5.137$

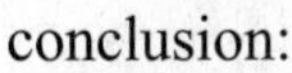

P-value = tcdf(-99,-5.137,99) = 7.000E-7 = 0.0000007

conclusion:

Reject H_o; there is sufficient evidence to conclude that $\mu_1 - \mu_2 < 0$. There is sufficient evidence to support the claim that the cents portions of the check amounts have a mean that is less than the cents portions of the credit card charges. One reason for the difference might be that check amounts include many more even dollar transactions (e.g., donations to church and charities) that have a cents portion of 0.

15. Let the supermodels be group 1, for which $n_1 = 9$.

$\bar{x}_1 - \bar{x}_2 = 70.0 - 63.2 = 6.8$

original claim: $\mu_1 - \mu_2 > 0$ inches

H_o: $\mu_1 - \mu_2 = 0$ inches

H_1: $\mu_1 - \mu_2 > 0$ inches

$\alpha = 0.01$ and df = 8

C.V. $t = t_\alpha = t_{0.01} = 2.896$

calculations:

$t_{\bar{x}_1 - \bar{x}_2} = (\bar{x}_1 - \bar{x}_2 - \mu_{\bar{x}_1 - \bar{x}_2})/s_{\bar{x}_1 - \bar{x}_2}$

$= (6.8 - 0)/\sqrt{(1.5)^2/9 + (2.7)^2/40}$

$= 6.8/0.6575 = 10.343$

P-value = tcdf(10.343,99,8) = 3.298E-6 = 0.000003

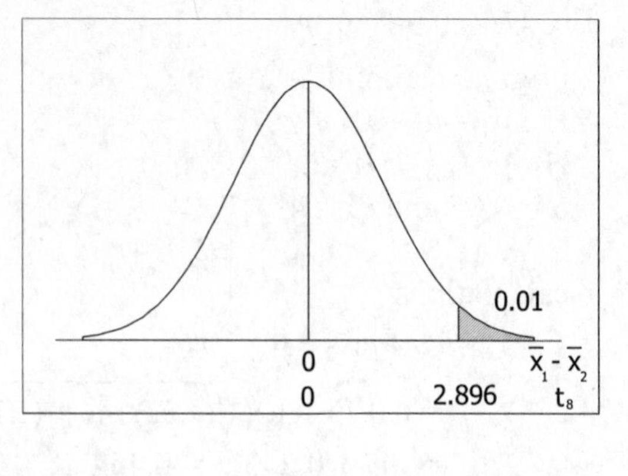

conclusion:

Reject H_o; there is sufficient evidence to conclude that $\mu_1 - \mu_2 > 0$. There is sufficient evidence to support the claim that the mean height of supermodels is greater than the mean height of women who are not supermodels.

17. Let the four-cylinder cars be group 1.

$\bar{x}_1 - \bar{x}_2 = 137.5 - 136.3 = 1.2$

$\alpha = 0.10$ and df = 11

$(\bar{x}_1 - \bar{x}_2) \pm t_{\alpha/2} s_{\bar{x}_1 - \bar{x}_2}$

$1.2 \pm 1.796\sqrt{(5.8)^2/13 + (9.7)^2/12}$

1.2 ± 5.800

$-4.6 < \mu_1 - \mu_2 < 7.0$ (feet)

Since the confidence interval includes zero, there could be no difference between the two population means. No; there does not appear to be a significant difference between the mean braking distance of four-cylinder and six-cylinder cars.

19. Let the menthol cigarettes be group 1.

$\bar{x}_1 - \bar{x}_2 = 0.87 - 0.92 = -0.05$

original claim: $\mu_1 - \mu_2 \neq 0$ mg

H_o: $\mu_1 - \mu_2 = 0$ mg

H_1: $\mu_1 - \mu_2 \neq 0$ mg

$\alpha = 0.05$ and df = 24

C.V. $t = \pm t_{\alpha/2} = t_{0.025} = \pm 2.064$

calculations:

$t_{\bar{x}_1 - \bar{x}_2} = (\bar{x}_1 - \bar{x}_2 - \mu_{\bar{x}_1 - \bar{x}_2})/s_{\bar{x}_1 - \bar{x}_2}$

$= (-0.05 - 0)/\sqrt{(0.24)^2/25 + (0.25)^2/25}$

$= -0.05/0.0693 = -0.721$

P-value = 2·tcdf(-99,-0.721,24) = 0.4776

0.025 0.025
-2.064 0 2.064 t_{24}

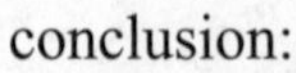

conclusion:

Do not reject H_o; there is not sufficient evidence to conclude that $\mu_1 - \mu_2 \neq 0$. There is not sufficient evidence to support the claim that menthol cigarettes and non-menthol cigarettes have different amounts of nicotine. No; menthol does not appear to have an effect on the nicotine content.

21. Let the high-interest mortgages be group 1.

$\bar{x}_1 - \bar{x}_2 = 594.8 - 785.2 = -190.4$

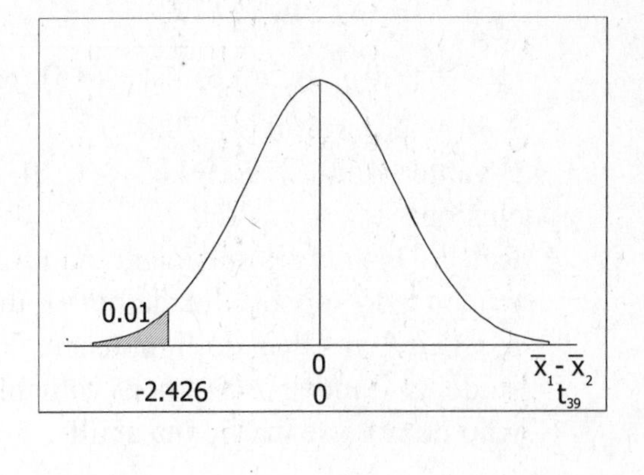

original claim: $\mu_1 - \mu_2 < 0$ FICO units

H_o: $\mu_1 - \mu_2 = 0$ FICO units

H_1: $\mu_1 - \mu_2 < 0$ FICO units

$\alpha = 0.01$ and df = 39

C.V. $t = -t_\alpha = -t_{0.01} = -2.426$

calculations:

$t_{\bar{x}_1 - \bar{x}_2} = (\bar{x}_1 - \bar{x}_2 - \mu_{\bar{x}_1 - \bar{x}_2})/s_{\bar{x}_1 - \bar{x}_2}$

$= (-190.4 - 0)/\sqrt{(12.2)^2/40 + (16.3)^2/40}$

$= -190.4/3.219 = -59.145$

P-value = tcdf(-99,-59.145,39) ≈ 0

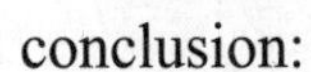
conclusion:

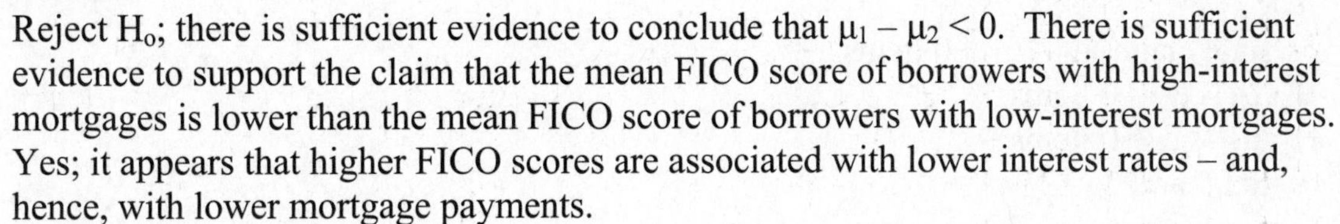
Reject H_o; there is sufficient evidence to conclude that $\mu_1 - \mu_2 < 0$. There is sufficient evidence to support the claim that the mean FICO score of borrowers with high-interest mortgages is lower than the mean FICO score of borrowers with low-interest mortgages. Yes; it appears that higher FICO scores are associated with lower interest rates – and, hence, with lower mortgage payments.

23. Let the unsuccessful applicants be group 1.

$\bar{x}_1 - \bar{x}_2 = 47.0 - 43.9 = 3.1$

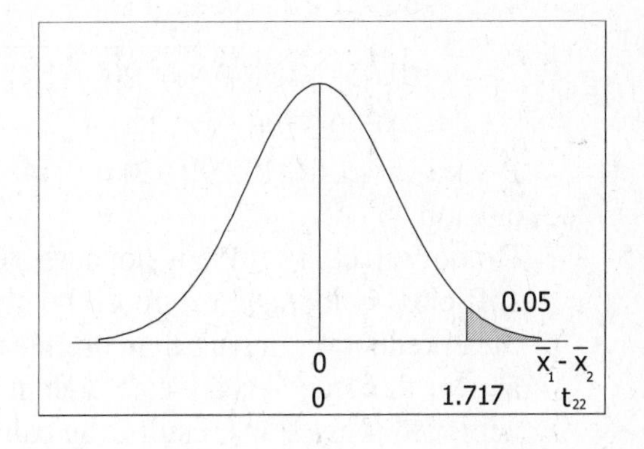

original claim: $\mu_1 - \mu_2 > 0$ years

H_o: $\mu_1 - \mu_2 = 0$ years

H_1: $\mu_1 - \mu_2 > 0$ years

$\alpha = 0.05$ and df = 22

C.V. $t = t_\alpha = t_{0.05} = 1.717$

calculations:

$t_{\bar{x}_1 - \bar{x}_2} = (\bar{x}_1 - \bar{x}_2 - \mu_{\bar{x}_1 - \bar{x}_2})/s_{\bar{x}_1 - \bar{x}_2}$

$= (3.1 - 0)/\sqrt{(7.2)^2/23 + (5.9)^2/30}$

$= 3.1/1.8478 = 1.678$

P-value = tcdf(1.678,99,22) = 0.0538

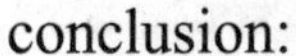
conclusion:

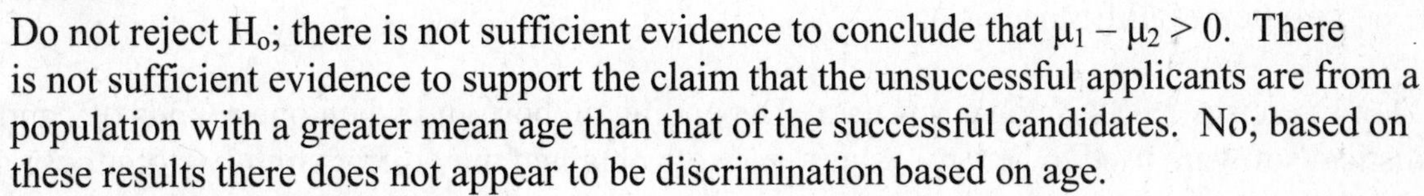
Do not reject H_o; there is not sufficient evidence to conclude that $\mu_1 - \mu_2 > 0$. There is not sufficient evidence to support the claim that the unsuccessful applicants are from a population with a greater mean age than that of the successful candidates. No; based on these results there does not appear to be discrimination based on age.

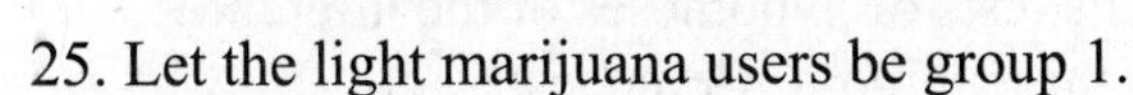
25. Let the light marijuana users be group 1.

$\bar{x}_1 - \bar{x}_2 = 53.3 - 51.3 = 2.0$

original claim: $\mu_1 - \mu_2 > 0$ items

H_o: $\mu_1 - \mu_2 = 0$ items

H_1: $\mu_1 - \mu_2 > 0$ items

$\alpha = 0.01$ and df = 63

C.V. $t = t_\alpha = t_{0.01} = 2.390$

0.01
0
0
2.390
$\bar{x}_1 - \bar{x}_2$
t_{63}

calculations:

$t_{\bar{x}_1-\bar{x}_2} = (\bar{x}_1-\bar{x}_2 - \mu_{\bar{x}_1-\bar{x}_2})/s_{\bar{x}_1-\bar{x}_2}$

$= (2.0 - 0)/\sqrt{(3.6)^2/64 + (4.5)^2/65}$

$= 2.0/0.7170 = 2.790$

P-value = tcdf(2.790,99,63) = 0.0035

conclusion:

Reject H_o; there is sufficient evidence to conclude that $\mu_1 - \mu_2 > 0$. There is sufficient evidence to support the claim that the heavy marijuana users have a lower mean number of recalled items than do light users. Yes; marijuana use should be of concern to college students – and an even more valuable study might be one comparing light users to those who do not use marijuana at all.

27. Let the magnet users be group 1.

$\bar{x}_1-\bar{x}_2 = 0.49 - 0.44 = 0.05$

original claim: $\mu_1 - \mu_2 > 0$ items

H_o: $\mu_1 - \mu_2 = 0$ items

H_1: $\mu_1 - \mu_2 > 0$ items

$\alpha = 0.05$ and df = 19

C.V. $t = t_\alpha = t_{0.05} = 1.729$

calculations:

$t_{\bar{x}_1-\bar{x}_2} = (\bar{x}_1-\bar{x}_2 - \mu_{\bar{x}_1-\bar{x}_2})/s_{\bar{x}_1-\bar{x}_2}$

$= (0.05 - 0)/\sqrt{(0.96)^2/20 + (1.4)^2/20}$

$= 0.05/0.3796 = 0.132$

P-value = tcdf(0.132,99,19) = 0.4483

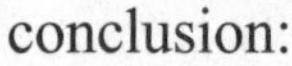

conclusion:

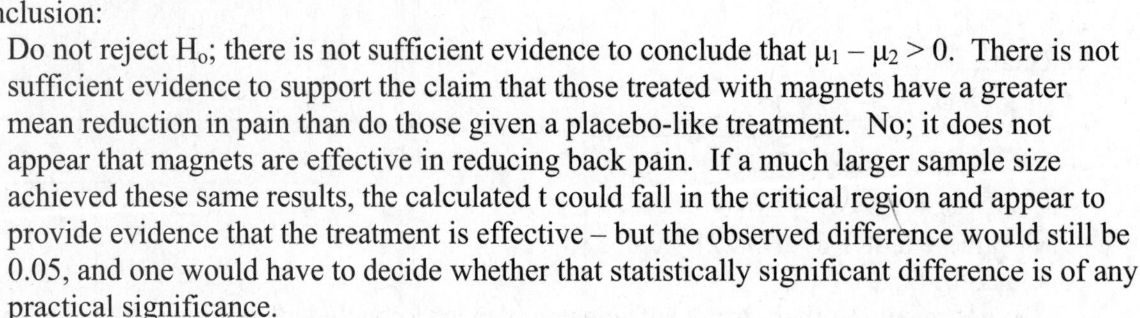

Do not reject H_o; there is not sufficient evidence to conclude that $\mu_1 - \mu_2 > 0$. There is not sufficient evidence to support the claim that those treated with magnets have a greater mean reduction in pain than do those given a placebo-like treatment. No; it does not appear that magnets are effective in reducing back pain. If a much larger sample size achieved these same results, the calculated t could fall in the critical region and appear to provide evidence that the treatment is effective – but the observed difference would still be 0.05, and one would have to decide whether that statistically significant difference is of any practical significance.

NOTE: Exercises 29-32 involve raw data. Depending on how and when one rounds off, and/or the statistical software used to get any values, answers obtained may differ slightly from those given in the text or in this manual. The final conclusions of the tests of hypotheses or the inferences made from the confidence intervals are not affected by these minor differences.

29. Let the recent winners be group 1.

group 1: recent (n=10)	group 2: former (n=10)
$\Sigma x = 187.6$	$\Sigma x = 201.6$
$\Sigma x^2 = 3532.04$	$\Sigma x^2 = 4083.94$
$\bar{x} = 18.76$	$\bar{x} = 20.16$
$s^2 = 1.4071$ (s=1.186)	$s^2 = 2.1871$ (s=1.479)

$\bar{x}_1-\bar{x}_2 = 18.76 - 20.16 = -1.40$

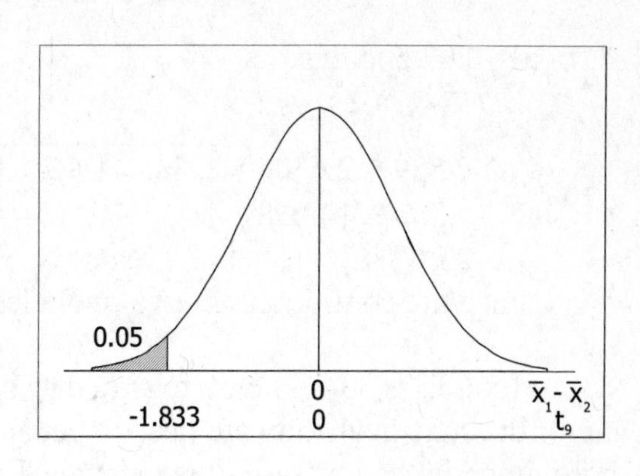

a. original claim: $\mu_1 - \mu_2 < 0$
   H_o: $\mu_1 - \mu_2 = 0$
   H_1: $\mu_1 - \mu_2 < 0$
   $\alpha = 0.05$ and df = 9
   C.V. $t = -t_\alpha = -t_{0.05} = -1.833$
   calculations:

$$t_{\overline{x}_1-\overline{x}_2} = (\overline{x}_1 - \overline{x}_2 - \mu_{\overline{x}_1-\overline{x}_2})/s_{\overline{x}_1-\overline{x}_2}$$
$$= (-1.40 - 0)/\sqrt{1.4071/10 + 2.1871/10}$$
$$= -1.40/0.5995$$
$$= -2.335$$

   P-value = tcdf(-99,-2.335,9) = 0.0222

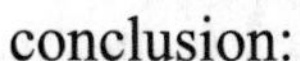

   conclusion:
   Reject H_o; there is sufficient evidence to conclude that $\mu_1 - \mu_2 < 0$. There is sufficient evidence to support the claim that recent winners have a lower mean BMI than winners from the 1920's and 1930's.

b. $\alpha = 0.10$ and df = 9

$$(\overline{x}_1 - \overline{x}_2) \pm t_{\alpha/2} s_{\overline{x}_1-\overline{x}_2}$$
$$-1.40 \pm 1.833\sqrt{1.4071/10 + 2.1871/10}$$
$$-1.40 \pm 1.0989$$
$$-2.50 < \mu_1 - \mu_2 < -0.30$$

31. Let the Popes be group 1.

group 1: Popes (n=24)	group 2: Rulers (n=14)
$\Sigma x = 315$	$\Sigma x = 318$
$\Sigma x^2 = 5981$	$\Sigma x^2 = 11722$
$\overline{x} = 13.125$	$\overline{x} = 22.714$
$s^2 = 80.2880$ (s=8.96)	$s^2 = 346.0659$ (s=18.60)

$\overline{x}_1 - \overline{x}_2 = 13.125 - 22.714 = -9.589$

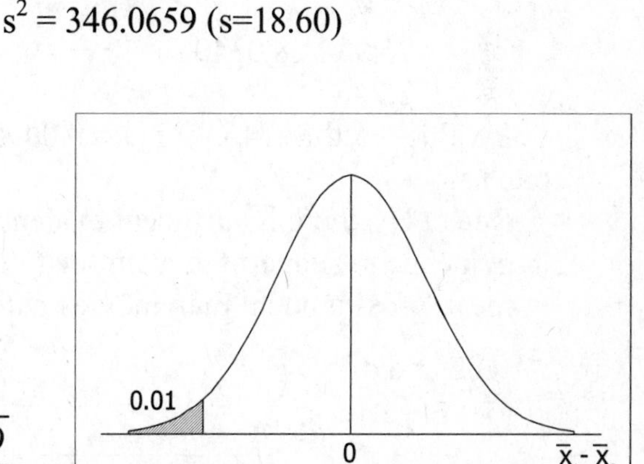

a. original claim: $\mu_1 - \mu_2 < 0$ years
   H_o: $\mu_1 - \mu_2 = 0$ years
   H_1: $\mu_1 - \mu_2 < 0$ years
   $\alpha = 0.01$ and df = 13
   C.V. $t = -t_\alpha = -t_{0.01} = -2.650$
   calculations:

$$t_{\overline{x}_1-\overline{x}_2} = (\overline{x}_1 - \overline{x}_2 - \mu_{\overline{x}_1-\overline{x}_2})/s_{\overline{x}_1-\overline{x}_2}$$
$$= (-9.589 - 0)/\sqrt{\frac{80.2880}{24} + \frac{346.0659}{14}}$$
$$= -9.589/5.2976$$
$$= -1.810$$

   P-value = tcdf(-99,-1.810,13) = 0.0467
   conclusion:
   Do not reject H_o; there is not sufficient evidence to conclude that $\mu_1 - \mu_2 < 0$. There is not sufficient evidence to support the claim that the mean longevity after election for popes is less than the mean longevity after coronation for British monarchs.

b. $\alpha = 0.02$ and df = 13

$(\bar{x}_1 - \bar{x}_2) \pm t_{\alpha/2} s_{\bar{x}_1 - \bar{x}_2}$

$-9.589 \pm 2.650\sqrt{80.2880/24 + 346.0659/14}$

-9.589 ± 14.038

$-23.6 < \mu_1 - \mu_2 < 4.4$ (years)

Since the confidence interval includes zero, the two population means could be equal.

NOTE: Exercises 33-36 involve large data sets. Depending on how and when one rounds off, and/or the statistical software used to get any values, answers obtained may differ slightly from those given in the text or in this manual. The final conclusions of the tests of hypotheses or the inferences made from the confidence intervals are not affected by these minor differences. This manual obtains the summary statistics from statistical software and then uses those (rounded) values to work the exercises by hand.

33. Let the PG & PG-13 movies be group 1. Statistical software yields the following summary.

group 1: $n = 23$ $\bar{x} = 149.391$ $s = 108.019$

group 2: $n = 12$ $\bar{x} = 72.417$ $s = 57.820$

$\bar{x}_1 - \bar{x}_2 = 76.974$

a. original claim: $\mu_1 - \mu_2 > 0$ million \$

H_o: $\mu_1 - \mu_2 = 0$ million \$

H_1: $\mu_1 - \mu_2 > 0$ million \$

$\alpha = 0.01$ and df = 11

C.V. $t = t_\alpha = t_{0.01} = 2.718$

calculations:

$t_{\bar{x}_1 - \bar{x}_2} = (\bar{x}_1 - \bar{x}_2 - \mu_{\bar{x}_1 - \bar{x}_2})/s_{\bar{x}_1 - \bar{x}_2}$

$= (76.974 - 0)/\sqrt{\dfrac{(108.019)^2}{23} + \dfrac{(57.820)^2}{12}}$

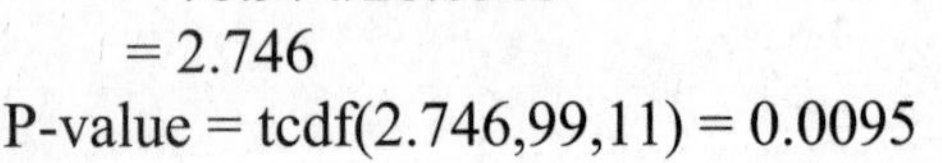

$= 76.974/28.0340$

$= 2.746$

P-value = tcdf(2.746,99,11) = 0.0095

conclusion:

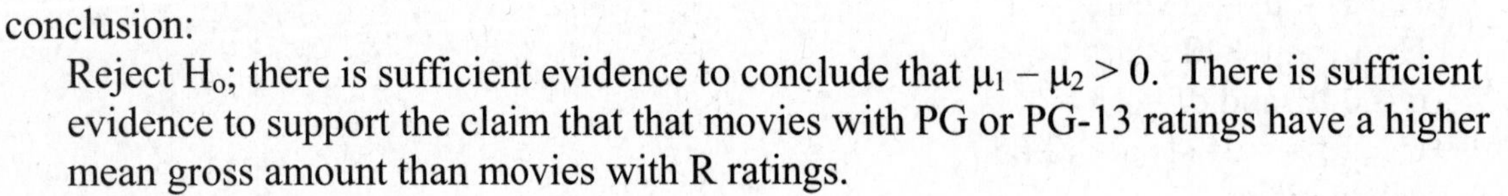

Reject H_o; there is sufficient evidence to conclude that $\mu_1 - \mu_2 > 0$. There is sufficient evidence to support the claim that that movies with PG or PG-13 ratings have a higher mean gross amount than movies with R ratings.

b. $\alpha = 0.02$ and df = 11

$(\bar{x}_1 - \bar{x}_2) \pm t_{\alpha/2} s_{\bar{x}_1 - \bar{x}_2}$

$76.974 \pm 2.718\sqrt{(108.019)^2/23 + (57.820)^2/12}$

76.974 ± 76.196

$0.8 < \mu_1 - \mu_2 < 153.2$ (million \$)

Since the confidence interval includes only positive values, it suggests that the mean gross amount is higher for PG & PG-13 rated movies than for R rated movies – i.e., that PG & PG-13 movies are a better investment than R movies in that they have a larger gross income.

35. Let the home voltages be group 1. Statistical software yields the following summary.

group 1: $n = 40$ $\bar{x} = 123.663$ $s = 0.240392$
group 2: $n = 40$ $\bar{x} = 124.663$ $s = 0.288842$
$\bar{x}_1 - \bar{x}_2 = -1.000$

original claim: $\mu_1 - \mu_2 = 0$ volts
H_o: $\mu_1 - \mu_2 = 0$ volts
H_1: $\mu_1 - \mu_2 \neq 0$ volts
$\alpha = 0.05$ and df = 39
C.V. $t = \pm t_{\alpha/2} = t_{0.025} = \pm 2.023$
calculations:

$$t_{\bar{x}_1-\bar{x}_2} = (\bar{x}_1 - \bar{x}_2 - \mu_{\bar{x}_1-\bar{x}_2})/s_{\bar{x}_1-\bar{x}_2}$$
$$= (-1.000 - 0)/\sqrt{\frac{(0.240392)^2}{40} + \frac{(0.288842)^2}{40}}$$
$$= -1.000/0.05942$$
$$= -16.830$$

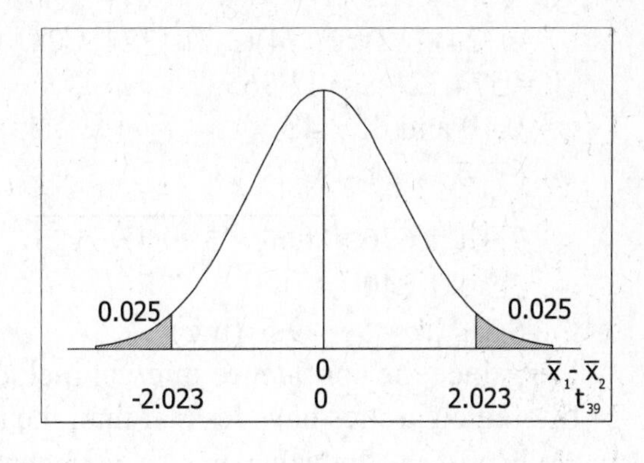

P-value = 2·tcdf(-99,-16.830,39) ≈ 0

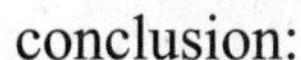

conclusion:

Reject H_o; there is sufficient evidence to reject the claim that $\mu_1 - \mu_2 = 0$. There is sufficient evidence to reject the claim that the samples of home voltages and generator voltages are from populations with the same mean. No; while the 1.000 volt difference is highly statistically significant (because of the consistency [low variability] of each source), it is likely of no practical significance or concern.

37. Let the children treated with low humidity be group 1.

$\bar{x}_1 - \bar{x}_2 = 0.98 - 1.09 = -0.11$ df = $df_1 + df_2 = 45 + 45 = 90$

$$s_p^2 = [(df_1)s_1^2 + (df_2)s_2^2]/[df_1 + df_2]$$
$$= [(45)(1.22)^2 + (45)(1.11)^2]/[45 + 45]$$
$$= 122.4225/90 = 1.36025$$

original claim: $\mu_1 - \mu_2 = 0$
H_o: $\mu_1 - \mu_2 = 0$
H_1: $\mu_1 - \mu_2 \neq 0$
$\alpha = 0.05$ and df = 90
C.V. $t = \pm t_{\alpha/2} = t_{0.025} = \pm 1.987$
calculations:

$$t_{\bar{x}_1-\bar{x}_2} = (\bar{x}_1 - \bar{x}_2 - \mu_{\bar{x}_1-\bar{x}_2})/s_{\bar{x}_1-\bar{x}_2}$$
$$= (-0.11 - 0)/\sqrt{1.36025/46 + 1.36025/46}$$
$$= -0.11/0.2432$$
$$= -0.452$$

P-value = 2·tcdf(-99,-0.452,90) = 0.6521

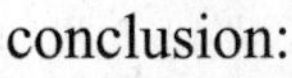

conclusion:

Do not reject H_o; there is not sufficient evidence to reject the claim that $\mu_1 - \mu_2 = 0$. There is not sufficient evidence to reject the claim that the two groups are from populations with the same mean. The results suggest that increasing the humidity does not have a significant effect on the treatment of croup.

*When $n_1 = n_2$, the calculated t statistic does not change at all. The only difference the assumption of equal standard deviations makes in this instance is to change the df from 45 to 90 and the P-value from 0.6532 to 0.6521. The conclusion is unaffected.

39. Let the unfiltered cigarettes be group 1.

$\bar{x}_1 - \bar{x}_2 = 21.1 - 13.2 = 7.9$ $\quad$ df = df$_1$ + df$_2$ = 24 + 24 = 48

$s_p^2 = [(df_1)s_1^2 + (df_2)s_2^2]/[df_1 + df_2]$

$= [(24)(3.2)^2 + (24)(3.7)^2]/[24 + 24]$

$= 574.32/48 = 11.965$

$\alpha = 0.10$ and df = 48

$(\bar{x}_1 - \bar{x}_2) \pm t_{\alpha/2} s_{\bar{x}_1 - \bar{x}_2}$

$7.9 \pm 1.676\sqrt{11.965/25 + 11.965/25}$

7.9 ± 1.640

$6.3 < \mu_1 - \mu_2 < 9.5$ (mg)

Yes; since the confidence interval includes only positive values, the results suggest that the filtered cigarettes have less tar than the unfiltered ones.

*When $n_1 = n_2$ the value of $s_{\bar{x}_1 - \bar{x}_2}$ is unchanged. The only difference the assumption of equal standard deviations makes in this instance is to change the df from 24 to 48 and the $t_{\alpha/2}$ from 1.711 to 1.676. This makes the interval slightly narrower, but the conclusion is unaffected.

41. Let the Popes be group 1.

group 1: Popes (n=24)	group 2: Rulers (n=14)
$\Sigma x = 315$	$\Sigma x = 2001$
$\Sigma x^2 = 5981$	$\Sigma x^2 = 2{,}901{,}433$
$\bar{x} = 13.125$	$\bar{x} = 142.929$
$s^2 = 80.2880$ (s=8.96)	$s^2 = 201187.1484$ (s=448.54)

$\bar{x}_1 - \bar{x}_2 = 13.125 - 142.929 = -129.804$

a. original claim: $\mu_1 - \mu_2 < 0$ years

H_o: $\mu_1 - \mu_2 = 0$ years

H_1: $\mu_1 - \mu_2 < 0$ years

$\alpha = 0.01$ and df = 13

C.V. $t = -t_\alpha = -t_{0.01} = -2.650$

calculations:

$t_{\bar{x}_1 - \bar{x}_2} = (\bar{x}_1 - \bar{x}_2 - \mu_{\bar{x}_1 - \bar{x}_2})/s_{\bar{x}_1 - \bar{x}_2}$

$= (-129.804 - 0)/\sqrt{\dfrac{80.2880}{24} + \dfrac{201187.1484}{14}}$

$= -129.804/119.8910 = -1.083$

P-value = tcdf(-99,-1.083,9) = 0.1536

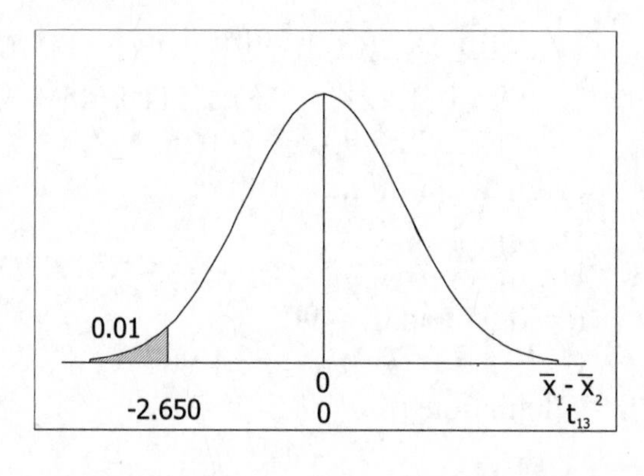

conclusion:

Do not reject H_o; there is not sufficient evidence to conclude that $\mu_1 - \mu_2 < 0$. There is not sufficient evidence to support the claim that the mean longevity after election for popes is less than the mean longevity after coronation for British monarchs.

b. $\alpha = 0.02$ and df = 13

$(\bar{x}_1 - \bar{x}_2) \pm t_{\alpha/2} s_{\bar{x}_1 - \bar{x}_2}$

$-129.804 \pm 2.650\sqrt{80.2880/24 + 201187.1484/14}$

-129.804 ± 317.711

$-447.5 < \mu_1 - \mu_2 < 187.9$ (years)

Since the confidence interval includes zero, the two population means could be equal.

*The conclusions from the test of hypothesis and the confidence remain the same, but the outlier has a dramatic effect on the calculations – greatly increasing the magnitude of the mean and standard deviation of group2, and the magnitude of the mean and standard deviation of the difference between the means. Although both $\bar{x}_1$-$\bar{x}_2$ and $s_{\bar{x}_1-\bar{x}_2}$ increase considerably, the calculated t statistic (which is a ratio) changes only from -1.810 to -1.083 – but the corresponding change in P-values from 0.0467 to 0.1536 is considerable. Since both $\bar{x}_1$-$\bar{x}_2$ and $s_{\bar{x}_1-\bar{x}_2}$ increase considerably, the change in the confidence interval is dramatic – with a shift in the center from -10 years to -129 years and an increase in width from 28 years to 635 years.

43. Let the treatment group be group 1.
$\bar{x}_1$-$\bar{x}_2$ = 0.049 – 0.000 = 0.049
original claim: $\mu_1 - \mu_2 = 0$
H_o: $\mu_1 - \mu_2 = 0$
H_1: $\mu_1 - \mu_2 \neq 0$
$\alpha = 0.05$ and df = 21
C.V. $t = \pm t_{\alpha/2} = t_{0.025} = \pm 2.080$
calculations:

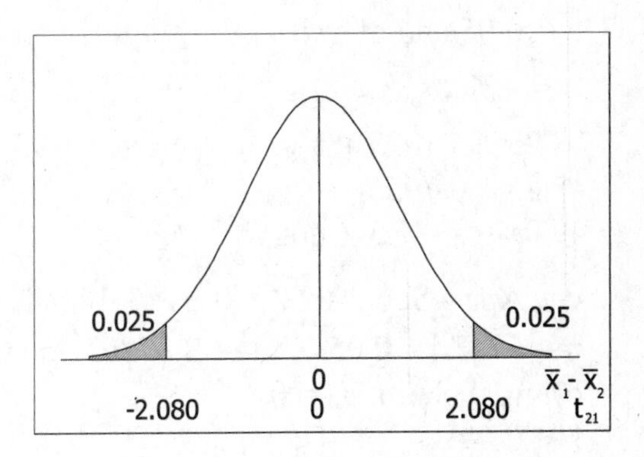

$t_{\bar{x}_1-\bar{x}_2} = (\bar{x}_1-\bar{x}_2 - \mu_{\bar{x}_1-\bar{x}_2})/s_{\bar{x}_1-\bar{x}_2}$
$= (0.049 - 0)/\sqrt{(0.015)^2/22 + (0)^2/22}$
$= 0.049/0.003198 = 15.322$
P-value = 2·tcdf(15.322,99,21) = 7.14E-13 ≈ 0
conclusion:
Reject H_o; there is sufficient evidence to reject the claim that $\mu_1 - \mu_2 = 0$ and conclude that $\mu_1 - \mu_2 \neq 0$ (in fact, that $\mu_1 - \mu_2 > 0$). There is sufficient evidence to reject the claim that the two sample groups come from populations with the same mean.
The fact that there was no variation in the second sample did not affect the calculations or present any special problems. Since there is no variation in x_2, it is really equivalent to the constant value zero – and the test is mathematically equivalent to the one-sample test H_o: $\mu_1 = 0$, for which $t = (\bar{x}_1 - 0)/s_{\bar{x}_1}$.

9-4 Inferences from Dependent Samples

1. The values of d = B–A are 6 -3 3 -13 6, for which n = 5 $\Sigma d = -1$ $\Sigma d^2 = 259$.
   a. $\bar{d} = (\Sigma d)/n = -1/5 = -0.2$ minutes
   b. $s_d^2 = [n(\Sigma d^2) - (\Sigma d)^2]/[n(n-1)] = [5(259) - (-1)^2]/[5(4)] = 1294/20 = 64.7$
   $s_d = 8.0$ minutes
   c. In general, μ_d represents the true mean of the differences from the population of matched pairs (which is mathematically equivalent to the true of the difference between the means of the two populations).

3. While the methods of this section can be used to produce numerical results, because there is paired data, those results will have no meaning. Because pulse rates and cholesterol levels measure different kinds of quantities and have different units, the differences between them have no meaningful interpretation and should not be calculated (either as matched pairs or as independent samples).

5. d = C-H: -8 -9 -8 -8
   summary: $n = 4$ $\Sigma d = -33$ $\Sigma d^2 = 273$
   a. $\bar{d} = (\Sigma d)/n = -33/4 = -8.3$ mpg
   b. $s_d^2 = [n(\Sigma d^2) - (\Sigma d)^2]/[n(n-1)] = [4(273) - (-33)^2]/[4(3)] = 3/12 = 0.25$
   $s_d = 0.5$ mpg
   c. $t_{\bar{d}} = (\bar{d} - \mu_{\bar{d}})/s_{\bar{d}}$
   $= (-8.25 - 0)/(0.5/\sqrt{4}) = -8.25/0.25 = -33.000$
   d. with df=3 and α=0.05, the critical values are $t = \pm t_{\alpha/2} = \pm t_{0.025} = \pm 3.182$

7. $\alpha = 0.05$ and df = 4
   $\bar{d} \pm t_{\alpha/2} \cdot s_{\bar{d}}$
   $-8.25 \pm 3.182 \cdot (0.5/\sqrt{4})$
   -8.25 ± 0.7955
   $-9.0 < \mu_d < 7.5$ (mpg)

9. d = Apr – Sept: -0.53 -0.24 1.18 -0.72 0.36
   $n = 5$ $\Sigma d = 0.05$ $\Sigma d^2 = 2.3789$ $\bar{d} = 0.1$ $s_d = 0.7711$
   original claim: $\mu_d = 0$
   H_o: $\mu_d = 0$
   H_1: $\mu_d \neq 0$
   $\alpha = 0.05$ and df = 4
   C.V. $t = \pm t_{\alpha/2} = \pm t_{0.025} = \pm 2.776$
   calculations
   $t_{\bar{d}} = (\bar{d} - \mu_{\bar{d}})/s_{\bar{d}}$
   $= (0.01 - 0)/(0.7711/\sqrt{5})$
   $= 0.01/0.3448 = 0.029$
   P-value = 2·tcdf(0.029,99,4) = 0.9783

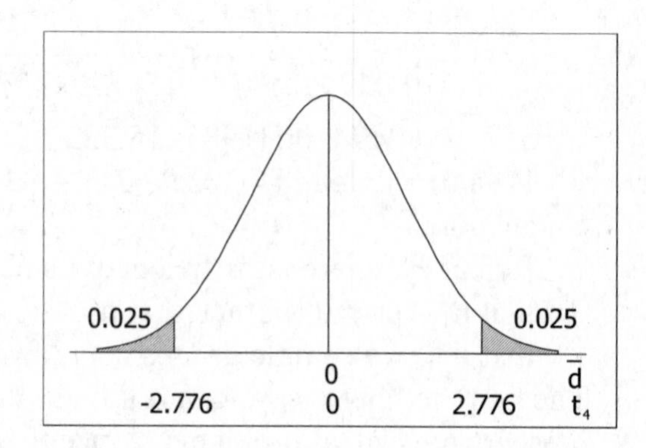

   conclusion:
   Do not reject H_o; there is not sufficient evidence to reject the claim that $\mu_d = 0$. There is not sufficient evidence to reject the claim that the mean change in BMI for all students is equal to 0. No; BMI does not appear to change during the freshman year.

11. d = actress - actor: -34 -9 -25 -14 -8 -16 -31 -12 2 -9 -21 -14 -9 -41 6
   $n = 15$ $\Sigma d = -235$ $\Sigma d^2 = 6003$ $\bar{d} = -15.67$ $s_d = 12.877$
   original claim: $\mu_d < 0$
   H_o: $\mu_d = 0$
   H_1: $\mu_d < 0$
   $\alpha = 0.05$ and df = 14
   C.V. $t = -t_\alpha = -t_{0.05} = -1.761$
   calculations
   $t_{\bar{d}} = (\bar{d} - \mu_{\bar{d}})/s_{\bar{d}}$
   $= (-15.67 - 0)/(12.877/\sqrt{15})$
   $= -15.67/5.3248 = -4.712$
   P-value = tcdf(-99,-4.712,14) = 0.0002

0.05
0
-1.761
0
$\bar{d}$
t_{14}

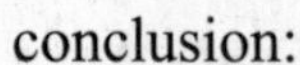
   conclusion:
   Reject H_o; there is sufficient evidence to conclude that $\mu_d < 0$. There is sufficient evidence to support the claim that best actresses are younger than best actors. Yes; the results

suggest that (unless acting ability actually peaks earlier in women than in men) actresses and actors are evaluated by different criteria.

13. d = 8am – 12am: -1.0 -2.4 -1.2 -1.6 -0.8 1.6
    $n = 6 \quad \Sigma d = -5.4 \quad \Sigma d^2 = 13.96 \quad \bar{d} = -0.90 \quad s_d = 1.349$
    $\alpha = 0.05$ and df = 5
    $\bar{d} \pm t_{\alpha/2} \cdot s_{\bar{d}}$
    $-0.90 \pm 2.571 \cdot (1.349/\sqrt{6})$
    -0.90 ± 1.4160
    $-2.32 < \mu_d < 0.52$ (°F)
    Yes; since the confidence interval includes 0, body temperature is basically the same at both times.

15. d = 6th – 13th: -4 -6 -3 1 -1 -7
    $n = 6 \quad \Sigma d = -20 \quad \Sigma d^2 = 112 \quad \bar{d} = -3.333 \quad s_d = 3.011$
    original claim: $\mu_d = 0$ admissions
    H_o: $\mu_d = 0$ admissions
    H_1: $\mu_d \neq 0$ admissions
    $\alpha = 0.05$ and df = 5
    C.V. $t = \pm t_{\alpha/2} = \pm t_{0.025} = \pm 2.571$
    calculations
    $t_{\bar{d}} = (\bar{d} - \mu_{\bar{d}})/s_{\bar{d}}$
    $= (-3.333 - 0)/(3.011/\sqrt{6})$
    $= -3.333/1.229 = -2.712$
    P-value = 2·tcdf(-99,-2.712,5) = 0.0422

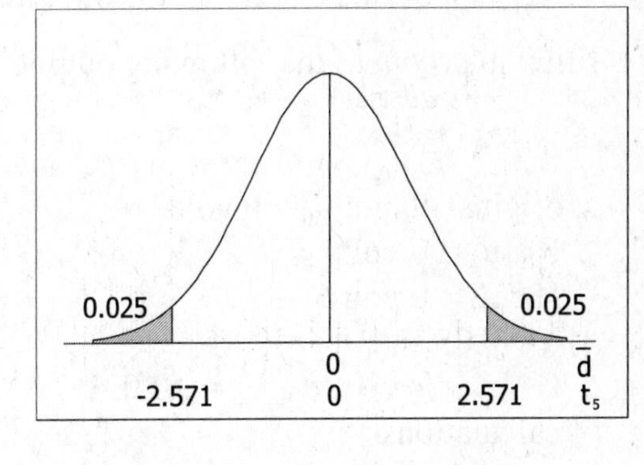

    conclusion:
    Reject H_o; there is sufficient evidence to reject the claim that $\mu_d = 0$ and conclude that $\mu_d \neq 0$ (in fact, that $\mu_d < 0$). There is sufficient evidence to reject the claim that the numbers of hospital admissions from motor vehicle crashes are not affected when the 13th of a month falls on a Friday – and to conclude that Friday the 13th does appear to be unlucky!

17. d = front – rear: -544 -224 1450 -2159 2789 3573 700 -777 2748
    $n = 9 \quad \Sigma d = 7556 \quad \Sigma d^2 = 36{,}299{,}976 \quad \bar{d} = 839.56 \quad s_d = 1935.081$
    $\alpha = 0.05$ and df = 8
    $\bar{d} \pm t_{\alpha/2} \cdot s_{\bar{d}}$
    $839.56 \pm 2.306 \cdot (1935.081/\sqrt{9})$
    839.56 ± 1487.4319
    $-647.9 < \mu_d < 2327.0$ (\$)
    Since the confidence interval includes 0, there is no significant mean difference between front repair costs and rear repair costs.

19. d = old – new: 1 2 3 2 4 3 4 2 2 2 2 3 2 2 2 3 2 2 2 2

$n = 20 \quad \Sigma d = 47 \quad \Sigma d^2 = 121 \quad \bar{d} = 2.35 \quad s_d = 0.745$

original claim: $\mu_d > 0$ mpg

H_o: $\mu_d = 0$ mpg

H_1: $\mu_d > 0$ mpg

$\alpha = 0.01$ and df = 19

C.V. $t = t_\alpha = t_{0.01} = 2.539$

calculations

$t_{\bar{d}} = (\bar{d} - \mu_{\bar{d}})/s_{\bar{d}}$

$= (2.35 - 0)/(0.745/\sqrt{20})$

$= 2.35/0.1666 = 14.104$

P-value = tcdf(14.104,99,19) = 9.093E-12 ≈ 0

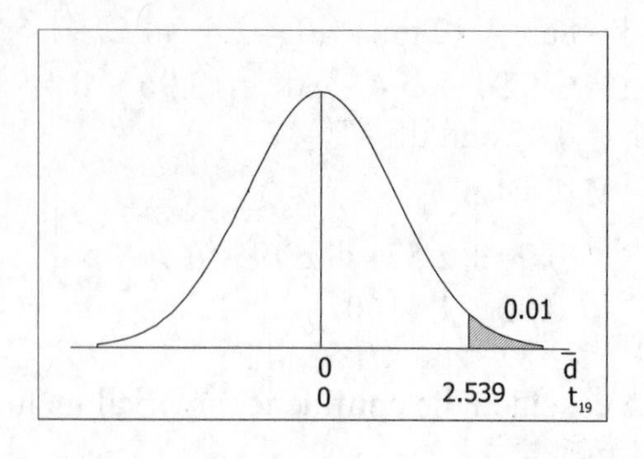

conclusion:

Reject H_o; there is sufficient evidence to conclude that $\mu_d > 0$. There is sufficient evidence to support the claim that the old ratings are higher than the new ratings.

21. Minitab produces the following output for a t test on the data column consisting of c2-c4.

```
Test of mu = 0 vs not = 0
Variable    N     Mean      StDev    SE Mean           95% CI              T      P
home-UPS   40  0.075000  0.449929  0.071140  (-0.068894, 0.218894)  1.05  0.298
```

a. original claim: $\mu_d = 0$ volts

H_o: $\mu_d = 0$ volts

H_1: $\mu_d \neq 0$ volts

$\alpha = 0.05$ and df = 39

C.V. $t = \pm t_{\alpha/2} = \pm t_{0.025} = \pm 2.023$

calculations

$t_{\bar{d}} = (\bar{d} - \mu_{\bar{d}})/s_{\bar{d}}$

$= (0.075 - 0)/(0.4499/\sqrt{40})$

$= 1.05$ [Minitab]

P-value = $2 \cdot P(t_{39} > 1.05) = 0.298$ [Minitab]

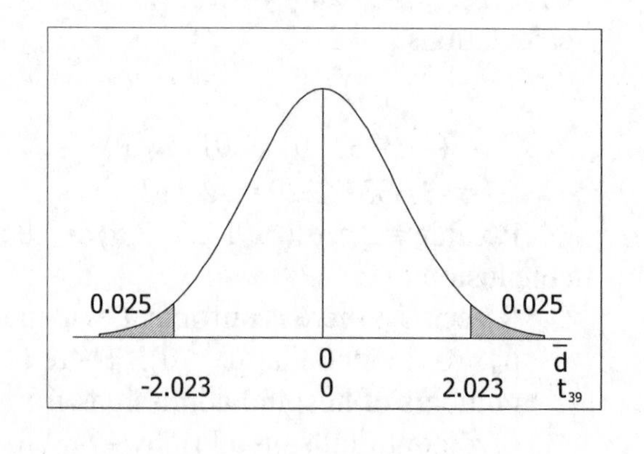

conclusion:

Do not reject H_o; there is not sufficient evidence to reject the claim that $\mu_d = 0$. There is not sufficient evidence to reject the claim that the paired sample values have differences that are from a population with a mean of 0 volts.

b. Even though the generator voltage amounts were recorded on the same day as the corresponding home voltage amounts, there is no physical justification for pairing the values. The gasoline-powered generator operates completely independently of the other sources, and so their data values represent independent samples.

23. Minitab produces the following output for a t interval on the data column consisting of c3-c4.

```
Variable    N    Mean     StDev   SE Mean       95% CI
pap-plas   62  7.51726  3.57036  0.45344  (6.61056, 8.42396)
```

$\alpha = 0.05$ and df = 61

$\bar{d} \pm t_{\alpha/2} \cdot s_{\bar{d}}$

$7.517 \pm 2.000 \cdot (3.570/\sqrt{62})$

$6.611 < \mu_d < 8.424$ (lbs) [Minitab]

Since the confidence interval includes only positive values, there discarded paper appears to weigh more than the discarded plastic.

25. The following values were used in the analysis. Different readings from the graph may produce slightly different values and slightly differ numbers in the test of hypothesis.

right hand	97	116	116	165	191
left hand	171	191	196	207	224
d = right – left	-74	-75	-80	-42	-33

$n = 5$ $\Sigma d = -304$ $\Sigma d^2 = 20354$ $\bar{d} = -60.8$ $s_d = 21.626$

original claim: $\mu_d = 0$ milliseconds

H_o: $\mu_d = 0$ milliseconds

H_1: $\mu_d \neq 0$ milliseconds

$\alpha = 0.05$ and df = 4

C.V. $t = \pm t_{\alpha/2} = \pm t_{0.025} = \pm 2.776$

calculations

$t_{\bar{d}} = (\bar{d} - \mu_{\bar{d}})/s_{\bar{d}}$

$= (-60.8 - 0)/(21.626/\sqrt{5})$

$= -60.8/9.6716 = -6.286$

P-value = 2·tcdf(-99,-6.286,4) = 0.0033

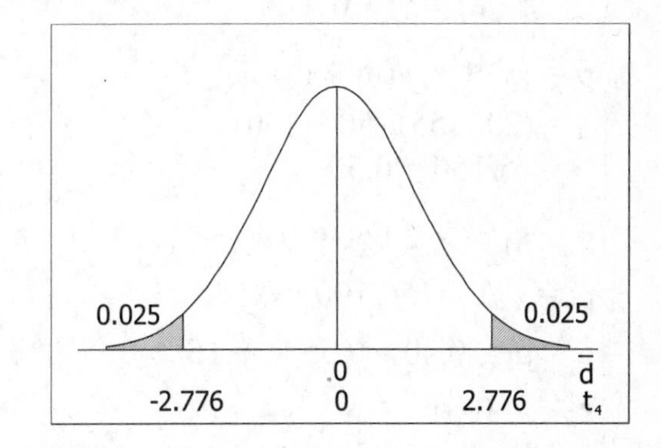

conclusion:

Reject H_o; there is sufficient evidence to reject the claim that $\mu_d = 0$ and conclude that $\mu_d \neq 0$ (in fact, that $\mu_d < 0$). There is sufficient evidence to reject the claim that there is no difference between the reaction times of the right and left hands.

Statistical Literacy and Critical Thinking

1. When a method is robust against departures from normality, it still performs well when the populations involved do have a normal distribution. When a method performs well, a test at the α level of significance will reject a true null hypothesis approximately α of the time and a confidence interval at the α level of significance will include the true value of the parameter approximately $1-\alpha$ of the time. The F test for comparing two population standard deviations (or variances) is not robust against departures from normality.

2. These results do not tell us anything about how the general population feels about the word *ginormous*. Since the sample is a voluntary response sample, it is not necessarily representative of the general population. No methods of statistics should be used on this sample data to make inferences about the general population.

3. The samples are independent. Since the Coke and Pepsi cans were selected independently of each, they are not related in any way and any pairing of the cans would be arbitrary and artificial.

4. The described procedure is not valid for at least two reasons. Since the states have different population sizes, the mean of the 50 state means is not necessarily the correct value for the mean of the entire country – a weighted mean should be used. The construction of confidence intervals to estimate population parameters assumes one is working with random samples – but the mean ages obtained for each state are not random selections, and they are presumably the true values (as determined from census or other records) and not sample values subject to sampling variability.

Chapter Quick Quiz

1. Let the California teachers be group 1. Since the original claim does not contain the equality, it must become the alternative hypothesis.
   original claim: $p_1 > p_2$ [i.e., $p_1 - p_2 > 0$]
   H_o: $p_1 - p_2 = 0$
   H_1: $p_1 - p_2 > 0$

2. $\bar{p} = (x_1 + x_2)/(n_1 + n_2)$
   $= (20 + 55)/(50 + 100)$
   $= 75/150 = 0.50$

3. $\hat{p}_1 = x_1/n_1 = 20/50 = 0.40$
   $\hat{p}_2 = x_2/n_2 = 55/100 = 0.55$
   $\hat{p}_1 - \hat{p}_2 = 0.40 - 0.55 = -0.15$
   $z_{\hat{p}_1-\hat{p}_2} = (\hat{p}_1-\hat{p}_2 - \mu_{\hat{p}_1-\hat{p}_2})/\sigma_{\hat{p}_1-\hat{p}_2}$
   $= (-0.15 - 0)/\sqrt{(0.5)(0.5)/50+(0.5)(0.5)/100}$
   $= -0.15/0.08660$
   $= -1.73$

4. P-value $= 2\cdot P(Z<-2.05)$
   $= 2\cdot(0.0202)$
   $= 0.0404$

5. A P-value of 0.0001 is small enough to reject the null hypothesis for any reasonable level of significance. The final conclusion is as follows.
   There is sufficient evidence to support the claim that $\mu_1 > \mu_2$.

6. This is a test involving matched pairs.
   original claim: $\mu_d = 0$
   H_o: $\mu_d = 0$ inches
   H_1: $\mu_d \neq 0$ inches

7. This is a test involving independent samples. Let the California voters be group 1.
   original claim $\mu_1 < \mu_2$ [i.e., $\mu_1 - \mu_2 < 0$]
   H_o: $\mu_1 - \mu_2 = 0$ years
   H_1: $\mu_1 - \mu_2 < 0$ years

8. Dependent, since the wives whose IQ's collected were married to husbands whose IQ's were collected.

9. A P-value of 0.0009 is small enough to reject the null hypothesis for any reasonable level of significance. The final conclusion is as follows.
   Reject H_o; there is sufficient evidence to conclude that $\mu_1 - \mu_2 \neq 0$. There is sufficient evidence to support the claim that the two populations gave different means.

10. True. In any test of hypothesis, the null hypothesis must contain the condition of equality.

Review Exercises

1. Let those treated with surgery be group 1.

   $\hat{p}_1 = x_1/n_1 = 67/73 = 0.918$ $\hat{p}_1 - \hat{p}_2 = 0.918 - 0.723 = 0.195$

   $\hat{p}_2 = x_2/n_2 = 60/83 = 0.723$

   $\bar{p} = (67+60)/(73+83)$

   $= 127/156 = 0.814$

   original claim: $p_1 - p_2 > 0$

   H_o: $p_1 - p_2 = 0$

   H_1: $p_1 - p_2 > 0$

   $\alpha = 0.01$

   C.V. $z = z_\alpha = z_{0.01} = 2.326$

   calculations:

   $z_{\hat{p}_1-\hat{p}_2} = (\hat{p}_1 - \hat{p}_2 - \mu_{\hat{p}_1-\hat{p}_2})/\sigma_{\hat{p}_1-\hat{p}_2}$

   $= (0.195 - 0)/\sqrt{(0.814)(0.186)/73 + (0.814)(0.186)/83} = 0.195/0.06242 = 3.12$

   P-value = $P(z>3.12) = 1 - 0.9991 = 0.0009$

   conclusion:

   Reject H_o; there is sufficient evidence to conclude that $p_1 - p_2 > 0$. There is sufficient evidence to support the claim that treatment of carpal tunnel syndrome with release surgery results in better outcomes than treatment with wrist splinting. In this context, surgery should generally be recommended instead of splinting.

2. Let those born to cocaine users be group 1.

   $\bar{x}_1 - \bar{x}_2 = 7.3 - 8.2 = -0.9$

   original claim: $\mu_1 - \mu_2 < 0$

   H_o: $\mu_1 - \mu_2 = 0$

   H_1: $\mu_1 - \mu_2 < 0$

   $\alpha = 0.05$ and df = 185

   C.V. $t = -t_\alpha = -t_{0.05} = -1.653$

   calculations:

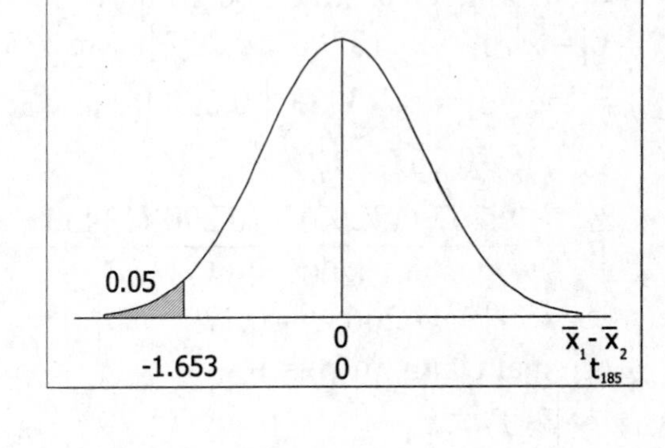

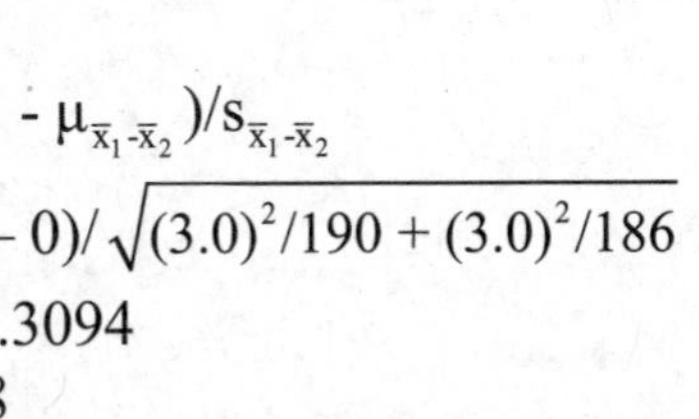

   $t_{\bar{x}_1-\bar{x}_2} = (\bar{x}_1 - \bar{x}_2 - \mu_{\bar{x}_1-\bar{x}_2})/s_{\bar{x}_1-\bar{x}_2}$

   $= (-0.9 - 0)/\sqrt{(3.0)^2/190 + (3.0)^2/186}$

   $= -0.9/0.3094$

   $= -2.908$

   P-value = tcdf(-99,-2.908,185) = 0.0020

   conclusion:

   Reject H_o; there is sufficient evidence to conclude that $\mu_1 - \mu_2 < 0$. There is sufficient evidence to support the claim that prenatal cocaine exposure is associated with lower scores on the test of object assembly.

3. d = regular – kiln: -5.75 -1.25 -1.00 -5.00 0.00 0.25 2.25 -0.50 0.75 -1.50 -0.25
 $n = 11 \quad \Sigma d = -12.00 \quad \Sigma d^2 = 68.8750 \quad \bar{d} = -1.091 \quad s_d = 2.3619$
 a. original claim: $\mu_d = 0$ cwt/acre
  H_o: $\mu_d = 0$ cwt/acre
  H_1: $\mu_d \neq 0$ cwt/acre
  $\alpha = 0.05$ and df = 10
  C.V. $t = \pm t_{\alpha/2} = \pm t_{0.025} = \pm 2.228$
  calculations

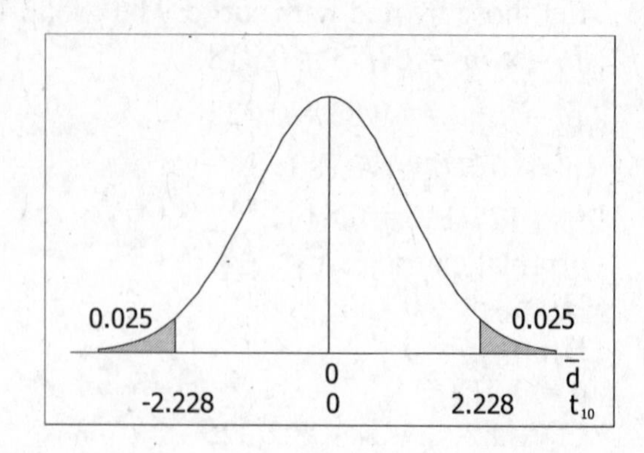

$$t_{\bar{d}} = (\bar{d} - \mu_{\bar{d}})/s_{\bar{d}}$$
$$= (-1.091 - 0)/(2.3619/\sqrt{11})$$
$$= -1.091/0.7121 = -1.532$$

  P-value = 2·tcdf(-99,-1.532,10) = 0.1565
  conclusion:
   Do not reject H_o; there is not sufficient evidence to reject the claim that $\mu_d = 0$. There is not sufficient evidence to reject the claim that there is no difference between the yields from the two types of seed.
 b. $\alpha = 0.05$ and df = 10
  $\bar{d} \pm t_{\alpha/2} \cdot s_{\bar{d}}$
  $-1.09 \pm 2.228 \cdot (2.3619/\sqrt{11})$
  -1.09 ± 1.587
  $-2.68 < \mu_d < 0.50$
 c. We cannot be 95% certain that either type of seed is better.

4. Let the blinded abstracts be group 1.
 $\hat{p}_1 = x_1/n_1 = x_1/13200 = 0.267$ [as any $3518 \leq x_1 \leq 3530$ gives $\hat{p}_1 = 26.7\%$, use 0.267 in all work]
 $\hat{p}_2 = x_2/n_2 = x_2/13433 = 0.290$ [as any $3889 \leq x_2 \leq 3902$ gives $\hat{p}_2 = 29.0\%$, use 0.290 in all work]
 $\hat{p}_1 - \hat{p}_2 = 0.267 - 0.290 = -0.023$

$$\bar{p} = \frac{(0.267)(13200) + (0.290)(13433)}{13200 + 13433}$$
$$= 7419.97/26633 = 0.279$$

 original claim: $p_1 - p_2 = 0$
 H_o: $p_1 - p_2 = 0$
 H_1: $p_1 - p_2 \neq 0$
 $\alpha = 0.01$
 C.V. $z = \pm z_{\alpha/2} = \pm z_{0.005} = \pm 2.575$
 calculations:

$$z_{\hat{p}_1 - \hat{p}_2} = (\hat{p}_1 - \hat{p}_2 - \mu_{\hat{p}_1 - \hat{p}_2})/\sigma_{\hat{p}_1 - \hat{p}_2}$$
$$= (-0.023 - 0)/\sqrt{(0.279)(0.721)/13200 + (0.279)(0.721)/13433} = -0.023/0.005494 = -4.19$$

  P-value = 2·P(z<-4.19) = 2·(0.0001) = 0.0002
 conclusion:
  Reject H_o; there is sufficient evidence to reject the claim that $p_1 - p_2 = 0$ and conclude that $p_1 - p_2 \neq 0$ (in fact, that $p_1 - p_2 < 0$). There is sufficient evidence to reject the claim that the acceptance rate is the same with or without blinding. When there is no blinding the reviewers know the names and institutions of the authors and may be (either subconsciously or knowingly) influenced by that knowledge.

5. Let the Rowling scores be group 1.

group 1: Rowling (n=12)
$\Sigma x = 969.0$
$\Sigma x^2 = 78487.82$
$\bar{x} = 80.75$
$s^2 = 21.915$ (s=4.681)

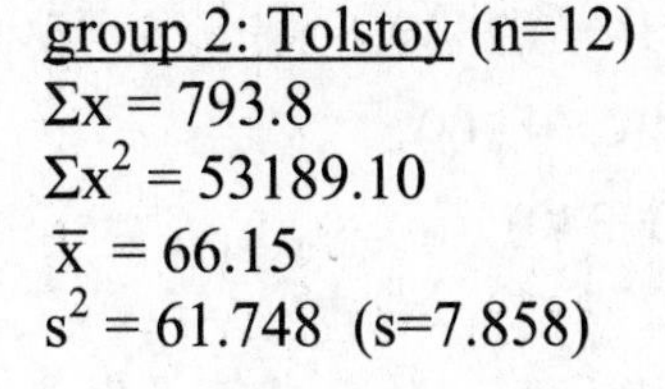
group 2: Tolstoy (n=12)
$\Sigma x = 793.8$
$\Sigma x^2 = 53189.10$
$\bar{x} = 66.15$
$s^2 = 61.748$ (s=7.858)

$\bar{x}_1 - \bar{x}_2 = 80.75 - 66.15 = 14.60$
original claim: $\mu_1 - \mu_2 > 0$
H_o: $\mu_1 - \mu_2 = 0$
H_1: $\mu_1 - \mu_2 > 0$
$\alpha = 0.05$ and df = 11
C.V. $t = t_\alpha = t_{0.05} = 1.796$
calculations:

$t_{\bar{x}_1-\bar{x}_2} = (\bar{x}_1 - \bar{x}_2 - \mu_{\bar{x}_1-\bar{x}_2})/s_{\bar{x}_1-\bar{x}_2}$
$= (14.60 - 0)/\sqrt{21.915/12 + 61.748/12}$
$= 14.60/2.6404$
$= 5.529$

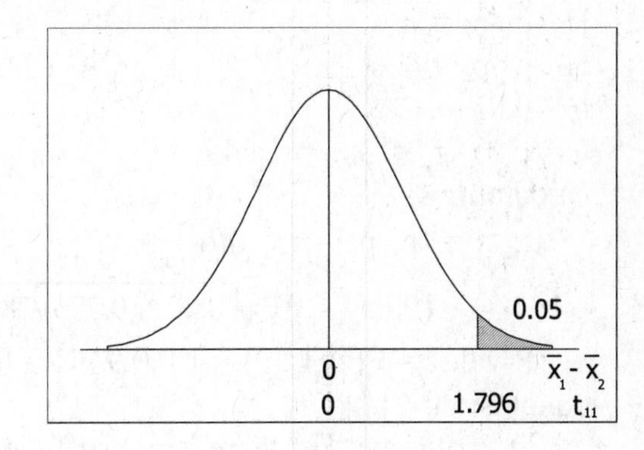

P-value = tcdf(5.529,99,11) = 0.0001
conclusion:

Reject H_o; there is sufficient evidence to conclude that $\mu_1 - \mu_2 > 0$. There is sufficient evidence to support the claim that *Harry Potter* is easier to read than *War and Peace*. Yes; this is the expected result.

6. d = before – after: 9 4 21 3 20 31 17 26 26 10 23 33
n = 12 $\Sigma d = 223$ $\Sigma d^2 = 5267$ $\bar{d} = 18.58$ $s_d = 10.104$

a. $\alpha = 0.01$ and df = 11
$\bar{d} \pm t_{\alpha/2} \cdot s_{\bar{d}}$
$18.58 \pm 3.106 \cdot (10.104/\sqrt{12})$
18.58 ± 9.05917
$9.5 < \mu_d < 27.6$ (mm Hg)

b. original claim: $\mu_d > 0$ mm Hg
H_o: $\mu_d = 0$ mm Hg
H_1: $\mu_d > 0$ mm Hg
$\alpha = 0.01$ [assumed] and df = 11
C.V. $t = t_\alpha = t_{0.01} = 2.718$
calculations

$t_{\bar{d}} = (\bar{d} - \mu_{\bar{d}})/s_{\bar{d}}$
$= (18.58 - 0)/(10.104/\sqrt{12})$
$= 18.58/2.9167$
$= 6.371$

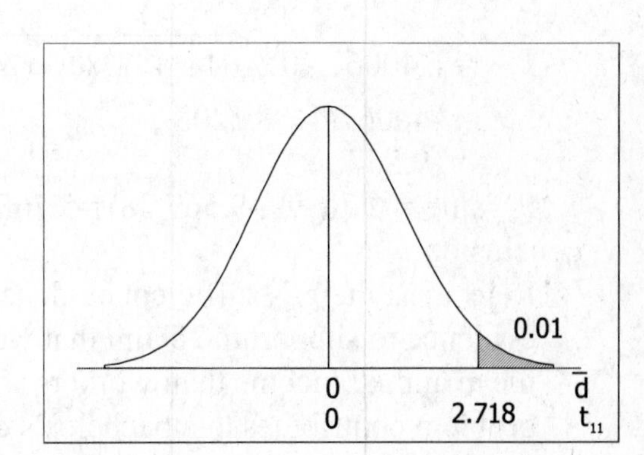

P-value = tcdf(6.371,99,11) = 2.643E-5 = 0.00003
conclusion:

Reject H_o; there is sufficient evidence to conclude that $\mu_d > 0$. There is sufficient evidence to support the claim that captopril is effective in lowering systolic blood pressure.

7. Let the men be group 1.

$\hat{p}_1 = x_1/n_1 = 71/280 = 0.254$ $\quad\quad$ $\hat{p}_1 - \hat{p}_2 = 0.254 - 0.200 = 0.054$

$\hat{p}_2 = x_2/n_2 = 68/340 = 0.200$

$\bar{p} = (71+68)/(280+340)$

$= 139/620 = 0.224$

original claim: $p_1 - p_2 > 0$

H_o: $p_1 - p_2 = 0$

H_1: $p_1 - p_2 > 0$

$\alpha = 0.05$

C.V. $z = z_\alpha = z_{0.05} = 1.645$

calculations:

$z_{\hat{p}_1-\hat{p}_2} = (\hat{p}_1 - \hat{p}_2 - \mu_{\hat{p}_1-\hat{p}_2})/\sigma_{\hat{p}_1-\hat{p}_2}$

$= (0.054 - 0)/\sqrt{(0.224)(0.776)/280 + (0.224)(0.776)/340} = 0.054/0.03366 = 1.59$

P-value = P(z>1.59) = 1 – 0.9441 = 0.0559

conclusion:

Do not reject H_o; there is not sufficient evidence to conclude that $p_1 - p_2 > 0$. There is not sufficient evidence to support the claim that the proportion of men who smoke is greater than the proportion of women who smoke.

8. Let those with a high school diploma be group 1.

$\bar{x}_1 - \bar{x}_2 = 37622 - 77689 = -40067$

original claim: $\mu_1 - \mu_2 < \$0$

H_o: $\mu_1 - \mu_2 = \$0$

H_1: $\mu_1 - \mu_2 < \$0$

$\alpha = 0.01$ and df = 38

C.V. $t = -t_\alpha = -t_{0.01} = -2.429$

calculations:

$t_{\bar{x}_1-\bar{x}_2} = (\bar{x}_1 - \bar{x}_2 - \mu_{\bar{x}_1-\bar{x}_2})/s_{\bar{x}_1-\bar{x}_2}$

$= (-40067 - 0)/\sqrt{(14115)^2/80 + (24227)^2/39}$

$= -40067/4188.1203$

$= -9.567$

P-value = tcdf(-99,-9.567,38) =5.76E-12 ≈ 0

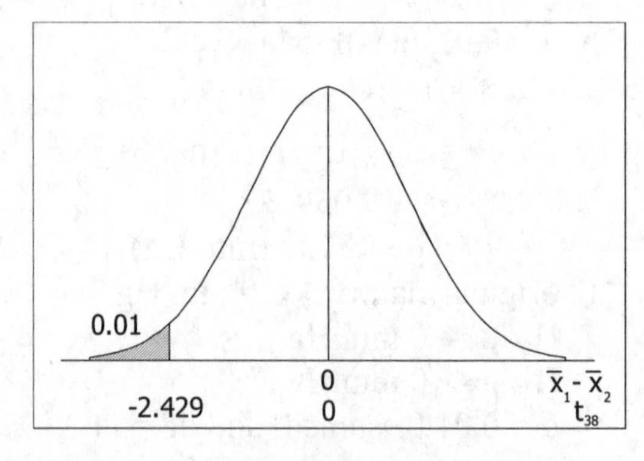

conclusion:

Reject H_o; there is sufficient evidence to conclude that $\mu_1 - \mu_2 < 0$. There is sufficient evidence to support the claim that workers with a only high school diploma have a lower mean annual income than workers with a bachelor's degree. Yes; since solving this problem contributes to a bachelor's degree, it contributes to a higher income.

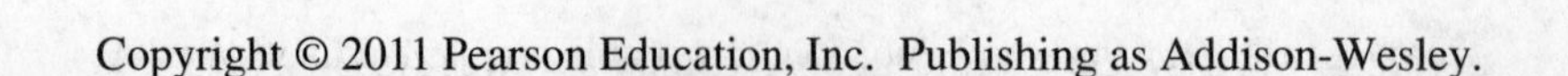

Cumulative Review Exercises

1. a. The samples are dependent, because the data are from married couples.
   b. male scores in numerical order: 8 8 9 14 14 15 16 19 21 25
   summary statistics: $n = 10$, $\Sigma x = 149$, $\Sigma x^2 = 2509$
   $\bar{x} = (\Sigma x)/n = (149)/10 = 14.9$ thousand words per day
   $\tilde{x} = (14+15)/2 = 14.5$ thousand words per day
   $M = 8,14$ thousand words per day [bi-modal]
   $R = 25 - 8 = 17$ thousand words per day
   $s^2 = [n(\Sigma x^2) - (\Sigma x)^2]/[n(n-1)] = [10(2509) - (149)^2]/[10(9)] = 2889/90 = 32.1$
   $s = \sqrt{32.1} = 5.666$, rounded to 5.7 thousand words per day
   c. Because differences are meaningful and there is a meaningful zero, the data are at the ratio level of measurement.

2. d = male – female: 0 13 -22 -7 -6 -8 -20 -1 -10 -7
   $n = 10$ $\Sigma d = -68$ $\Sigma d^2 = 1352$ $\bar{d} = -6.8$ $s_d = 9.942$
   original claim: $\mu_d < 0$
   H_o: $\mu_d = 0$
   H_1: $\mu_d < 0$
   $\alpha = 0.05$ and df = 9
   C.V. $t = t_\alpha = t_{0.05} = -1.833$
   calculations
   $t_{\bar{d}} = (\bar{d} - \mu_{\bar{d}})/s_{\bar{d}}$
   $= (-6.8 - 0)/(9.942/\sqrt{10})$
   $= -6.8/3.1440 = -2.163$
   P-value = tcdf(-99,-2.163,9) = 0.0294

   0.05 | -1.833 | 0 | 0 | $\bar{d}$ | t_9

   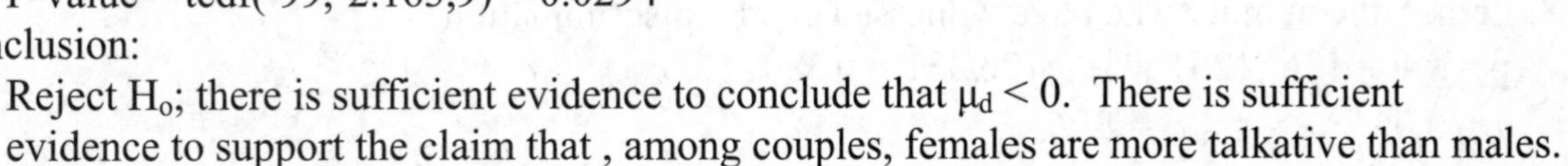
   conclusion:
   Reject H_o; there is sufficient evidence to conclude that $\mu_d < 0$. There is sufficient evidence to support the claim that , among couples, females are more talkative than males.

3. Let the males be group 1.

   group 1: males (n=10)
   $\Sigma x = 149$
   $\Sigma x^2 = 2509$
   $\bar{x} = 14.9$
   $s^2 = 32.100$ (s=5.67)

   group 2: females (n=10)
   $\Sigma x = 217$
   $\Sigma x^2 = 5472$
   $\bar{x} = 21.7$
   $s^2 = 84.789$ (s=9.208)

   $\bar{x}_1 - \bar{x}_2 = 14.9 - 21.7 = -6.8$
   original claim: $\mu_1 - \mu_2 = 0$
   H_o: $\mu_1 - \mu_2 = 0$
   H_1: $\mu_1 - \mu_2 \neq 0$
   $\alpha = 0.05$ and df = 9
   C.V. $t = \pm t_{\alpha/2} = t_{0.025} = \pm 2.262$
   calculations:
   $t_{\bar{x}_1-\bar{x}_2} = (\bar{x}_1 - \bar{x}_2 - \mu_{\bar{x}_1-\bar{x}_2})/s_{\bar{x}_1-\bar{x}_2}$
   $= (-6.8 - 0)/\sqrt{32.100/10 + 84.789/10}$
   $= -6.8/3.4189 = -1.989$
   P-value = 2·tcdf(-99,-1.989,9) = 0.0779

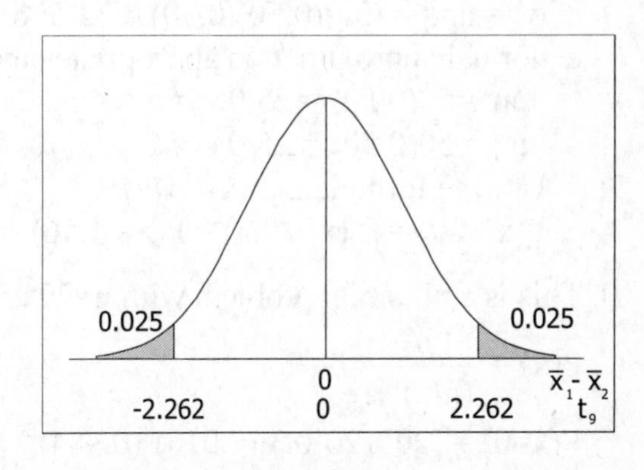

conclusion:

Do not reject H_o; there is not sufficient evidence to reject the claim that $\mu_1 - \mu_2 = 0$. There is not sufficient evidence to reject the claim that the two samples are from populations with the same mean.

4. σ unknown (and assuming the distribution is approximately normal), use t with df=9

$\alpha = 0.05$, $t_{df,\alpha/2} = t_{9,0.025} = 2.262$

$\bar{x} \pm t_{\alpha/2} \cdot s/\sqrt{n}$

$14.9 \pm 2.262(5.666)/\sqrt{10}$

14.9 ± 4.0527

$10.8 < \mu < 19.0$ (thousand words per day)

5. The requested frequency distribution is given at the right. The word count is in thousands of words per day.

word count	f
6 – 9	3
10 – 13	0
14 – 17	4
18 – 21	2
22 – 25	1
	10

6. For each part, use the formula $z = (x - \mu)/\sigma$.

Let x be the word count in thousands of words per day.

a. $P(x>17) = P(z>0.33) = 1 - 0.6293 = 0.3707$

b. The $\bar{x}$'s are normally distributed with $\mu_{\bar{x}} = \mu = 15$ and $\sigma_{\bar{x}} = \sigma/\sqrt{n} = 6/\sqrt{9} = 2$.

$P(\bar{x}>17) = P(z>1.00) = 1 - 0.8413 = 0.1587$

c. From Table A-2, the z score with 0.9000 [closest entry is 0.8997]below it is 1.28.

$x = \mu + z\sigma = 15 + (1.28)(6) = 22.68$, or 22,680 words per day

7. $\alpha = 0.10$, $z_{\alpha/2} = z_{0.05} = 1.645$ and $E = 0.025$; $\hat{p}$ unknown, use $\hat{p}=0.5$

$n = [(z_{\alpha/2})^2\hat{p}\hat{q}]/E^2$

$= [(1.645)^2(0.5)(0.5)]/(0.025)^2 = 1082.41$, rounded up to 1083

8. Let x = the number who have witnessed gender discrimination.

$\hat{p} = x/n = 126/(126+205) = 126/331 = 0.3807$

$\alpha = 0.05$, $z_{\alpha/2} = z_{0.025} = 1.96$

$\hat{p} \pm z_{\alpha/2}\sqrt{\hat{p}\hat{q}/n}$

$0.3807 \pm 1.96\sqrt{(0.3807)(0.6193)/331}$

0.3807 ± 0.0523

$0.328 < p < 0.433$ or $32.8\% < p < 43.3\%$

9. This is a binomial problem with n=50 and p=0.50.

a. $\mu = np = (50)(0.50) = 25.0$

b. $\sigma^2 = npq = (50)(0.50)(0.50) = 12.5$; $\sigma = 3.536$, rounded to 3.5

c. normal approximation appropriate since

$np = 50(0.50) = 25.0 \geq 5$

$nq = 50(0.50) = 25.0 \geq 5$

Use the formula $z = (x - \mu)/\sigma$.

$P(x \geq 20) = P_c(x>19.5) = P(z>-1.56) = 1 - 0.0594 = 0.9406$

10. This is a binomial problem with n=20 and p=0.016

$$P(x) = \frac{n!}{(n-x)!x!}p^x q^{n-x}$$

$P(x=0) = [20!/(20!0!)](0.016)^0(0.984)^{20} = [1](1)(0.984)^{20} = 0.724$

No; since $0.724 \geq 0.05$ such an event is not unusual.

Chapter 10

Correlation and Regression

10-2 Correlation

1. a. r = the correlation in the sample. In this context, r is the linear correlation coefficient computed using the chosen paired (points in Super Bowl, number of new cars sold) values for the randomly selected years in the sample.
   b. ρ = the correlation in the population. In this context, ρ is the linear correlation coefficient computed using all the paired (points in Super Bowl, number of new cars sold) values for every year there has been a Super Bowl.
   c. Since there is no relationship between the number of points scored in a Super Bowl and the number of new cars sold that year, the estimated value of r is 0.

3. Correlation is the existence of a relationship between two variables – so that knowing the value of one of the variables allows a researcher to make a reasonable inference about the value of the other. Correlation measures only association and not causality. If there is an association between two variables, it may or may not be cause-and-effect – and if it is cause-and-effect, there is nothing in the mathematics of correlation analysis to identify which variable is the cause and which is the effect.

5. From Table A-5 for n = 62 [closest entry is n=60], C.V. = ±0.254. Therefore r = 0.758 indicates a significant (positive) linear correlation. Yes; there is sufficient evidence to support the claim that there is a linear correlation between the weight of discarded garbage and the household size.

7. From Table A-5 for n = 40, C.V. = ±0.312. Therefore r = -0.202 does not indicate a significant linear correlation. No; there is not sufficient evidence to support the claim that there is a linear correlation between the heights and pulse rates of women.

9. a. Excel produces the following scatterplot.

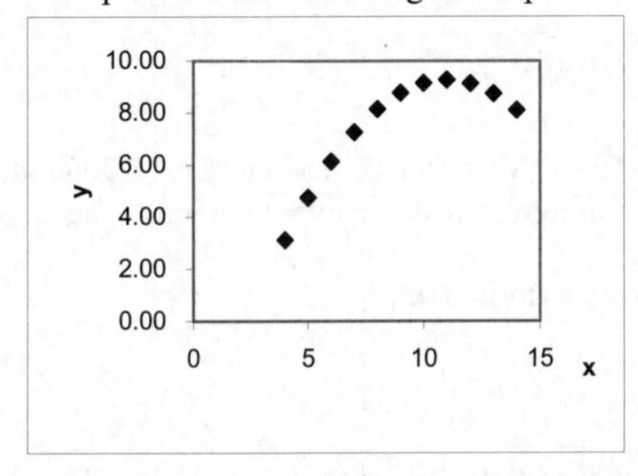

x	y	xy	x^2	y^2
10	9.14	91.40	100	83.5396
8	8.14	65.12	64	66.2596
13	8.74	113.62	169	76.3876
9	8.77	78.93	81	76.9129
11	9.26	101.86	121	85.7476
14	8.10	113.40	196	65.61
6	6.13	36.78	36	37.5769
4	3.10	12.40	16	9.61
12	9.13	109.56	144	83.3569
7	7.26	50.82	49	52.7076
5	4.74	23.70	25	22.4676
99	82.51	797.59	1001	660.1763

b. See the preceding chart on the right, where n = 11.

$n(\Sigma xy) - (\Sigma x)(\Sigma y) = 11(797.59) - (99)(82.51) = 605.00$
$n(\Sigma x^2) - (\Sigma x)^2 = 11(1001) - (99)^2 = 1210$
$n(\Sigma y^2) - (\Sigma y)^2 = 11(660.1763) - (82.51)^2 = 454.0392$
$r = [n(\Sigma xy) - (\Sigma x)(\Sigma y)]/[\sqrt{n(\Sigma x^2) - (\Sigma x)^2}\sqrt{n(\Sigma y^2) - (\Sigma y)^2}]$
$= 605.00/[\sqrt{1210}\sqrt{454.0392}] = 0.816$

From Table A-5 for n = 11, C.V. = ±0.602. Therefore r = 0.816 indicates a significant (positive) linear correlation. Yes; there is sufficient evidence to support the claim that there is a linear correlation between the two variables.

c. The scatterplot indicates that the relationship between the variables is quadratic, not linear.

NOTE: In addition to the value of n, calculation of r requires five sums: Σx, Σy, Σx^2, Σy^2 and Σxy. As the sums can usually be found conveniently using a calculator and without constructing a chart as in exercises 9 and 10, the remaining exercises give only the values of the sums and do not show a chart. In addition, calculation of r involves three subcalculations.

(1) $n(\Sigma xy) - (\Sigma x)(\Sigma y)$ determines the sign of r. If large values of x are associated with large values of y, it will be positive. If large values of x are associated with small values of y, it will be negative. If not, a mistake has been made.
(2) $n(\Sigma x^2) - (\Sigma x)^2$ cannot be negative. If it is, a mistake had been made.
(3) $n(\Sigma y^2) - (\Sigma y)^2$ cannot be negative. If it is, a mistake had been made.

Finally, r must be between -1 and 1 inclusive. If not, a mistake has been made. If this or any of the previous mistakes occurs, stop immediately and find the error – continuing is a waste of effort.

11. The following table and summary statistics apply to all parts of this exercise.

x: 1 1 1 2 2 2 3 3 3 10
y: 1 2 3 1 2 3 1 2 3 10

using all the points: n =10 $\Sigma x = 28$ $\Sigma y = 28$ $\Sigma xy = 136$ $\Sigma x^2 = 142$ $\Sigma y^2 = 142$
without the outlier: n = 9 $\Sigma x = 18$ $\Sigma y = 18$ $\Sigma xy = 36$ $\Sigma x^2 = 42$ $\Sigma y^2 = 42$

a. There appears to be a strong positive linear correlation, with r close to 1.

b. $n(\Sigma xy) - (\Sigma x)(\Sigma y) = 10(136) - (28)(28) = 576$
$n(\Sigma x^2) - (\Sigma x)^2 = 10(142) - (28)^2 = 636$
$n(\Sigma y^2) - (\Sigma y)^2 = 10(142) - (28)^2 = 636$
$r = [n(\Sigma xy) - (\Sigma x)(\Sigma y)]/[\sqrt{n(\Sigma x^2) - (\Sigma x)^2}\sqrt{n(\Sigma y^2) - (\Sigma y)^2}]$
$= 576/[\sqrt{636}\sqrt{636}] = 0.906$

From Table A-5 for n = 10, assuming $\alpha = 0.05$, C.V. = ±0.632. Therefore r = 0.906 indicates a significant (positive) linear correlation. This agrees with the interpretation of the scatterplot.

c. There appears to be no linear correlation, with r close to 0.

$n(\Sigma xy) - (\Sigma x)(\Sigma y) = 9(36) - (18)(18) = 0$
$n(\Sigma x^2) - (\Sigma x)^2 = 9(42) - (18)^2 = 54$
$n(\Sigma y^2) - (\Sigma y)^2 = 9(42) - (18)^2 = 54$
$r = [n(\Sigma xy) - (\Sigma x)(\Sigma y)]/[\sqrt{n(\Sigma x^2) - (\Sigma x)^2}\sqrt{n(\Sigma y^2) - (\Sigma y)^2}]$
$= 0/[\sqrt{54}\sqrt{54}] = 0$

From Table A-5 for n = 9 assuming $\alpha = 0.05$, C.V. = ±0.666. Therefore r = 0 does not indicate a significant linear correlation. This agrees with the interpretation of the scatterplot.

d. The effect of a single pair of values can be dramatic, changing the conclusion entirely.

NOTE: In each of exercises 13-28 the first variable listed is designated x, and the second variable listed is designated y. In correlation problems the designation of x and y is arbitrary – so long as a person remains consistent after making the designation. In each test of hypothesis, the C.V. and test statistic are given in terms of t using the P-value Method. The usual t formula written for r is

$$t_r = (r - \mu_r)/s_r, \text{ where } \mu_r = \rho = 0 \text{ and } s_r = \sqrt{(1-r^2)/(n-2)} \text{ and df} = n-2.$$

Performing the test using the t statistic allows the calculation of exact P-values. <u>For the r method, the C.V. in terms of r is given in brackets and indicated on the accompanying graph – and the test statistic is simply r.</u> The two methods are mathematically equivalent and always agree.

The scatterplots for the following exercises were generated by Minitab. Scatterplots produced by other statistical software, while the x and y scales may be slightly different, will produce the same visual impression as to how closely the data cluster around a straight line.

13. a. $n = 6$
$\Sigma x = 742.7$
$\Sigma y = 6.50$
$\Sigma xy = 1067.910$
$\Sigma x^2 = 118115.51$
$\Sigma y^2 = 9.7700$

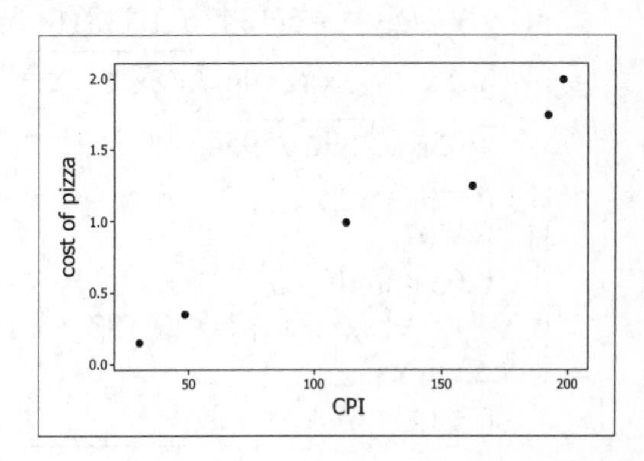

b. $n(\Sigma xy) - (\Sigma x)(\Sigma y) = 6(1067.910) - (742.7)(6.50) = 1579.910$
$n(\Sigma x^2) - (\Sigma x)^2 = 6(118115.51) - (742.7)^2 = 157{,}089.77$
$n(\Sigma y^2) - (\Sigma y)^2 = 6(9.7700) - (6.50)^2 = 16.3700$

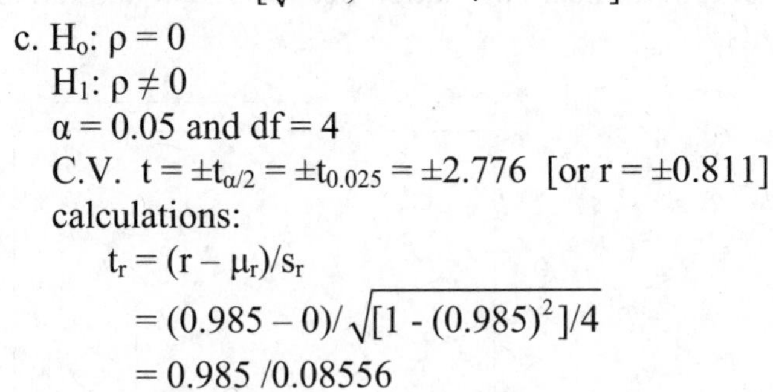

$r = [n(\Sigma xy) - (\Sigma x)(\Sigma y)]/[\sqrt{n(\Sigma x^2) - (\Sigma x)^2}\sqrt{n(\Sigma y^2) - (\Sigma y)^2}]$
$= 1579.910/[\sqrt{157{,}089.77}\sqrt{16.3700}] = 0.985$

c. $H_o: \rho = 0$
$H_1: \rho \neq 0$
$\alpha = 0.05$ and df = 4
C.V. $t = \pm t_{\alpha/2} = \pm t_{0.025} = \pm 2.776$ [or $r = \pm 0.811$]
calculations:
$t_r = (r - \mu_r)/s_r$
$= (0.985 - 0)/\sqrt{[1 - (0.985)^2]/4}$
$= 0.985/0.08556$
$= 11.504$
P-value = 2·tcdf(11.504,99,4) = 0.0003

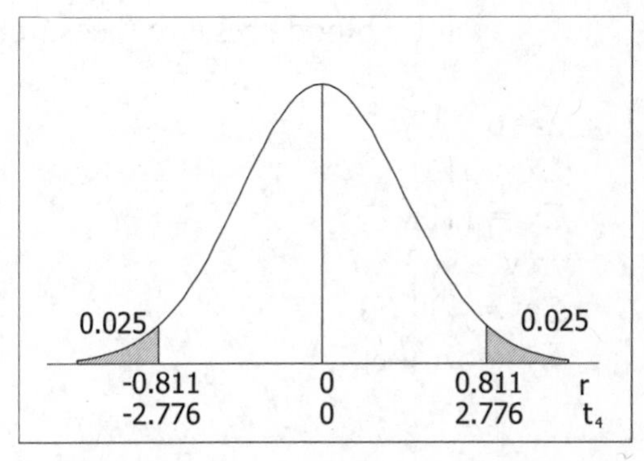

conclusion:
Reject H_o; there is sufficient evidence to conclude that $\rho \neq 0$ (in fact, that $\rho > 0$). Yes; there is sufficient evidence to support the claim of a linear correlation between the CPI and the cost of a slice of pizza.

15. a. $n = 5$
$\Sigma x = 455$
$\Sigma y = 816$
$\Sigma xy = 74937$
$\Sigma x^2 = 41923$
$\Sigma y^2 = 134362$

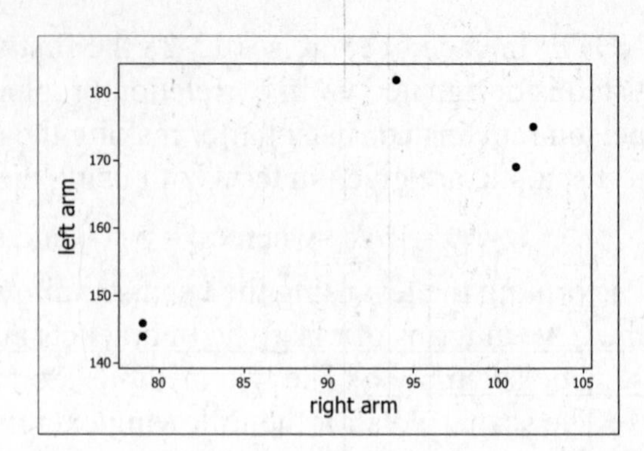

b. $n(\Sigma xy) - (\Sigma x)(\Sigma y) = 5(74937) - (455)(816) = 3405$
$n(\Sigma x^2) - (\Sigma x)^2 = 5(41923) - (255)^2 = 2590$
$n(\Sigma y^2) - (\Sigma y)^2 = 5(134362) - (816)^2 = 5954$
$r = [n(\Sigma xy) - (\Sigma x)(\Sigma y)]/[\sqrt{n(\Sigma x^2) - (\Sigma x)^2}\sqrt{n(\Sigma y^2) - (\Sigma y)^2}]$
$= 3405/[\sqrt{2590}\sqrt{5954}] = 0.867$

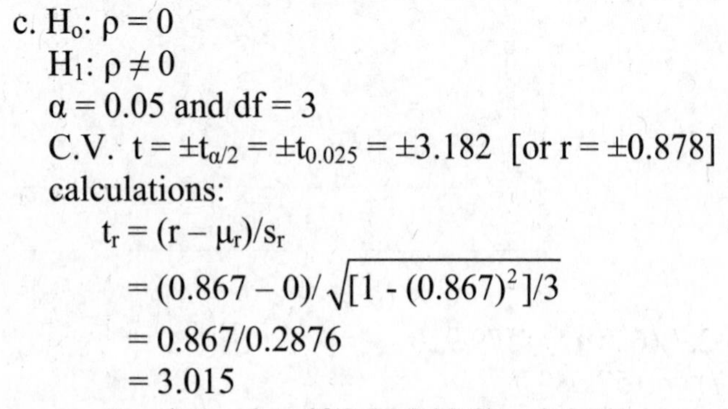

c. H_o: $\rho = 0$
H_1: $\rho \neq 0$
$\alpha = 0.05$ and df = 3
C.V. $t = \pm t_{\alpha/2} = \pm t_{0.025} = \pm 3.182$ [or $r = \pm 0.878$]
calculations:
$t_r = (r - \mu_r)/s_r$
$= (0.867 - 0)/\sqrt{[1 - (0.867)^2]/3}$
$= 0.867/0.2876$
$= 3.015$
P-value = 2·tcdf(3.015,99,3) = 0.0570

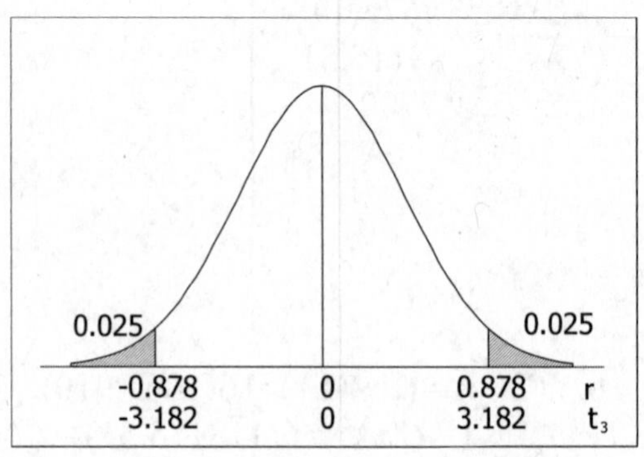

conclusion:
Do not reject H_o; there is not sufficient evidence to conclude that $\rho \neq 0$. No; there is not sufficient evidence to support the claim of a linear correlation between right and left arm systolic blood pressure measurements.

17. a. $n = 6$
$\Sigma x = 51.0$
$\Sigma y = 1108$
$\Sigma xy = 9639.0$
$\Sigma x^2 = 439.00$
$\Sigma y^2 = 214482$

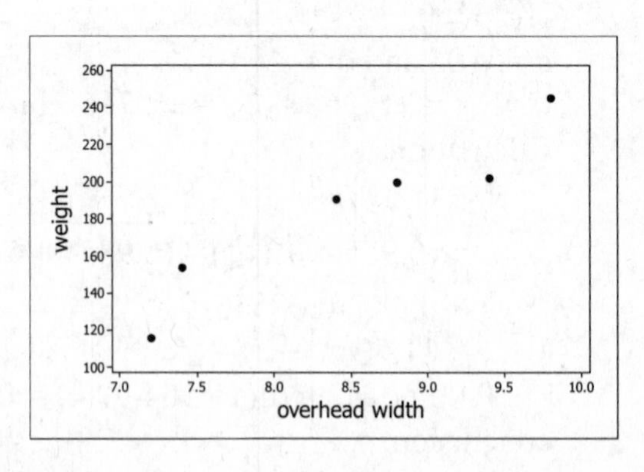

b. $n(\Sigma xy) - (\Sigma x)(\Sigma y) = 6(9639.0) - (51.0)(1108) = 1326.0$
$n(\Sigma x^2) - (\Sigma x)^2 = 6(439.00) - (51.0)^2 = 33.00$
$n(\Sigma y^2) - (\Sigma y)^2 = 6(214482) - (1108)^2 = 59{,}228$
$r = [n(\Sigma xy) - (\Sigma x)(\Sigma y)]/[\sqrt{n(\Sigma x^2) - (\Sigma x)^2}\sqrt{n(\Sigma y^2) - (\Sigma y)^2}]$
$= 1326.0/[\sqrt{33.00}\sqrt{59{,}228}] = 0.948$

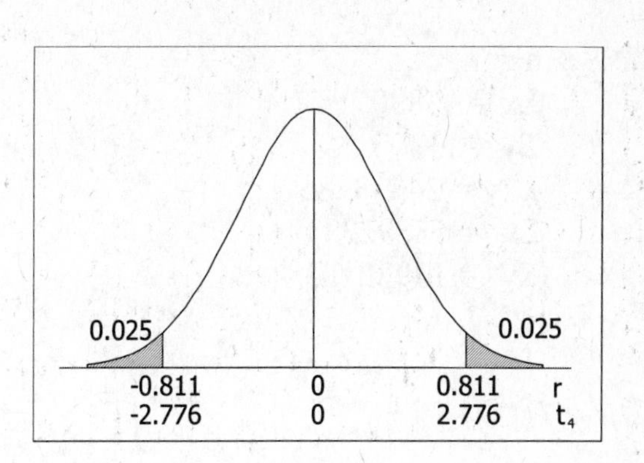

c. H_o: $\rho = 0$
H_1: $\rho \neq 0$
$\alpha = 0.05$ and df = 4
C.V. $t = \pm t_{\alpha/2} = \pm t_{0.025} = \pm 2.776$ [or $r = \pm 0.811$]
calculations:

$t_r = (r - \mu_r)/s_r$
$= (0.948 - 0)/\sqrt{[1 - (0.948)^2]/4}$
$= 0.948/0.1584$
$= 5.986$

P-value = 2·tcdf(5.986,99,4) = 0.0039
conclusion:

Reject H_o; there is sufficient evidence to conclude that $\rho \neq 0$ (in fact, that $\rho > 0$). Yes; there is sufficient evidence to support the claim of a linear correlation between the overhead widths of seals from photographs and the weights of the seals.

19. a. n = 7
$\Sigma x = 1908$
$\Sigma y = 4832$
$\Sigma xy = 1340192$
$\Sigma x^2 = 523336$
$\Sigma y^2 = 3661094$

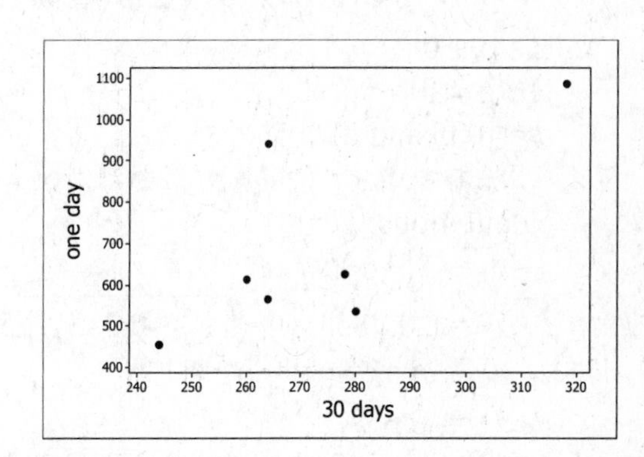

b. $n(\Sigma xy) - (\Sigma x)(\Sigma y) = 7(1340192) - (1908)(4832) = 161{,}888$
$n(\Sigma x^2) - (\Sigma x)^2 = 7(523336) - (1908)^2 = 22{,}888$
$n(\Sigma y^2) - (\Sigma y)^2 = 7(3661094) - (4832)^2 = 2{,}279{,}434$
$r = [n(\Sigma xy) - (\Sigma x)(\Sigma y)]/[\sqrt{n(\Sigma x^2) - (\Sigma x)^2}\sqrt{n(\Sigma y^2) - (\Sigma y)^2}]$
$= 161{,}888/[\sqrt{22{,}888}\sqrt{2{,}279{,}434}] = 0.709$

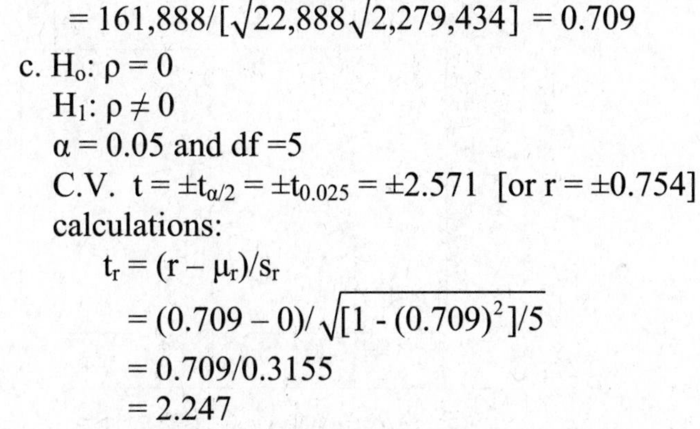
c. H_o: $\rho = 0$
H_1: $\rho \neq 0$
$\alpha = 0.05$ and df = 5
C.V. $t = \pm t_{\alpha/2} = \pm t_{0.025} = \pm 2.571$ [or $r = \pm 0.754$]
calculations:

$t_r = (r - \mu_r)/s_r$
$= (0.709 - 0)/\sqrt{[1 - (0.709)^2]/5}$
$= 0.709/0.3155$
$= 2.247$

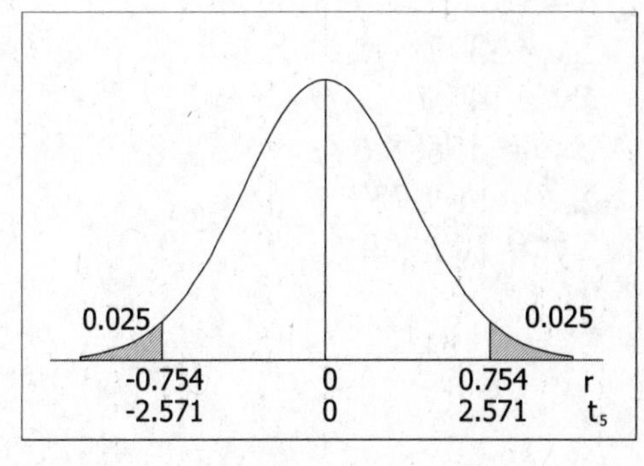

P-value = 2·tcdf(2.247,99,5) = 0.0746
conclusion:

Do not reject H_o; there is not sufficient evidence to conclude that $\rho \neq 0$. No; there is not sufficient evidence to support the claim of a linear correlation between the costs of tickets purchased 30 days in advance and those purchased one day in advance.

21. a. $n = 7$
$\Sigma x = 16890$
$\Sigma y = 11303$
$\Sigma xy = 24833485$
$\Sigma x^2 = 53892334$
$\Sigma y^2 = 23922183$

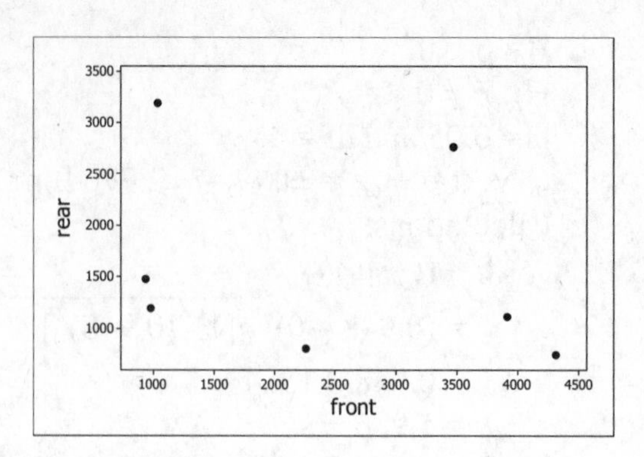

b. $n(\Sigma xy) - (\Sigma x)(\Sigma y) = 7(24833)(485) - (16890)(11303) = -17{,}073{,}275$
$n(\Sigma x^2) - (\Sigma x)^2 = 7(53892334) - (16890)^2 = 91{,}974{,}238$
$n(\Sigma y^2) - (\Sigma y)^2 = 7(23922183) - (11303)^2 = 39{,}697{,}472$
$r = [n(\Sigma xy) - (\Sigma x)(\Sigma y)]/[\sqrt{n(\Sigma x^2) - (\Sigma x)^2}\sqrt{n(\Sigma y^2) - (\Sigma y)^2}]$
$= -17{,}073{,}275/[\sqrt{91{,}974{,}238}\sqrt{39{,}697{,}472}] = -0.283$

c. H_o: $\rho = 0$
H_1: $\rho \neq 0$
$\alpha = 0.05$ and df = 5
C.V. $t = \pm t_{\alpha/2} = \pm t_{0.025} = \pm 2.571$ [or $r = \pm 0.754$]
calculations:
$t_r = (r - \mu_r)/s_r$
$= (-0.283 - 0)/\sqrt{[1 - (-0.283)^2]/5}$
$= -0.283/0.4290$
$= -0.659$

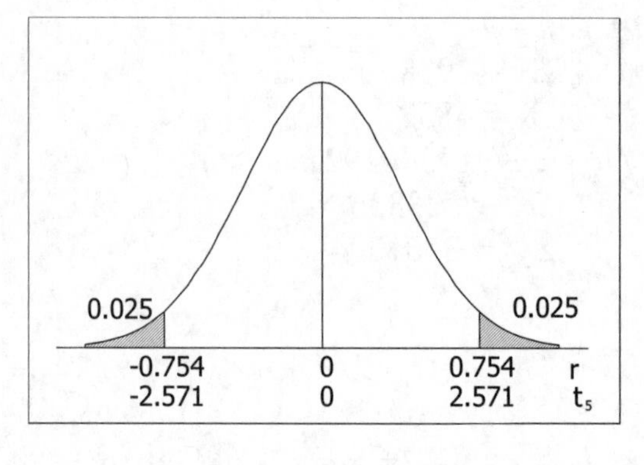

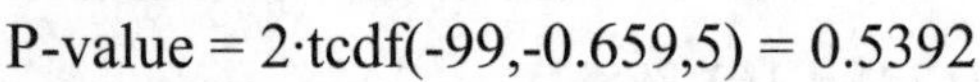
P-value = 2·tcdf(-99,-0.659,5) = 0.5392

conclusion:
Do not reject H_o; there is not sufficient evidence to conclude that $\rho \neq 0$. No; there is not sufficient evidence to support the claim of a linear correlation between the repair costs from full-front crashes and full-rear crashes.

23. a. $n = 10$
$\Sigma x = 3377$
$\Sigma y = 141.7$
$\Sigma xy = 47888.6$
$\Sigma x^2 = 1143757$
$\Sigma y^2 = 2008.39$

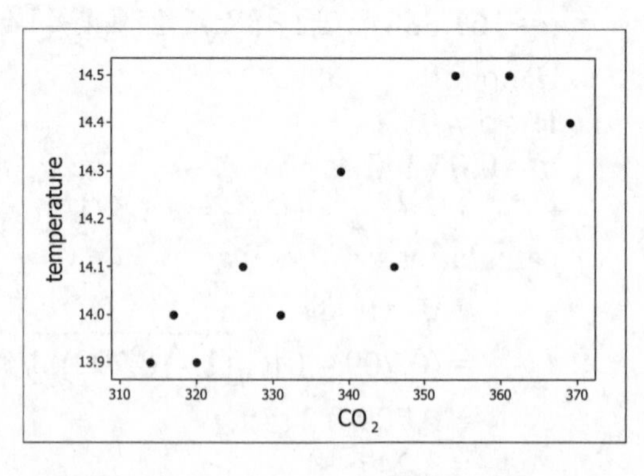

b. $n(\Sigma xy) - (\Sigma x)(\Sigma y) = 10(47888.6) - (3377)(141.7) = 365.1$
$n(\Sigma x^2) - (\Sigma x)^2 = 10(1143757) - (3377)^2 = 33{,}441$
$n(\Sigma y^2) - (\Sigma y)^2 = 10(2008.39 - (141.7)^2 = 5.01$
$r = [n(\Sigma xy) - (\Sigma x)(\Sigma y)]/[\sqrt{n(\Sigma x^2) - (\Sigma x)^2}\sqrt{n(\Sigma y^2) - (\Sigma y)^2}]$
$= 365.1/[\sqrt{33{,}441}\sqrt{5.01}] = 0.892$

c. H_o: $\rho = 0$
H_1: $\rho \neq 0$
$\alpha = 0.05$ and df = 8
C.V. $t = \pm t_{\alpha/2} = \pm t_{0.025} = \pm 2.306$ [or $r = \pm 0.632$]
calculations:
$t_r = (r - \mu_r)/s_r$
$= (0.892 - 0)/\sqrt{[1 - (0.892)^2]/8}$
$= 0.892/0.1598$
$= 5.581$
P-value = 2·tcdf(5.581,99,8) = 0.0005
conclusion:
Reject H_o; there is sufficient evidence to conclude that $\rho \neq 0$ (in fact, that $\rho > 0$). Yes; there is sufficient evidence to support the claim of a linear correlation between global temperature and the concentration of CO_2.

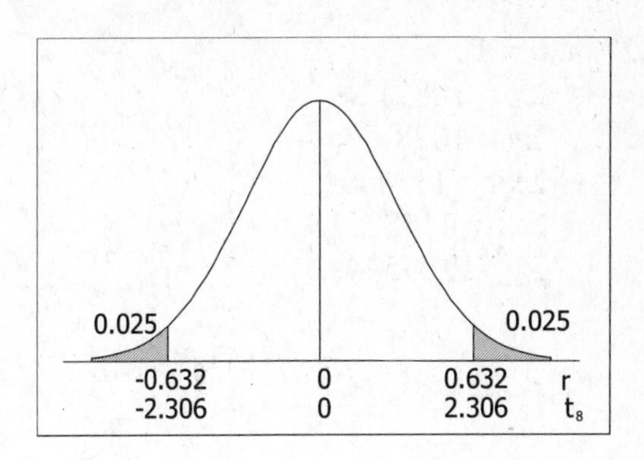

25. a. n = 7
$\Sigma x = 154$
$\Sigma y = 3.531$
$\Sigma xy = 118.173$
$\Sigma x^2 = 86016$
$\Sigma y^2 = 1.807253$

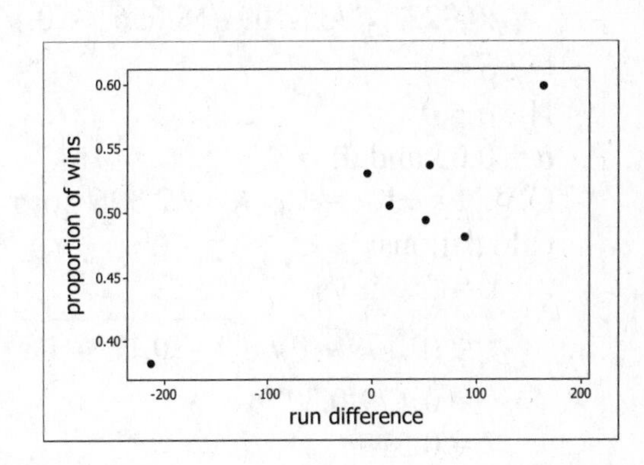

b. $n(\Sigma xy) - (\Sigma x)(\Sigma y) = 7(118.173) - (154)(3.531) = 283.437$
$n(\Sigma x^2) - (\Sigma x)^2 = 7(86016) - (152)^2 = 578{,}396$
$n(\Sigma y^2) - (\Sigma y)^2 = 7(1.807253) - (3.531)^2 = 0.182810$
$r = [n(\Sigma xy) - (\Sigma x)(\Sigma y)]/[\sqrt{n(\Sigma x^2) - (\Sigma x)^2}\sqrt{n(\Sigma y^2) - (\Sigma y)^2}]$
$= 283.437/[\sqrt{578{,}396}\sqrt{0.182810}] = 0.872$

c. H_o: $\rho = 0$
H_1: $\rho \neq 0$
$\alpha = 0.05$ and df = 5
C.V. $t = \pm t_{\alpha/2} = \pm t_{0.025} = \pm 2.571$ [or $r = \pm 0.754$]
calculations:
$t_r = (r - \mu_r)/s_r$
$= (0.872 - 0)/\sqrt{[1 - (0.872)^2]/5}$
$= 0.872/0.2192$
$= 3.977$
P-value = 2·tcdf(3.977,99,5) = 0.0106
conclusion:
Reject H_o; there is sufficient evidence to conclude that $\rho \neq 0$ (in fact, that $\rho > 0$). Yes; there is sufficient evidence to support the claim of a linear correlation between a team's proportion of wins and its difference between numbers of runs scored and runs allowed.

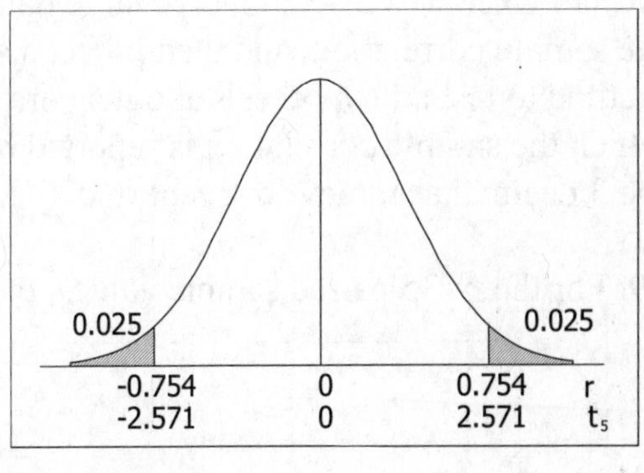

27. a. $n = 10$
$\Sigma x = 10821$
$\Sigma y = 1028$
$\Sigma xy = 1114491$
$\Sigma x^2 = 11782515$
$\Sigma y^2 = 107544$

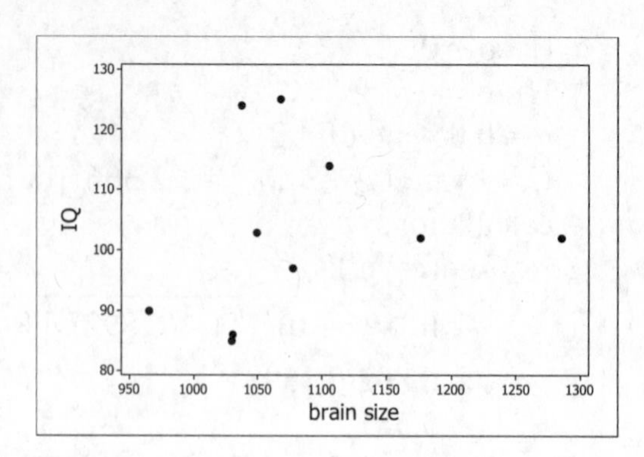

b. $n(\Sigma xy) - (\Sigma x)(\Sigma y) = 8(1114491) - (10821)(1028) = 20{,}922$
$n(\Sigma x^2) - (\Sigma x)^2 = 10(11782515) - (10821)^2 = 731{,}109$
$n(\Sigma y^2) - (\Sigma y)^2 = 10(107544) - (1028)^2 = 18{,}656$
$r = [n(\Sigma xy) - (\Sigma x)(\Sigma y)]/[\sqrt{n(\Sigma x^2) - (\Sigma x)^2}\sqrt{n(\Sigma y^2) - (\Sigma y)^2}]$
$= 20{,}922/[\sqrt{731{,}109}\sqrt{18{,}656}] = 0.179$

c. $H_o: \rho = 0$
$H_1: \rho \neq 0$
$\alpha = 0.05$ and $df = 8$
C.V. $t = \pm t_{\alpha/2} = \pm t_{0.025} = \pm 2.306$ [or $r = \pm 0.632$]
calculations:
$t_r = (r - \mu_r)/s_r$
$= (0.179 - 0)/\sqrt{[1 - (0.179)^2]/8}$
$= 0.179/0.3478$
$= 0.515$

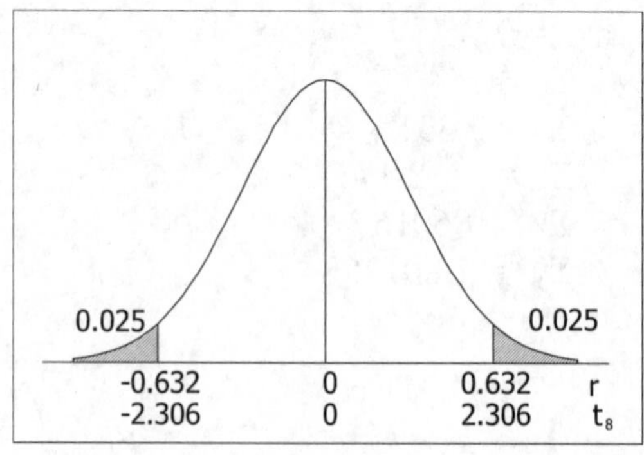

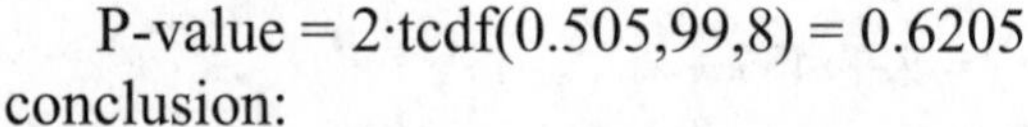
P-value = 2·tcdf(0.505,99,8) = 0.6205

conclusion:
Do not reject H_o; there is not sufficient evidence to conclude that $\rho \neq 0$. No; there is not sufficient evidence to support the claim of a linear correlation between brain size and IQ score. No; it does not appear that people with larger brains are more intelligent.

NOTE: Exercises 29-32 involve large data sets from Appendix B. Use statistical software to find the sample correlation, and then proceed as usual using that value. Those using the P-value method to test an hypothesis about a correlation will be limited by the degree of accuracy with which the sample correlation is reported by the statistical software. This manual proceeds using the 3 decimal accuracy for r reported by Minitab as if it were the exact sample value.

29. For the n=35 paired sample values, the Minitab regression of c3 on c4 yields $r = 0.744$.

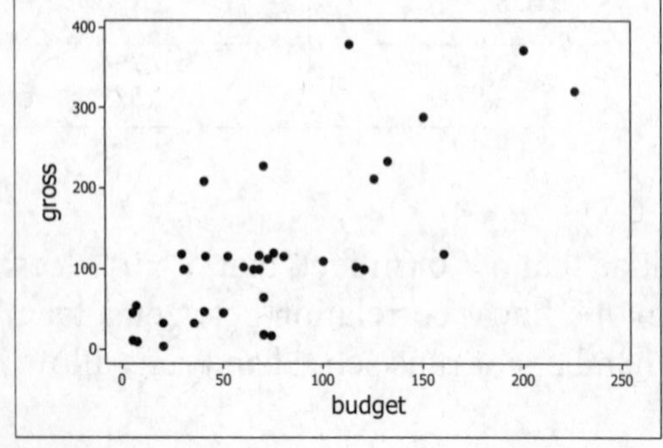

H_o: $\rho = 0$
H_1: $\rho \neq 0$
$\alpha = 0.05$ and df = 33
C.V. $t = \pm t_{\alpha/2} = \pm t_{0.025} = \pm 2.035$ [or $r = \pm 0.335$]
calculations:

$t_r = (r - \mu_r)/s_r$
$= (0.744 - 0)/\sqrt{[1 - (0.744)^2]/33}$
$= 0.744/0.1163$
$= 6.396$

P-value = 2·tcdf(6.396 ,99,33) = 3.018E-7 = 0.0000003

conclusion:

Reject H_o; there is sufficient evidence to conclude that $\rho \neq 0$ (in fact, that $\rho > 0$). Yes; there is sufficient evidence to support the claim of a linear correlation between a movie's budget amount and the amount that movie grosses.

31. For the n=56 paired sample values, the Minitab regression of c1 on c2 yields r = 0.319.

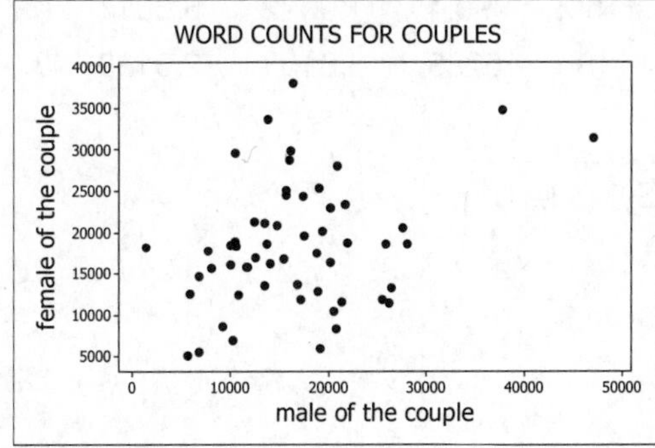

H_o: $\rho = 0$
H_1: $\rho \neq 0$
$\alpha = 0.05$ and df = 54
C.V. $t = \pm t_{\alpha/2} = \pm t_{0.025} = \pm 2.009$ [or $r = \pm 0.254$]
calculations:

$t_r = (r - \mu_r)/s_r$
$= (0.319 - 0)/\sqrt{[1 - (0.319)^2]/54}$
$= 0.319/0.1290$
$= 2.473$

P-value = 2·tcdf(2.473,99,54) = 0.0166

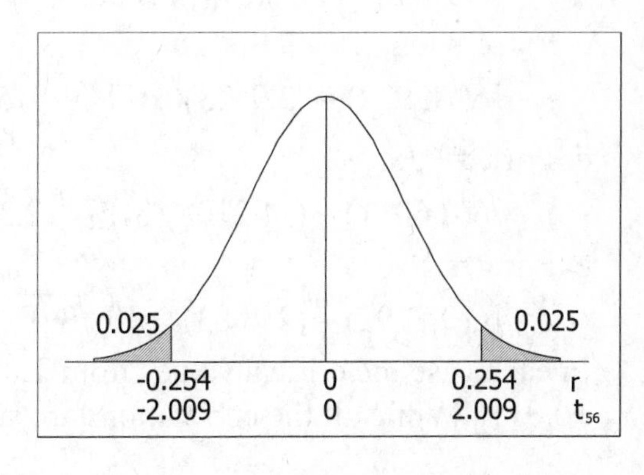

conclusion:

Reject H_o; there is sufficient evidence to conclude that $\rho \neq 0$ (in fact, that $\rho > 0$). Yes; there is sufficient evidence to support the claim of a linear correlation between the numbers of words spoken by men and women who are a couple.

33. A significant linear correlation indicates that the factors are associated, not that there is a cause-and-effect relationship. Even if there is a cause-and-effect relationship, correlation analysis cannot identify which factor is the cause and which factor is the effect.

35. A significant linear correlation between group averages indicates nothing about the relationship between the individual scores – which may be uncorrelated, correlated in the opposite direction, or have different correlations in each of the groups.

37. The following table gives the values for y, x, x^2, log x, $\sqrt{x}$ and 1/x. The rows at the bottom of the table give the sum of the values (Σv), the sum of the squares of the values (Σv^2), the sum of each value times the corresponding y value (Σvy), and the quantity $n\Sigma v^2 - (\Sigma v)^2$ needed in subsequent calculations.

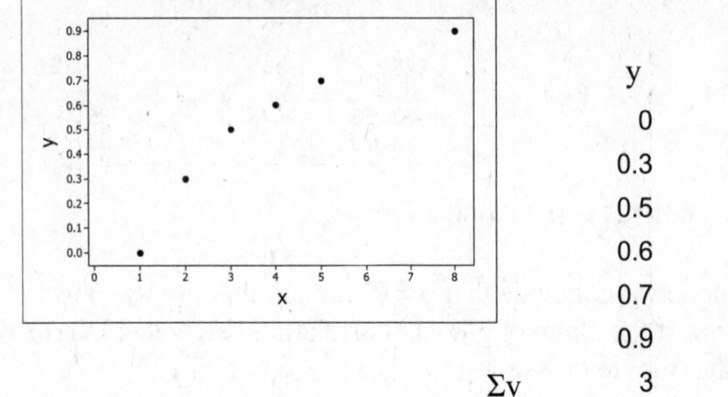

	y	x	x^2	log x	$\sqrt{x}$	1/x
	0	1	1	0	1.0000	1.0000
	0.3	2	4	0.3010	1.4142	0.5000
	0.5	3	9	0.4771	1.7321	0.3333
	0.6	4	16	0.6021	2.0000	0.2500
	0.7	5	25	0.6990	2.2361	0.2000
	0.9	8	64	0.9031	2.8284	0.1250
Σv	3	23	119	2.9823	11.2108	2.4083
Σv^2	2	119	5075	1.9849	23.0000	1.4792
Σvy		15.2	90.4	1.9922	6.6011	0.7192
$n\Sigma v^2 - (\Sigma v)^2$	3	185	16289	3.0153	12.3189	3.0753

In general, $r = [n(\Sigma vy) - (\Sigma v)(\Sigma y)]/[\sqrt{n(\Sigma v^2) - (\Sigma v)^2}\sqrt{n(\Sigma y^2) - (\Sigma y)^2}]$

a. For v = x,

$r = [6(15.2) - (23)(3)]/[\sqrt{185}\sqrt{3}] = 0.9423$

b. For v = x^2,

$r = [6(90.4) - (119)(3)]/[\sqrt{16289}\sqrt{3}] = 0.8387$

c. For v = log x,

$r = [6(1.9922) - (2.9823)(3)]/[\sqrt{3.0153}\sqrt{3}] = 0.9996$

d. For v = $\sqrt{x}$,

$r = [6(6.6011) - (11.2108)(3)]/[\sqrt{12.3189}\sqrt{3}] = 0.9827$

e. For v = 1/x,

$r = [6(0.7192) - (2.4083)(3)]/[\sqrt{3.0753}\sqrt{3}] = -0.9580$

In each case the critical values from Table A-5 for testing significance at the 0.05 level are ±0.811. While all the correlations are significant, the largest value for r occurs in part (c).

10-3 Regression

1. The symbol $\hat{y}$ represents the predicted cholesterol level. The predictor variable x represents weight. The response variable y represents cholesterol level.

3. Since s_y and s_x must be non-negative, the regression line has a slope (which is equal to $r \cdot s_y/s_x$) with the same sign as r. If r is positive, the slope of the regression line is positive and the regression line rises as it goes from left to right. If r is negative, the slope of the regression line is negative and the regression line fall as it goes from left to right.

5. For n=62, C.V. = ±0.254. Since r = 0.759 > 0.254, use the regression line for prediction.
   $\hat{y} = 0.445 + 0.119x$
   $\hat{y}_{50} = 0.455 + 0.119(50) = 6.4$ people

7. For n=40, C.V. = ±0.312. Since r = 0.202 < 0.312, use the mean for prediction.
   $\hat{y} = \bar{y}$
   $\hat{y}_{70} = 76.3$ beats/minute

9. Excel produces the following scatterplot.

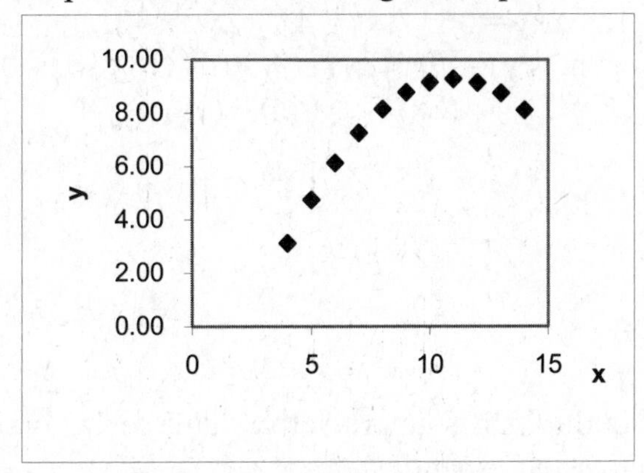

x	y	xy	x^2	y^2
10	9.14	91.40	100	83.5396
8	8.14	65.12	64	66.2596
13	8.74	113.62	169	76.3876
9	8.77	78.93	81	76.9129
11	9.26	101.86	121	85.7476
14	8.10	113.40	196	65.61
6	6.13	36.78	36	37.5769
4	3.10	12.40	16	9.61
12	9.13	109.56	144	83.3569
7	7.26	50.82	49	52.7076
5	4.74	23.70	25	22.4676
99	82.51	797.59	1001	660.1763

See the chart above at the right, where n = 11.
   $\bar{x} = (\Sigma x)/n = 99/11 = 9.0$ $\quad n(\Sigma xy) - (\Sigma x)(\Sigma y) = 11(797.59) - (99)(82.51) = 605.00$
   $\bar{y} = (\Sigma y)/n = 82.52/11 = 7.50$ $\quad n(\Sigma x^2) - (\Sigma x)^2 = 11(1001) - (99)^2 = 1210$
 $b_1 = [n(\Sigma xy) - (\Sigma x)(\Sigma y)]/[n(\Sigma x^2) - (\Sigma x)^2]$
   $= 605.00/1210 = 0.500$
 $b_o = \bar{y} - b_1\bar{x}$
   $= 7.50 - 0.500(9.0) = 3.00$
 $\hat{y} = b_o + b_1x$
 $\hat{y} = 3.00 + 0.500x$
 The scatterplot indicates that the relationship between the variables is quadratic, not linear.

NOTE: In addition to the value of n, calculations associated with regression involve five sums: Σx, Σy, Σx^2, Σy^2 and Σxy. As the sums can usually be found conveniently using a calculator, the remaining exercises give only the values of the sums without constructing a chart as in exercises 9 and 10,. In addition, the calculations typically involve the following subcalculations.
  (1) $n(\Sigma xy) - (\Sigma x)(\Sigma y)$ determines the sign of the slope of the regression line. If large values of x are associated with large values of y, it will be positive. If large values of x are associated with small values of y, it will be negative. If not, a mistake has been made.
  (2) $n(\Sigma x^2) - (\Sigma x)^2$ cannot be negative. If it is, a mistake had been made.
  (3) $n(\Sigma y^2) - (\Sigma y)^2$ cannot be negative. If it is, a mistake had been made.
If any of these mistakes occurs, stop immediately and find the error – continuing is wasted effort.

11. a. using all the points: $n = 10$ $\Sigma x = 28$ $\Sigma y = 28$ $\Sigma xy = 136$ $\Sigma x^2 = 142$ $\Sigma y^2 = 142$

$\bar{x} = (\Sigma x)/n = 28/10 = 2.8$

$\bar{y} = (\Sigma y)/n = 28/10 = 2.8$

$n(\Sigma xy) - (\Sigma x)(\Sigma y) = 10(136) - (28)(28) = 576$

$n(\Sigma x^2) - (\Sigma x)^2 = 10(142) - (28)^2 = 636$

$b_1 = [n(\Sigma xy) - (\Sigma x)(\Sigma y)]/[n(\Sigma x^2) - (\Sigma x)^2]$
$= 576/636 = 0.906$

$b_o = \bar{y} - b_1\bar{x}$
$= 2.8 - 0.906(2.8) = 0.264$

$\hat{y} = b_o + b_1x$

$\hat{y} = 0.264 + 0.906x$

b. without the outlier: $n = 9$ $\Sigma x = 18$ $\Sigma y = 18$ $\Sigma xy = 36$ $\Sigma x^2 = 42$ $\Sigma y^2 = 42$

$\bar{x} = (\Sigma x)/n = 18/9 = 2.0$

$\bar{y} = (\Sigma y)/n = 18/9 = 2.0$

$n(\Sigma xy) - (\Sigma x)(\Sigma y) = 9(36) - (18)(18) = 0$

$n(\Sigma x^2) - (\Sigma x)^2 = 9(42) - (18)^2 = 54$

$b_1 = [n(\Sigma xy) - (\Sigma x)(\Sigma y)]/[n(\Sigma x^2) - (\Sigma x)^2]$
$= 0/54 = 0$

$b_o = \bar{y} - b_1\bar{x}$
$= 2.0 - 0(2.0) = 2.0$

$\hat{y} = b_o + b_1x$

$\hat{y} = 2.0 + 0x$ [or simply $\hat{y} = 2.0$, for any x]

c. The results are very different – without the outlier, x has no predictive value for y. A single outlier can have a dramatic effect on the regression equation.

NOTE: For exercises 13-26, the exact summary statistics (i.e., without any rounding) are given on the right. While the intermediate calculations on the left are presented rounded to various degrees of accuracy, the entire unrounded values were preserved in the calculator until the end. When finding a predicted value, always verify that it is reasonable for the story problem and consistent with the given data points used to find the regression equation. The final prediction is made either using the regression equation $\hat{y} = b_o + b_1x$ or the sample mean $\bar{y}$. Refer back to the corresponding test for a significant linear correlation in the previous section (the exercise numbers are the same), and use $\hat{y} = b_o + b_1x$ only if there is a significant linear correlation.

13. $\bar{x} = 123.78$

$\bar{y} = 1.08$

$b_1 = [n(\Sigma xy) - (\Sigma x)(\Sigma y)]/[n(\Sigma x^2) - (\Sigma x)^2]$
$= 1579.910/157{,}089.77 = 0.0101$

$b_o = \bar{y} - b_1\bar{x}$
$= 1.08 - 0.0101(123.78) = -0.162$

$\hat{y} = b_o + b_1x$
$= -0.162 + 0.0101x$

$\hat{y}_{182.5} = -0.162 + 0.0101(182.5) = \$1.67$ [\$1.68 using rounded values]

$n = 6$
$\Sigma x = 742.7$
$\Sigma y = 6.50$
$\Sigma x^2 = 118115.51$
$\Sigma y^2 = 9.7700$
$\Sigma xy = 1067.910$

15. $\bar{x} = 91.0$

$\bar{y} = 163.2$

$b_1 = [n(\Sigma xy) - (\Sigma x)(\Sigma y)]/[n(\Sigma x^2) - (\Sigma x)^2]$
$= 3405/2590 = 1.315$

$b_o = \bar{y} - b_1\bar{x}$
$= 163.2 - 1.315(91.0) = 43.56$

$\hat{y} = b_o + b_1x$
$= 43.6 + 1.31x$

$\hat{y}_{100} = \bar{y} = 163.2$ mm Hg [no significant correlation]

$n = 5$
$\Sigma x = 455$
$\Sigma y = 816$
$\Sigma x^2 = 41923$
$\Sigma y^2 = 134362$
$\Sigma xy = 74937$

17. $\bar{x} = 8.50$
$\bar{y} = 184.67$
$b_1 = [n(\Sigma xy) - (\Sigma x)(\Sigma y)]/[n(\Sigma x^2) - (\Sigma x)^2]$
$= 1326.0/33.00 = 40.18$
$b_o = \bar{y} - b_1\bar{x}$
$= 184.67 - 40.18(8.50) = -156.87$
$\hat{y} = b_o + b_1x$
$= -156.9 + 40.2x$
$\hat{y}_{9.0} = -156.9 + 40.2(9.0) = 204.8$ kg

$n = 6$
$\Sigma x = 51.0$
$\Sigma y = 1108$
$\Sigma x^2 = 439.00$
$\Sigma y^2 = 214482$
$\Sigma xy = 9639.0$

19. $\bar{x} = 272.57$
$\bar{y} = 690.29$
$b_1 = [n(\Sigma xy) - (\Sigma x)(\Sigma y)]/[n(\Sigma x^2) - (\Sigma x)^2]$
$= 161{,}888/22{,}888 = 7.07$
$b_o = \bar{y} - b_1\bar{x}$
$= 690.29 - 7.07(272.57) = -1237.62$
$\hat{y} = b_o + b_1x = -1237.6 + 7.07x$
$\hat{y}_{300} = \bar{y} = \$690.3$ [no significant correlation]

$n = 7$
$\Sigma x = 1908$
$\Sigma y = 4832$
$\Sigma x^2 = 523336$
$\Sigma y^2 = 3661094$
$\Sigma xy = 1340192$

21. $\bar{x} = 2412.86$
$\bar{y} = 1614.71$
$b_1 = [n(\Sigma xy) - (\Sigma x)(\Sigma y)]/[n(\Sigma x^2) - (\Sigma x)^2]$
$= -17{,}073{,}275/91{,}974{,}238 = -0.186$
$b_o = \bar{y} - b_1\bar{x}$
$= 1614.71 - (-0.186)(2412.86) = 2062.62$
$\hat{y} = b_o + b_1x = 2062.6 - 0.186x$
$\hat{y}_{4594} = \bar{y} = \$1614.7$ [no significant correlation]
The result does not compare very well to the actual repair cost of $982.

$n = 7$
$\Sigma x = 16890$
$\Sigma y = 11303$
$\Sigma x^2 = 53892334$
$\Sigma y^2 = 23922183$
$\Sigma xy = 24833485$

23. $\bar{x} = 337.70$
$\bar{y} = 14.17$
$b_1 = [n(\Sigma xy) - (\Sigma x)(\Sigma y)]/[n(\Sigma x^2) - (\Sigma x)^2]$
$= 365.1/33{,}441 = 0.0109$
$b_o = \bar{y} - b_1\bar{x}$
$= 14.17 - 0.0109(337.70) = 10.48$
$\hat{y} = b_o + b_1x = 10.5 + 0.0109x$
$\hat{y}_{370.9} = 10.5 + 0.0109(370.9) = 14.5$ °C
Yes; in this instance the predicted temperature is equal to the actual temperature of 14.5 °C.

$n = 10$
$\Sigma x = 3377$
$\Sigma y = 141.7$
$\Sigma x^2 = 1143757$
$\Sigma y^2 = 2008.39$
$\Sigma xy = 47888.6$

25. $\bar{x} = 22.00$
$\bar{y} = 0.504$
$b_1 = [n(\Sigma xy) - (\Sigma x)(\Sigma y)]/[n(\Sigma x^2) - (\Sigma x)^2]$
$= 283.437/578{,}396 = 0.000490$
$b_o = \bar{y} - b_1\bar{x}$
$= 0.504 - 0.000490(22.00) = 0.494$
$\hat{y} = b_o + b_1x = 0.494 + 0.000490x$
$\hat{y}_{52} = 0.494 + 0.000490(52) = 0.519$
Yes; the predicted proportion is reasonably close to the actual proportion of 0.543.

$n = 7$
$\Sigma x = 154$
$\Sigma y = 3.531$
$\Sigma x^2 = 86016$
$\Sigma y^2 = 1.807253$
$\Sigma xy = 118.173$

27. $\bar{x} = 1082.10$ $\quad n = 10$
$\bar{y} = 102.80$ $\quad \Sigma x = 10821$
$b_1 = [n(\Sigma xy) - (\Sigma x)(\Sigma y)]/[n(\Sigma x^2) - (\Sigma x)^2]$ $\quad \Sigma y = 1028$
$= 20{,}922/731{,}109 = 0.0286$ $\quad \Sigma x^2 = 11782515$
$b_o = \bar{y} - b_1\bar{x}$ $\quad \Sigma y^2 = 107544$
$= 102.80 - 0.0286(1082.10) = 71.83$ $\quad \Sigma xy = 1114491$
$\hat{y} = b_o + b_1x = 71.8 - 0.0286x$
$\hat{y}_{1275} = \bar{y} = 102.8$ [no significant correlation]

NOTE: Exercises 29-32 involve large data sets from Appendix B. Use statistical software to find the regression equation. When finding a predicted value, always verify that it is reasonable for the story problem and consistent with the given data points used to find the regression equation. The final prediction is made either using the regression equation $\hat{y} = b_o + b_1x$ or the sample mean $\bar{y}$. Refer back to the corresponding test for a significant linear correlation in the previous section (the exercise numbers are the same), and use $\hat{y} = b_o + b_1x$ only if there is a significant linear correlation. If there is no significant linear correlation, use statistical software to find the mean of the response variable (i.e., the y variable) and use that for the predicted value.

29. For the n=35 paired sample values, the Minitab regression of c4 on c3 yields
```
gross = 20.6 + 1.38 budget
```
$\hat{y} = 20.6 + 1.38x$
$\hat{y}_{120} = 20.6 + 1.38(120) = 186.2$ million \$

31. For the n=56 paired sample values, the Minitab regression of c2 on c1 yields
```
1F = 13439 + 0.302 1M
```
$\hat{y} = 13439 + 0.302x$
$\hat{y}_{6000} = 13439 + 0.302(6000) = 15{,}248$ words per day

33. If H_o:ρ=0 is true, there is no linear correlation between x and y and $\hat{y} = \bar{y}$ is the appropriate prediction for y for any x.
If Ho:β_1=0 is true, then the true regression line is $y = \beta_o + 0x = \beta_o$ and the best estimate for β_o is $b_o = \bar{y} - 0\bar{x} = \bar{y}$, producing the line $\hat{y} = \bar{y}$.
Since both hypotheses imply precisely the same result, they are equivalent.

35. Refer to the table at the right, where
x = the pulse rate
y = the systolic blood pressure
$\hat{y} = 71.68 + 0.5956x$
= the value predicted by the regression equation
$y - \hat{y}$ = the residuals for the regression line

x	y	$\hat{y}$	$y-\hat{y}$
68	125	112.181	12.819
64	107	109.798	-2.798
88	126	124.093	1.907
72	110	114.563	-4.563
64	110	109.798	0.202
72	107	114.563	-7.563
428	685	684.997	0.003

The residual plot below was obtained by plotting the predictor variable (pulse rate) on the horizontal axis and the corresponding residual from the table on the vertical axis. The scatterplot shows the original (x,y) = (pulse,systolic) pairs.

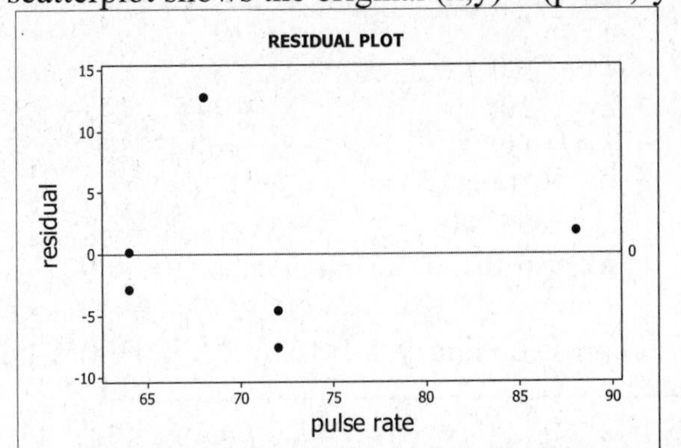

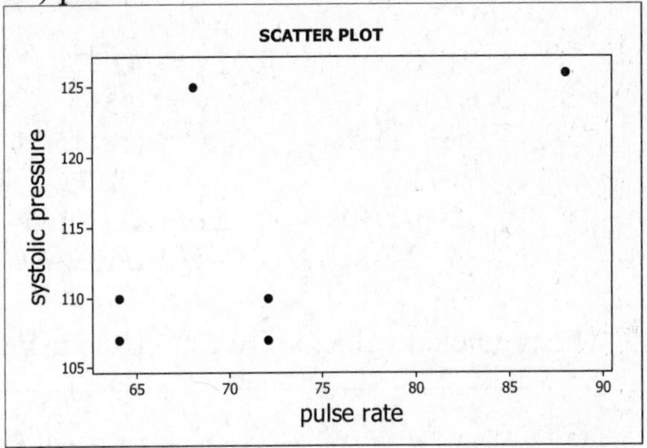

The residual plot seems to suggest that the regression equation is a good model – because the residuals are randomly scattered around the zero line, with no obvious pattern or change in variability. The scatterplot suggests that the regression equation is not a good model – because the points do not appear to fit a straight line pattern.

10-4 Variation and Prediction Intervals

1. In general, s measures the spread of the data around some reference. For a set of y values in one dimension, s_y measures the spread of the y values around $\overline{y}$. For ordered pairs (x,y) in two dimensions, s_y measures the spread of the points around the line $y = \overline{y}$. For ordered pairs (x,y), s_e measures the spread of the points around the regression line $\hat{y} = b_o + b_1x$.

3. By providing a range of values instead of a single point, a prediction interval gives an indication of the accuracy of the prediction. A confidence interval is an interval estimate of a parameter – i.e., of a conceptually fixed, although unknown, value. A prediction interval is an interval estimate of a random variable – i.e., of a value from a distribution of values.

5. The coefficient of determination is $r^2 = (0.873)^2 = 0.762$.
   The portion of the total variation in y explained by the regression is $r^2 = 0.762 = 76.2\%$

7. The coefficient of determination is $r^2 = (-0.865)^2 = 0.748$.
   The portion of the total variation in y explained by the regression is $r^2 = 0.748 = 74.8\%$.

9. Since the slope of the regression line $b_1 = r\cdot(s_y/s_x)$ is negative, r must be negative.
   Since $r^2 = 65.0\% = 0.650$, $r = -\sqrt{0.650} = -0.806$.
   For n=32 [closest entry is n=30], Table A-5 gives C.V. = ±0.361.
   Since $-0.806 < -0.361$, there is sufficient evidence to support the claim of a linear correlation between the weights of cars and their highway fuel consumption amounts.

11. The given point estimate is $\hat{y} = 27.028$ mpg.

NOTE: The following summary statistics apply to exercises 13-16 and 17-20. They are all that is necessary to use the chapter formulas to work the problems.

exercise #13	exercise #14	exercise #15	exercise #16
$n = 6$	$n = 6$	$n = 6$	$n = 10$
$\Sigma x = 742.7$	$\Sigma x = 742.77$	$\Sigma x = 51.0$	$\Sigma x = 3377$
$\Sigma y = 6.50$	$\Sigma y = 6.35$	$\Sigma y = 1108$	$\Sigma y = 141.7$
$\Sigma x^2 = 118115.51$	$\Sigma x^2 = 118115.51$	$\Sigma x^2 = 439.00$	$\Sigma x^2 = 1143757$
$\Sigma y^2 = 9.7700$	$\Sigma y^2 = 9.2175$	$\Sigma y^2 = 214482$	$\Sigma y^2 = 2008.39$
$\Sigma xy = 1067.910$	$\Sigma xy = 1036.155$	$\Sigma xy = 9639.0$	$\Sigma xy = 47888.6$
see also 10.2-3 #13	see also 10.2-3 #14	see also 10.2-3 #17	see also 10.2-3 #23

13. The predicted values were calculated using the regression line $\hat{y} = -0.161601 + 0.0100574x$.

x	y	$\hat{y}$	$\bar{y}$	$\hat{y}-\bar{y}$	$(\hat{y}-\bar{y})^2$	$y-\hat{y}$	$(y-\hat{y})^2$	$y-\bar{y}$	$(y-\bar{y})^2$
30.2	0.15	0.142	1.083	-0.940	0.886	0.008	0.000	-0.930	0.871
48.3	0.35	0.324	1.083	-0.760	0.576	0.026	0.001	-0.730	0.538
112.3	1.00	0.968	1.083	-0.120	0.013	0.032	0.001	-0.080	0.007
162.2	1.25	1.470	1.083	0.386	0.149	-0.220	0.048	0.167	0.028
191.9	1.75	1.768	1.083	0.685	0.469	-0.018	0.000	0.667	0.444
197.8	2.00	1.828	1.083	0.744	0.554	0.172	0.030	0.917	0.840
742.7	6.50	6.500	6.500	0.000	2.648	0.000	0.080	0.000	2.728

a. The explained variation is $\Sigma(\hat{y}-\bar{y})^2 = 2.648$
b. The unexplained variation is $\Sigma(y-\hat{y})^2 = 0.080$
c. The total variation is $\Sigma(y-\bar{y})^2 = 2.728$
d. $r^2 = \Sigma(\hat{y}-\bar{y})^2/\Sigma(y-\bar{y})^2 = 2.648/2.728 = 0.971$
e. $s_e^2 = \Sigma(y-\hat{y})^2/(n-2) = 0.080/4 = 0.020$; $s_e = \sqrt{0.020} = 0.141$

NOTE: A table such as the one in the preceding exercise organizes the work and provides all the values needed to discuss variation. In such a table, the following must always be true (except for minor discrepancies due to rounding) and can be used as a check before proceeding.
(1) $\Sigma y = \Sigma\hat{y} = \Sigma\bar{y}$
(2) $\Sigma(\hat{y}-\bar{y}) = \Sigma(y-\hat{y}) = \Sigma(y-\bar{y}) = 0$
(3) $\Sigma(y-\hat{y})^2 + \Sigma(\hat{y}-\bar{y})^2 = \Sigma(y-\bar{y})^2$

15. The predicted values were calculated using the regression line $\hat{y} = -156.879 + 40.1818x$.

x	y	$\hat{y}$	$\bar{y}$	$\hat{y}-\bar{y}$	$(\hat{y}-\bar{y})^2$	$y-\hat{y}$	$(y-\hat{y})^2$	$y-\bar{y}$	$(y-\bar{y})^2$
7.2	116	132.43	184.67	-52.20	2728.67	-16.43	269.94	-68.70	4715.11
7.4	154	140.47	184.67	-44.20	1953.67	13.53	183.16	-30.70	940.44
9.8	245	236.9	184.67	52.24	2728.60	8.097	65.567	60.33	3640.11
9.4	202	220.83	184.67	36.16	1307.78	-18.83	354.57	17.33	300.44
8.8	200	196.72	184.67	12.05	145.303	3.279	10.753	15.33	235.11
8.4	191	180.65	184.67	-4.02	16.15	10.35	107.16	6.33	40.11
51.0	1108	1108.00	1108.00	0.00	8880.17	0.00	991.15	0.00	9871.33

a. The explained variation is $\Sigma(\hat{y}-\bar{y})^2 = 8880.17$
b. The unexplained variation is $\Sigma(y-\hat{y})^2 = 991.15$
c. The total variation is $\Sigma(y-\bar{y})^2 = 9871.33$

d. $r^2 = \Sigma(\hat{y}-\bar{y})^2/\Sigma(y-\bar{y})^2 = 8880.17/9871.33 = 0.900$

e. $s_e^2 = \Sigma(y-\hat{y})^2/(n-2) = 991.15/4 = 247.7875$; $s_e = \sqrt{247.7875} = 15.74$

17. a. $\hat{y} = -0.161601 + 0.0100574x$

$\hat{y}_{187.1} = -0.161601 + 0.0100574(187.1) = 1.7201$, rounded to \$1.72

b. preliminary calculations for n = 6

$\bar{x} = (\Sigma x)/n = 742.7/6 = 123.783$

$n\Sigma x^2 - (\Sigma x)^2 = 6(118115.51) - (742.7)^2 = 157{,}089.77$

$\alpha = 0.05$ and df = n–2 = 4

$$\hat{y} \pm t_{\alpha/2}s_e\sqrt{1 + 1/n + n(x_o - \bar{x})^2/[n\Sigma x^2 - (\Sigma x)^2]}$$

$$\hat{y}_{187.1} \pm t_{0.025}(0.141)\sqrt{1 + 1/6 + 6(187.1-123.783)^2/[157089.77]}$$

$1.7201 \pm (2.776)(0.141)\sqrt{1.31979}$

1.7201 ± 0.4450

$1.27 < y_{187.1} < 2.17$ (dollars)

19. a. $\hat{y} = -156.879 + 40.1818x$

$\hat{y}_{9.0} = -156.879 + 40.1818(9.0) = 204.757$, rounded to 204.8 kg

b. preliminary calculations for n = 6

$\bar{x} = (\Sigma x)/n = 51.0/6 = 8.50$

$n\Sigma x^2 - (\Sigma x)^2 = 6(439.00) - (51.0)^2 = 33.00$

$\alpha = 0.05$ and df = n–2 = 4

$$\hat{y} \pm t_{\alpha/2}s_e\sqrt{1 + 1/n + n(x_o - \bar{x})^2/[n\Sigma x^2 - (\Sigma x)^2]}$$

$$\hat{y}_{187.1} \pm t_{0.025}(15.74)\sqrt{1 + 1/6 + 6(9.0-8.50)^2/[33.00]}$$

$204.757 \pm (2.776)(15.74)\sqrt{1.21212}$

204.757 ± 48.110

$156.6 < y_{9.0} < 252.9$ (kg)

Exercises 21–24 refer to the chapter problem of Table 10-1. Use the following, which are calculated and/or discussed in the text,

$n = 6 \quad \Sigma x = 6.50 \quad \Sigma x^2 = 9.7700 \quad \hat{y} = 0.034560 + 0.945021x \quad s_e = 0.122987$

and the additional values

$\bar{x} = (\Sigma x)/n = 6.50/6 = 1.083333 \qquad n\Sigma x^2 - (\Sigma x)^2 = 6(9.7700) - (6.50)^2 = 16.3700$

NOTE: Using a slightly different regression equation for $\hat{y}$ or a slightly different value for s_e may result in slightly different values in exercises 21-24.

21. $\hat{y}_{2.10} = 0.034560 + 0.945021(2.10) = 2.019$

$\alpha = 0.01$ and df = n–2 = 4

$$\hat{y} \pm t_{\alpha/2}s_e\sqrt{1 + 1/n + n(x_o - \bar{x})^2/[n\Sigma x^2 - (\Sigma x)^2]}$$

$$\hat{y}_{2.10} \pm t_{0.005}(0.122987)\sqrt{1 + 1/6 + 6(2.10-1.083333)^2/[16.3700]}$$

$2.019 \pm (4.604)(0.122987)\sqrt{1.545510}$

2.019 ± 0.704

$1.32 < y_{2.10} < 2.72$ (dollars)

23. $\hat{y}_{0.50} = 0.034560 + 0.945021(0.50) = 0.507$

$\alpha = 0.05$ and df = n–2 = 4

$$\hat{y} \pm t_{\alpha/2}s_e\sqrt{1 + 1/n + n(x_o - \bar{x})^2/[n\Sigma x^2 - (\Sigma x)^2]}$$

$$\hat{y}_{0.50} \pm t_{0.25}(0.122987)\sqrt{1 + 1/6 + 6(0.50 - 1.083333)^2/[16.3700]}$$

$0.507 \pm (2.776)(0.122987)\sqrt{1.291387}$

0.507 ± 0.388

$0.12 < y_{0.50} < 0.89$ (dollars)

25. Use the following, which are calculated and/or discussed in the text,

$n = 6 \quad \Sigma x = 6.50 \quad \Sigma x^2 = 9.7700 \quad \hat{y} = 0.034560 + 0.945021x \quad s_e = 0.122987$

and the additional values

$\bar{x} = (\Sigma x)/n = 6.50/6 = 1.083333 \qquad \Sigma x^2 - (\Sigma x)^2/n = 9.7700 - (6.50)^2/6 = 2.728333$

a. $\alpha = 0.05$ and df = n–2 = 4

$$b_o \pm t_{\alpha/2}s_e\sqrt{1/n + (\bar{x})^2/[\Sigma x^2 - (\Sigma x)^2/n]}$$

$0.034560 \pm t_{0.025}(0.122987)\sqrt{1/6 + (1.083333)^2/[2.728333]}$

$0.034560 \pm (2.776)(0.122987)\sqrt{0.596823}$

0.034560 ± 0.263755

$-0.229 < \beta_o < 0.298$ (dollars)

b. $\alpha = 0.05$ and df = n–2 = 4

$$b_1 \pm t_{\alpha/2}s_e / \sqrt{\Sigma x^2 - (\Sigma x)^2/n}$$

$0.945021 \pm t_{0.025}(0.122987)/\sqrt{2.728333}$

$0.945021 \pm (2.776)(0.122987)/\sqrt{2.728333}$

0.945021 ± 0.206695

$0.738 < \beta_1 < 1.152$ (dollars/dollar)

NOTE: The confidence interval for $\beta_o = y_o$ may also be found as the confidence interval [as distinguished from the prediction interval, see exercise #26] for x = 0.

$\hat{y}_0 = 0.034560 + 0.945021(0) = 0.034560$

$\alpha = 0.05$ and df = n–2 = 4

$$\hat{y} \pm t_{\alpha/2}s_e\sqrt{1/n + n(x_o - \bar{x})^2/[n\Sigma x^2 - (\Sigma x)^2]}$$ modifies to become

$$\hat{y}_0 \pm t_{\alpha/2}s_e\sqrt{1/n + n(\bar{x})^2/[n\Sigma x^2 - (\Sigma x)^2]}$$

$0.034560 \pm t_{0.025}(0.122987)\sqrt{1/6 + (1.083333)^2/[2.728333]}$

$0.034560 \pm (2.776)(0.122987)\sqrt{0.596823}$

0.034560 ± 0.263755

$-0.229 < \beta_o < 0.298$ (dollars)

10-5 Rank Correlation

1. No. When there is a significant linear correlation between two sets of raw data, the linear regression line can be used for prediction using the raw scores. But a significant linear correlation between two sets of ranks does not necessarily mean that there is a linear correlation between the two sets of raw data that produced the ranks, and so the linear regression line should not be used for prediction using the raw scores. Consider, for example, (x,y) pairs that have a perfect quadratic relationship (x,x^2): there is a perfect correlation $r_s = 1.00$ between the ranks, but there is not a linear relationship between the raw scores – and the plot of the (x,y) ordered pairs of raw data is not a straight line.

3. In general, r_s refers to the rank correlation in the sample and ρ_s refers the rank correlation in the population. In this context, r_s is the rank correlation for five recent years between the amount spent on new cars and the amount spent on clothing, and ρ_s is that rank correlation for all years in the population of interest – that population being determined by the details and intention of the study. The subscript "s" is necessary to distinguish rank correlation from the linear correlation of section 10-2. The "s" stands for "Spearman" – and rank correlation is sometimes referred to as Spearman's rank correlation. The subscript "s" used to designate Spearman's rank correlation is not related to the lower case "s" used to designate standard deviation.

5. The scatterplot includes n=6 (x,y) points. The critical values are $r_s = \pm 0.886$.
   Since the points can be joined to form a constantly ascending line from right to left, the ranks of both the x variable and the y variable will be in ascending order from 1 to n. This means that $r_s = 1$.
   Since $1 > 0.866$, reject H_o:$\rho_s = 0$ and conclude $\rho_s \neq 0$ (in fact, that $\rho_s > 0$). There is sufficient evidence to support the claim that there is a correlation between distance and time.

7. a. From Table A-6, the critical values are ± 0.521.
   b. From Table A-6, the critical values are ± 0.521.
   c. From Formula 10-7, the critical values are $\pm z_{\alpha/2}/\sqrt{n\text{-}1} = \pm 1.960/\sqrt{99} = \pm 0.197$.
   d. From Formula 10-7, the critical values are $\pm z_{\alpha/2}/\sqrt{n\text{-}1} = \pm\ 2.575/\sqrt{64} = \pm 0.322$.

NOTE: The rank correlation is correctly calculated using the ranks in the Pearson product moment correlation Formula 10-1 to produce

$$r_s = [n\Sigma R_x R_y - (\Sigma R_x)(\Sigma R_y)]/[\sqrt{n(\Sigma R_x^2) - (\Sigma R_x)^2}\sqrt{n(\Sigma R_y^2) - (\Sigma R_y)^2}]$$

Since $\Sigma Rx = \Sigma Ry = 1 + 2 + \ldots + n = n(n+1)/2$ [always]
and $\Sigma R_x^2 = \Sigma R_y^2 = 1^2 + 2^2 + \ldots + n^2 = n(n+1)(2n+1)/6$ [when there are no ties in the ranks]
it can be shown by algebra that the above formula can be shortened to

$$r_s = 1 - [6(\Sigma d^2)]/[n(n^2\text{-}1)]$$ when there are no ties in the ranks.

This manual calculates $d = R_x - R_y$, thus preserving the sign of d. This convention means that $\Sigma d = 0$ must always be true and provides a check for the assigning and differencing of the ranks – and it must also be true that $\Sigma R_x = \Sigma R_y = n(n+1)/2$.

Unless there is an excessive number of ties, the difference between using Formula 10-1 and the shortcut formula is usually trivial – in fact, some textbooks give only the shortcut formula and ignore the fact that it does not give the exact answer when there are ties. A related item ignored by virtually every textbook is the fact that the critical values in Table A-6 are not exact when there are ties.

In addition, some exercises in this section appeared previously in a parametric context in section 10-2. Because comparison of the parametric and non-parametric results are usually informative and provide statistical insights, we list those pairings.

this section	#15	#16	#17	#18
section 10-2	#17	#26	#31	#32

9. The following table summarizes the calculations.

R_x	R_y	d	d^2
1	6	-5	25
3	4	-1	1
4	5	-1	1
7	1	6	36
5	3	2	4
6	2	4	16
2	7	-5	25
28	28	0	108

$r_s = 1 - [6(\Sigma d^2)]/[n(n^2-1)]$
$= 1 - [6(108)]/[7(48)]$
$= 1 - 1.929$
$= -0.929$

H_o: $\rho_s = 0$
H_1: $\rho_s \neq 0$
$\alpha = 0.05$
C.V. $r_s = \pm 0.786$
calculations:
$r_s = -0.929$

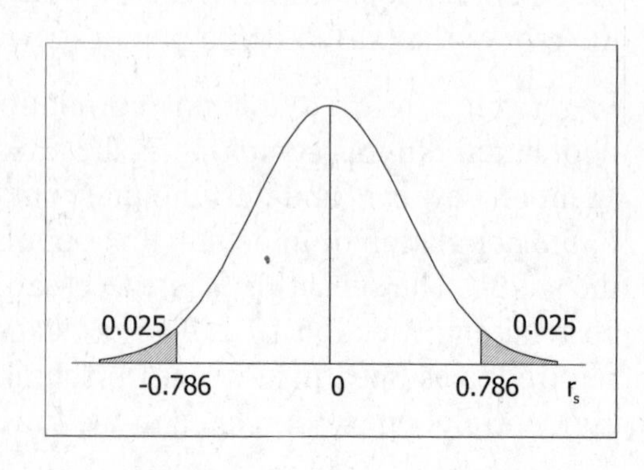

conclusion:

Reject H_o; there is sufficient evidence to reject the claim that $\rho_s = 0$ and to conclude that $\rho_s \neq 0$ (in fact, that $\rho_s < 0$). There is sufficient evidence to support the claim that there is a correlation between the rankings of the two judges. In this case the correlation is negative, and so the judges rank very differently – in general, the first judge gave high ranks to those bands that received low ranks from the second judge, and vice-versa.

11. The following table summarizes the calculations.

R_x	R_y	d	d^2
1	6	-5	25
2	2	0	0
3	10	-7	49
4	4	0	0
5	12	-7	49
6	9	-3	9
7	8	-1	1
8	7	1	1
9	1	8	64
10	5	5	25
11	3	8	64
12	11	1	1
78	78	0	288

$r_s = 1 - [6(\Sigma d^2)]/[n(n^2-1)]$
$= 1 - [6(288)]/[12(143)]$
$= 1 - 1.007$
$= -0.007$

H_o: $\rho_s = 0$
H_1: $\rho_s \neq 0$
$\alpha = 0.05$
C.V. $r_s = \pm 0.587$
calculations:
$r_s = -0.007$

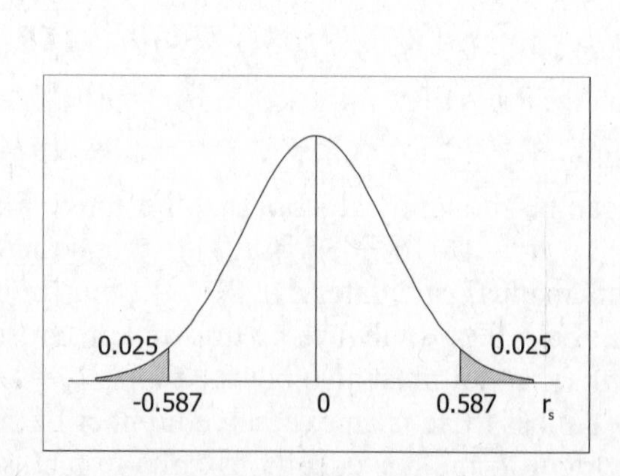

conclusion:

Do not reject H_o; there is not sufficient evidence to reject the claim that $\rho_s = 0$. There is not sufficient evidence to support the claim that there is a correlation between the conviction rate and the recidivism rate. No; recidivism rates do not appear to be related to conviction rates.

13. The following table summarizes the calculations.
Let x be the quality. Give the highest quality rank 1 and give the highest cost rank 1.

x	R_x	y	R_y
74	1	27	3.5
71	2	30	2
68	3	38	1
65	4	23	5.5
63	5	20	7.5
62	6	13	10.5
59	7	27	3.5
57	8.5	23	5.5
57	8.5	14	9
53	10	13	10.5
51	11	20	7.5
	66		66

$\Sigma R_xR_x = 505.50$
$\Sigma R_yR_y = 504.00$
$\Sigma R_xR_y = 468.25$

H_o: $\rho_s = 0$
H_1: $\rho_s \neq 0$
$\alpha = 0.05$
C.V. $r_s = \pm 0.618$
calculations:

$$r_s = [n\Sigma R_xR_y - (\Sigma R_x)(\Sigma R_y)]/ [\sqrt{n(\Sigma R_x^2) - (\Sigma R_x)^2} \sqrt{n(\Sigma R_y^2) - (\Sigma R_y)^2}]$$
$$= [11(468.25) - (66)(66)]/ [\sqrt{11(505.50) - (66)^2} \sqrt{11(504.00) - (66)^2}]$$
$$= [794.75]/[\sqrt{1204.50}\sqrt{1188.00}] = 0.664$$

conclusion:
Reject H_o; there is sufficient evidence to reject the claim that $\rho_s = 0$ and conclude that $\rho_s \neq 0$ (in fact, that $\rho_s > 0$). There is sufficient evidence to support the claim that there is a correlation between the quality and price of LCD TV's. Yes; based on these results you can expect to get higher quality by purchasing a more expensive LCD TV.

15. The following table summarizes the calculations.
Let x be the overhead width

x	R_x	y	R_y	d	d^2
7.2	6	116	6	0	0
7.4	5	154	5	0	0
9.8	1	245	1	0	0
9.4	2	202	2	0	0
8.8	3	200	3	0	0
8.4	4	191	4	0	0
	21		21	0	0

$r_s = 1 - [6(\Sigma d^2)]/[n(n^2-1)]$
$= 1 - [6(0)]/[6(35)]$
$= 1 - 0$
$= 1$

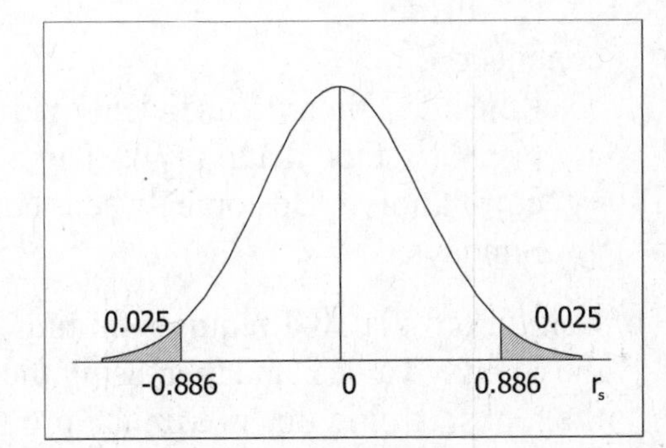

H_o: $\rho_s = 0$
H_1: $\rho_s \neq 0$
$\alpha = 0.05$
C.V. $r_s = \pm 0.886$
calculations:
$r_s = 1$
conclusion:
Reject H_o; there is sufficient evidence to reject the claim that $\rho_s = 0$ and to conclude that $\rho_s \neq 0$ (in fact, that $\rho_s > 0$). There is sufficient evidence to support the claim that there is a correlation between the overhead width of seals and the weights of the seals.

17. There are n=56 male-female pairs in the first two columns of Data Set 4.
    From Formula 10-7, the critical values are $\pm z_{\alpha/2}/\sqrt{n-1} = \pm 1.960/\sqrt{55} = \pm 0.264$.
    The Minitab correlation command on the ranked data yields the following.

```
Pearson correlation of men and women = 0.231
```

H_o: $\rho_s = 0$
H_1: $\rho_s \neq 0$
$\alpha = 0.05$
C.V. $r_s = \pm 0.264$
calculations:
    $r_s = 0.231$

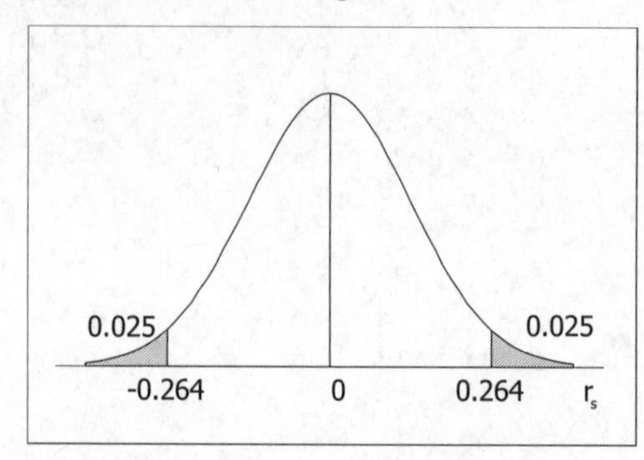

conclusion:
    Do not reject H_o; there is not sufficient evidence to reject the claim that $\rho_s = 0$. There is not sufficient evidence to conclude that there is a correlation between the numbers of words spoken by men and women who are in a couple relationship.

19. There are n=52 Monday-Sunday pairs in Data Set 14.
    From Formula 10-7, the critical values for α=0.05 are $\pm z_{\alpha/2}/\sqrt{n-1} = \pm 1.960/\sqrt{51} = \pm 0.274$.
    a. Using the shortcut method requires finding the ranks for the Monday values, finding the ranks for the Sunday values, subtracting to find the difference in ranks for each pair, and then summing the squares of the differences. Doing this in Minitab produces the following.

```
Sum of d-squared = 14415
```

Using the shortcut formula

$$r_s = 1 - [6(\Sigma d^2)]/[n(n^2\text{-}1)]$$
$$= 1 - [6(14415)]/[52(2703)]$$
$$= 1 - 0.615$$
$$= 0.385$$

H_o: $\rho_s = 0$
H_1: $\rho_s \neq 0$
$\alpha = 0.05$ [assumed]
C.V. $r_s = \pm 0.274$
calculations:
    $r_s = 0.385$
conclusion:
    Reject H_o; there is sufficient evidence to reject the claim that $\rho_s = 0$ and conclude that $\rho_s \neq 0$ (in fact, that $\rho_s > 0$). There is sufficient evidence to conclude that there is a correlation in Boston between the amounts of rainfall on Mondays and the following Sundays.

b. Using Formula 10-1 requires finding the ranks for the Monday values, finding the ranks for the Sunday values, and then using the regular command for Pearson's linear correlation on the ranks. Doing this in Minitab produces the following.

```
Pearson correlation of Monday and Sunday = 0.109
```

H_o: $\rho_s = 0$
H_1: $\rho_s \neq 0$
$\alpha = 0.05$ [assumed]
C.V. $r_s = \pm 0.274$
calculations:
$r_s = 0.109$

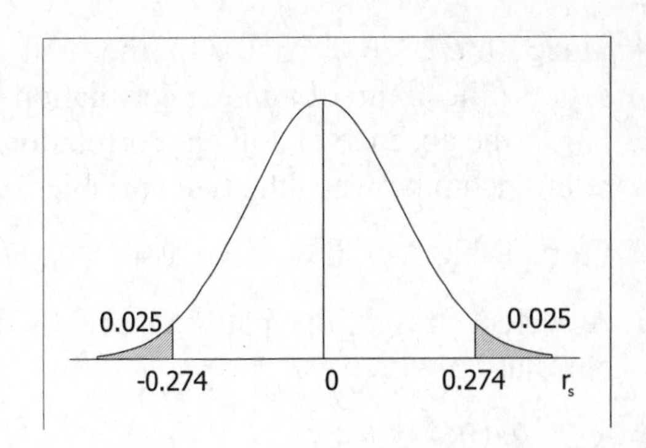

conclusion:
Do not reject H_o; there is not sufficient evidence to reject the claim that $\rho_s = 0$. There is not sufficient evidence to conclude that there is a correlation in Boston between the amounts of rainfall on Mondays and the following Sundays.

c. In this instance there is a substantial difference between the two tests – caused by the high number of individual ties, and especially by the many (0,0) pairs. The two conclusions are different. The second method (using Formula 10-1, the Pearson linear correlation formula, on the ranks) always gives the correct answer and is the preferred method. The first method (using the d^2 values) exists only as a computational shortcut for Formula 10-1 – a shortcut that always gives the same answer when there are no ties, but that fails to give the correct answer when there are ties.

Statistical Literacy and Critical Thinking

1. Section 9-4 deals with making inferences about the mean of the differences between matched pairs and requires that each member of the pair have the same unit of measurement. Section 10-2 deals with making inference about the relationship between the members of the pairs and does not require that each member of the pair have the same unit of measurement.

2. Yes; since 0.963 > 0.279 (the C.V. from Table A-5), there is sufficient evidence to support the claim of a linear correlation between chest size and weight. No; the conclusion is only that larger chest sizes are associated with larger weights – not that there is a cause and effect relationship, and not that the direction of any cause and effect relationship can be identified.

3. No; a perfect positive correlation means only that larger values of one variable are associated with larger values of the other variable and that the value of one of the variables can be perfectly predicted from the value of the other. A perfect correlation does not imply equality between the paired values, or even that the paired values have the same unit of measurement.

4. No; a value of r=0 suggests only that there is no linear relationship between the two variables, but the two variables may be related in some other manner.

Chapter Quick Quiz

1. If the calculation indicate that r = 2.650, then an error has been made. For any set of data, it must be true that $-1 \leq r \leq 1$.

2. Since 0.989 > 0.632 (the C.V. from Table A-5), there is sufficient evidence to support the claim of a linear correlation between the two variables.

3. True.

4. Since -0.632 < 0.099 < 0.632 (the C.V.'s from Table A-5), there is not sufficient evidence to support the claim of a linear correlation between the two variables.
5. False; the absence of a linear correlation does not preclude the existence of another type of relationship between the two variables.
6. From Table A-5, C.V. = ±0.514.
7. A perfect straight line pattern that falls from left to right describes a perfect negative correlation with r = -1.
8. $\hat{y}_{10} = 2(10) - 5 = 15$
9. The proportion of the variation in y that is explained by the linear relationship between x and y is $r^2 = (0.400)^2 = 0.160$, or 16%.
10. False; the conclusion is only that larger amounts of salt consumption are associated with higher measures of blood pressure – not that there is a cause and effect relationship, and not that the direction of any cause and effect relationship can be identified.

Review Exercises

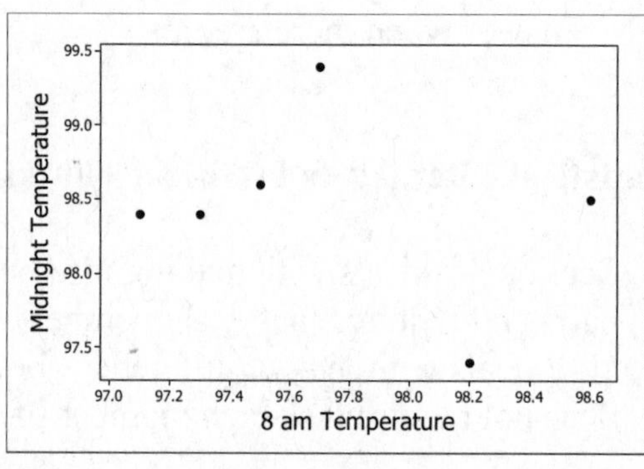

1. These are the necessary summary statistics.
   $n = 6$
   $\Sigma x = 586.4$
   $\Sigma y = 590.7$
   $\Sigma x^2 = 57312.44$
   $\Sigma y^2 = 58156.45$
   $\Sigma xy = 57730.62$
   $n(\Sigma x^2) - (\Sigma x)^2 = 6(57312.44) - (586.4)^2 = 9.68$
   $n(\Sigma y^2) - (\Sigma y)^2 = 6(58156.45) - (590.7)^2 = 12.21$
   $n(\Sigma xy) - (\Sigma x)(\Sigma y) = 6(57730.62) - (586.4)(590.7) = -2.76$

   a. The scatterplot is given above at the right. The scatterplot suggests that there is not a linear relationship between the two variables.

   b. $r = [n(\Sigma xy) - (\Sigma x)(\Sigma y)]/[\sqrt{n(\Sigma x^2) - (\Sigma x)^2}\sqrt{n(\Sigma y^2) - (\Sigma y)^2}]$
   $= -2.76/[\sqrt{9.68}\sqrt{12.21}] = -0.254$

   H_o: $\rho = 0$
   H_1: $\rho \neq 0$
   $\alpha = 0.05$ [assumed] and df = 4
   C.V. $t = \pm t_{\alpha/2} = \pm t_{0.025} = \pm 2.776$ [or $r = \pm 0.811$]
   calculations:
   $t_r = (r - \mu_r)/s_r$
   $= (-0.254 - 0)/\sqrt{[1 - (-0.254)^2]/4}$
   $= -0.254/0.4836 = -0.525$
   P-value = 2·tcdf(-99,-0.525,4) = 0.6274

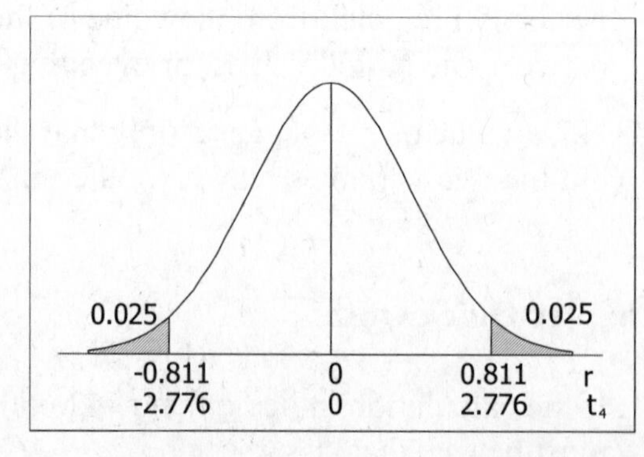

   conclusion:
   Do not reject H_o; there is not sufficient evidence to conclude that $\rho \neq 0$. No; there is not sufficient evidence to support the claim of a linear correlation between the 8 am and midnight temperatures.

c. $\bar{x} = (\Sigma x)/n = 586.4/6 = 97.73$ $\bar{y} = (\Sigma y)/n = 590.7/6 = 98.45$

$b_1 = [n(\Sigma xy) - (\Sigma x)(\Sigma y)]/[n(\Sigma x^2) - (\Sigma x)^2]$
$= -2.76/9.68 = -0.2851$

$b_o = \bar{y} - b_1\bar{x}$
$= 98.45 - (-0.2851)(97.73) = 126.32$

$\hat{y} = b_o + b_1 x$
$= 126.32 - 0.2851x$

d. $\hat{y}_{98.3} = \bar{y} = 98.45$ °F [no significant correlation]

2. a. Yes. Assuming $\alpha = 0.05$, Table A-5 indicates C.V. $= \pm 0.312$. Since $0.522 > 0.312$, there is sufficient evidence to support a claim of a linear correlation between heights and weights of males.

b. $r^2 = (0.522)^2 = 0.272$, or 27.2%

c. $\hat{y} = -139 + 4.55x$

d. $\hat{y}_{72} = -139 + 4.55(72) = 188.6$ lbs

3. These are the necessary summary statistics.

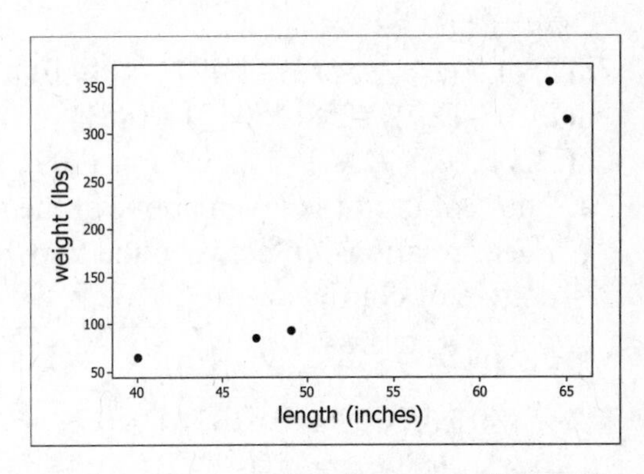

$n = 5$
$\Sigma x = 265$
$\Sigma y = 917$
$\Sigma x^2 = 14531$
$\Sigma y^2 = 247049$
$\Sigma xy = 54572$
$n(\Sigma x^2) - (\Sigma x)^2 = 5(14531) - (265)^2 = 2430$
$n(\Sigma y^2) - (\Sigma y)^2 = 5(247049) - (917)^2 = 394356$
$n(\Sigma xy) - (\Sigma x)(\Sigma y) = 5(54572) - (265)(917) = 29855$

a. The scatterplot is given above at the right. The scatterplot suggests that there is a linear relationship between the two variables.

b. $r = [n(\Sigma xy) - (\Sigma x)(\Sigma y)]/[\sqrt{n(\Sigma x^2) - (\Sigma x)^2}\sqrt{n(\Sigma y^2) - (\Sigma y)^2}]$
$= 29855/[\sqrt{2430}\sqrt{394356}] = 0.964$

H_o: $\rho = 0$
H_1: $\rho \neq 0$
$\alpha = 0.05$ [assumed] and df = 3
C.V. $t = \pm t_{\alpha/2} = \pm t_{0.025} = \pm 3.182$ [or $r = \pm 0.878$]

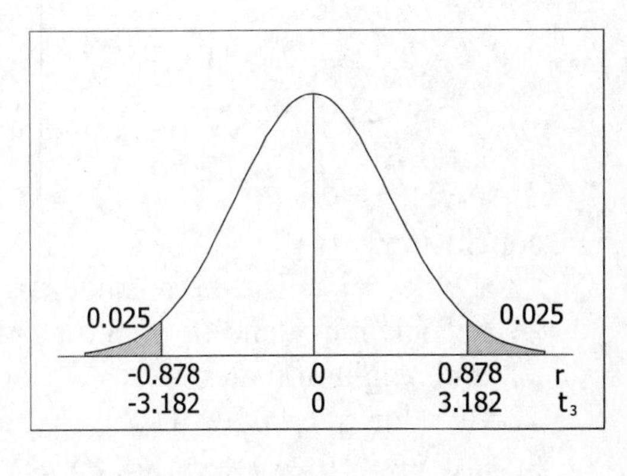

calculations:
$t_r = (r - \mu_r)/s_r$
$= (0.964 - 0)/\sqrt{[1 - (0.964)^2]/3}$
$= 0.964/0.1526$
$= 6.319$

P-value = 2·tcdf(6.319,99,3) = 0.0080

conclusion:
Reject H_o; there is sufficient evidence to conclude that $\rho \neq 0$ (in fact, that $\rho > 0$). Yes; there is sufficient evidence to support the claim of a linear correlation between the lengths and weights of bears.

c. $\bar{x} = (\Sigma x)/n = 265/5 = 53.0$ $\bar{y} = (\Sigma y)/n = 917/5 = 183.4$

$b_1 = [n(\Sigma xy) - (\Sigma x)(\Sigma y)]/[n(\Sigma x^2) - (\Sigma x)^2]$
$= 29855/2430 = 12.286$

$b_o = \bar{y} - b_1\bar{x}$
$= 183.4 - (12.286)(53.0) = -467.8$

$\hat{y} = b_o + b_1 x$
$= -467.8 + 12.286x$

d. $\hat{y}_{72} = -467.8 + 12.286(72) = 416.8$ lbs

4. These are the necessary summary statistics, where x = leg and y = height.

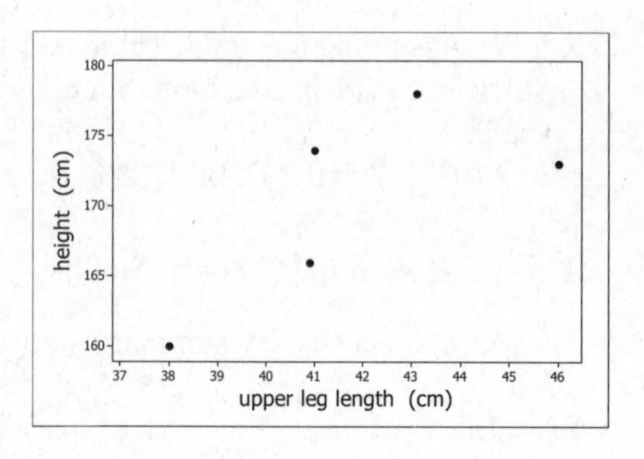

$n = 5$
$\Sigma x = 209.0$
$\Sigma y = 851$
$\Sigma x^2 = 8771.42$
$\Sigma y^2 = 145045$
$\Sigma xy = 35633.2$
$n(\Sigma x^2) - (\Sigma x)^2 = 5(8771.42) - (209.0)^2 = 176.10$
$n(\Sigma y^2) - (\Sigma y)^2 = 5(145045) - (851)^2 = 1024$
$n(\Sigma xy) - (\Sigma x)(\Sigma y) = 5(35633.2) - (209.0)(851) = 307.0$

a. The scatterplot is given above at the right. The scatterplot suggests that there may be a linear relationship between the two variables, but only a formal test determine that with any degree of conficence.

b. $r = [n(\Sigma xy) - (\Sigma x)(\Sigma y)]/[\sqrt{n(\Sigma x^2) - (\Sigma x)^2}\sqrt{n(\Sigma y^2) - (\Sigma y)^2}]$
$= 307.0/[\sqrt{176.10}\sqrt{1024}] = 0.723$

H_o: $\rho = 0$
H_1: $\rho \neq 0$
$\alpha = 0.05$ [assumed] and df = 3
C.V. $t = \pm t_{\alpha/2} = \pm t_{0.025} = \pm 3.182$ [or $r = \pm 0.878$]
calculations:

$t_r = (r - \mu_r)/s_r$
$= (0.723 - 0)/\sqrt{[1 - (0.723)^2]/3}$
$= 0.723/0.3989$
$= 1.812$

P-value = 2·tcdf(1.812,99,3) = 0.1676

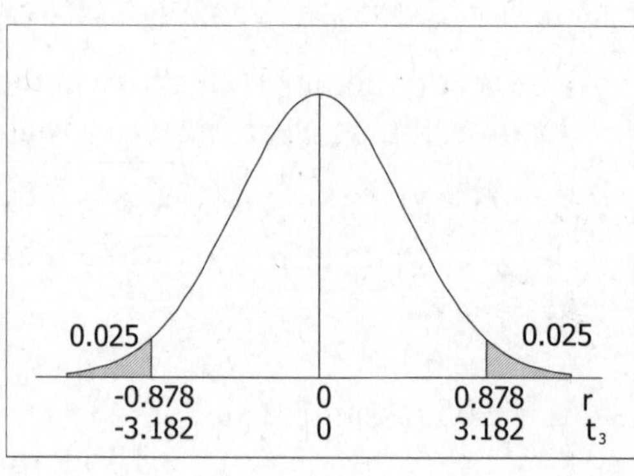

conclusion:

Do not reject H_o; there is not sufficient evidence to conclude that $\rho \neq 0$. No; there is not sufficient evidence to support a claim of a linear correlation between upper leg length and height of males.

c. $\bar{x} = (\Sigma x)/n = 209.0/5 = 41.80$ $\bar{y} = (\Sigma y)/n = 170.2$

$b_1 = [n(\Sigma xy) - (\Sigma x)(\Sigma y)]/[n(\Sigma x^2) - (\Sigma x)^2] = 307.0/176.10 = 1.743$
$b_o = \bar{y} - b_1\bar{x} = 170.2 - (1.743)(41.80) = 97.329$
$\hat{y} = b_o + b_1 x = 97.33 + 1.743x$

d. $\hat{y}_{45} = \bar{y} = 170.2$ cm [no significant correlation]

5. The following table summarizes the calculations.

R_x	R_y	d	d^2
1	1	0	0
2	2	0	0
3	5	-2	4
4	4	0	0
5	7	-2	4
6	6	0	0
7	3	4	16
8	8	0	0
36	36	0	24

$r_s = 1 - [6(\Sigma d^2)]/[n(n^2-1)]$
$= 1 - [6(24)]/[8(63)]$
$= 1 - 0.286$
$= 0.714$

H_o: $\rho_s = 0$
H_1: $\rho_s \neq 0$
$\alpha = 0.05$
C.V. $r_s = \pm 0.738$
calculations:
$r_s = 0.714$
conclusion:
Do not reject H_o; there is not sufficient evidence to reject the claim that $\rho_s = 0$. There is not sufficient evidence to support the claim that there is a correlation between the rankings of the students and the ranking of the magazine.

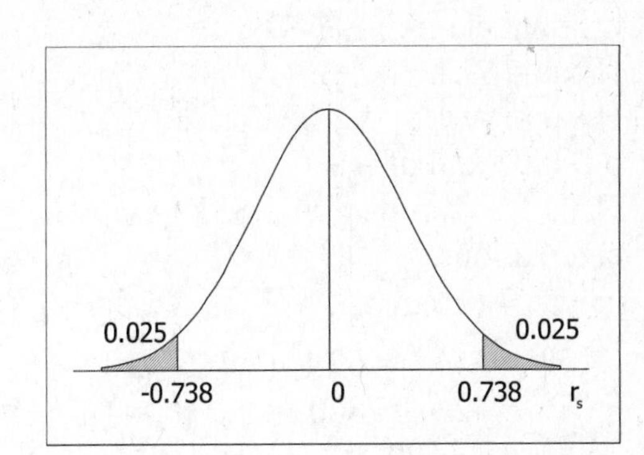

Cumulative Review Exercises

The following summary statistics apply to exercises 1-6. The ordered heights are as follows.
1877: 62 64 65 65 66 66 67 68 68 71
recent: 62 63 66 68 68 69 69 71 72 73
Let the 1877 heights be group 1.

group 1: 1877 (n=10)
$\Sigma x = 662$
$\Sigma x^2 = 43{,}880$
$\bar{x} = 66.2$
$s^2 = 6.178$ (s=2.486)

group 2: recent (n=10)
$\Sigma x = 681$
$\Sigma x^2 = 46{,}493$
$\bar{x} = 68.1$
$s^2 = 12.989$ (s=3.604)

$\bar{x}_1 - \bar{x}_2 = 66.2 - 68.1 = -1.9$

1. For 1877: $\bar{x} = 66.2$ inches $\tilde{x} = (66+66)/2 = 66.0$ inches $s = 2.5$ inches
For recent: $\bar{x} = 68.1$ inches $\tilde{x} = (68+69)/2 = 68.5$ inches $s = 3.6$ inches

2. original claim: $\mu_1 - \mu_2 < 0$
H_o: $\mu_1 - \mu_2 = 0$
H_1: $\mu_1 - \mu_2 < 0$
$\alpha = 0.05$ and df = 9
C.V. $t = -t_\alpha = -t_{0.05} = -1.833$
calculations:
$t_{\bar{x}_1-\bar{x}_2} = (\bar{x}_1 - \bar{x}_2 - \mu_{\bar{x}_1-\bar{x}_2})/s_{\bar{x}_1-\bar{x}_2}$
$= (-1.9 - 0)/\sqrt{6.178/10 + 12.989/10}$
$= -1.9/1.3844$
$= -1.372$
P-value = tcdf(-99,-1.372,9) = 0.1016

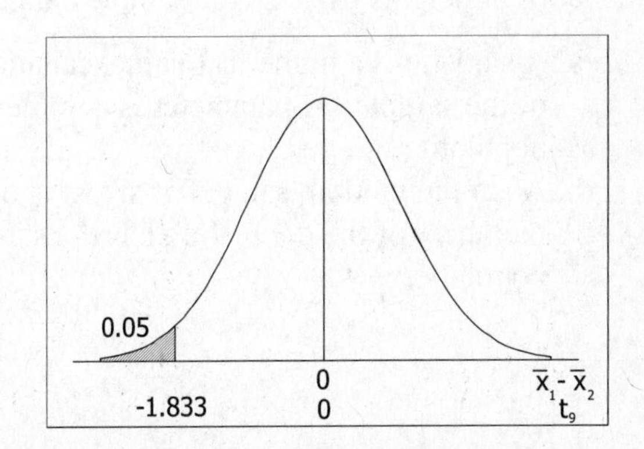

conclusion:

Do not reject H_o; there is not sufficient evidence to conclude that $\mu_1 - \mu_2 < 0$. There is not sufficient evidence to support the claim that the males in 1877 had a mean height that is less than the mean height of males today.

3. original claim: $\mu < 69.1$
   H_o: $\mu = 69.1$
   H_1: $\mu < 69.1$
   $\alpha = 0.05$ and df = 9
   C.V. $t = -t_\alpha = -t_{0.05} = -1.833$
   calculations:
   $t_{\bar{x}} = (\bar{x} - \mu)/s_{\bar{x}}$
   $= (66.2 - 69.1)/(2.486/\sqrt{10})$
   $= -2.9/0.7860$
   $= -3.690$

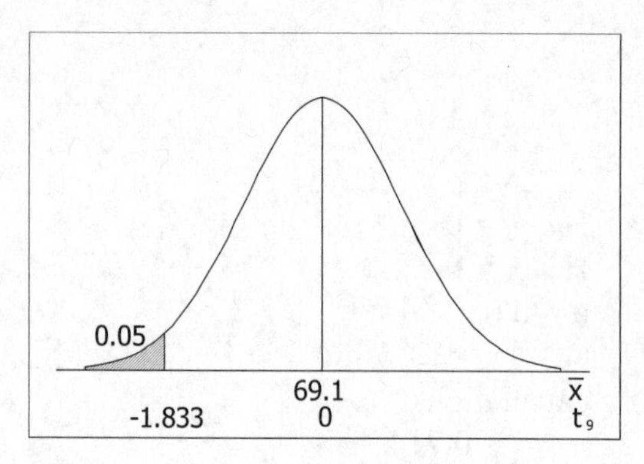

P-value = $P(t_9 < -3.690)$ = tcdf(-99,-3.690,9) = 0.0025
   conclusion:

Reject H_o; there is sufficient evidence to conclude that $\mu < 69.1$. There is sufficient evidence to support the claim that the men from 1877 have a mean height that is less than 69.1 inches.

4. σ unknown (and assuming the distribution is approximately normal), use t with df=9
   $\alpha = 0.05$, $t_{df,\alpha/2} = t_{9,0.05} = 2.262$
   $\bar{x} \pm t_{\alpha/2} \cdot s/\sqrt{n}$
   $66.2 \pm 2.262(2.486)/\sqrt{10}$
   66.2 ± 1.8
   $64.4 < \mu < 68.0$ (inches)

5. $\alpha = 0.05$ and df = 9
   $(\bar{x}_1 - \bar{x}_2) \pm t_{\alpha/2} s_{\bar{x}_1 - \bar{x}_2}$
   $-1.9 \pm 2.262\sqrt{6.178/10 + 12.989/10}$
   -1.9 ± 3.1
   $-5.0 < \mu_1 - \mu_2 < 1.2$ (inches)
   Yes; the confidence interval includes the value 0. Since the confidence interval includes the value 0, we cannot reject the notion that the two populations may have the same mean.

6. It would not be appropriate to test for a linear correlation between heights from 1877 and current heights because the sample data are not matched pairs, as required for that test.

7. a. A statistic is a numerical value, calculated from sample data, that describes a characteristic of the sample. A parameter is a numerical value that describes a characteristic of the population.
   b. A simple random sample of size n is one chosen in such a way that every group of n members of the population has the same chance of being selected as the sample from that population.

c. A voluntary response sample is one in which the respondents themselves decide whether or not to be included. Such samples are generally unsuited for making inferences about populations because they are not likely to be representative of the population. In general, those with a strong interest in the topic are more likely to make the effort to include themselves in the sample – and the sample will contain an over-representation of persons with a strong interest in the topic, and an under-representation of persons with little or no interest in the topic.

8. Yes; since 40 is (40-26)/5 = 2.8 standard deviation from the mean, it is considered an outlier. In general, any observation more than 2 standard deviations from the mean (which typically accounts for the most extreme 5% of the observations) is considered an outlier.

9. a. Use $\mu = 26$ and $\sigma = 5$.

$$P(x>28) = P(z>0.40) = 1 - 0.6554 = 0.3446$$

b. Use $\mu_{\bar{x}} = \mu = 26$ and $\sigma_{\bar{x}} = \sigma/\sqrt{n} = 5/\sqrt{16} = 1.25$.

$$P(\bar{x}>28) = P(z>1.60) = 1 - 0.9452 = 0.0548$$

10. For independent events, $P(G_1 \text{ and } G_2 \text{ and } G_3 \text{ and } G_4) = P(G_1)\cdot P(G_2)\cdot P(G_3)\cdot P(G_4)$
$= (0.12)(0.12)(0.12)(0.12)$
$= 0.000207$

Because the probably of getting four green-eyed persons by random selection is so small, it appears that the researcher (either knowingly or unknowingly) did not make the selections at random from the population.

Chapter 11

Chi-Square and Analysis of Variance

11-2 Goodness-of-Fit

1. When digits are randomly generated, they should form a uniform distribution – i.e., a distribution in which each of the digits is equally likely. The test for goodness-of-fit is a test of the hypothesis that the sample data fit the uniform distribution.

3. O represents the observed frequencies. The twelve values for O are 5, 8, 7, 9, 13, 17, 11, 10, 10, 12, 8, 10. E represents the expected frequencies. The twelve values for E dependent on the expected (i.e., the hypothesized) distribution. If the hypothesized distribution is that weddings occur in different months with the same frequency, we expect the $\Sigma O = 120$ observations to be equally distributed among the 12 months so that each of the twelve values for E is $120/12 = 12$.

NOTE FOR EXERCISES 5 AND 6: The principle that the null hypothesis must contain the equality still applies. When the original claim is given in words, it will reduce either to a statement that the observed results fit or match (i.e., are equal to) some specified distribution or to a statement that the observed results are different from some specified distribution. In the first case the original claim is the null hypothesis, in the second case the original claim is the alternative hypothesis.

5. original claim: the actual outcomes agree with the expected frequencies
   H_o: the actual outcomes agree with the expected frequencies
   H_1: at least one outcome is not as expected
   $\alpha = 0.05$ and df = 9
   C.V. $\chi^2 = \chi^2_\alpha = \chi^2_{0.05} = 16.919$
   calculations:
   $\chi^2 = \Sigma[(O - E)^2/E] = 8.185$
   P-value = χ^2cdf (8.185,99,9) = 0.5156
   conclusion:
   Do not reject Ho; there is not sufficient evidence to reject the claim that the actual outcomes agree with the expected frequencies. There is no reason to say the slot machine is not functioning as expected.

NOTES FOR THE REMAINING EXERCISES IN THIS SECTION:

(1) In goodness-of-fit problems, always verify that $\Sigma E = \Sigma O$ before proceeding. If these sums are not equal, then an error has been made and further calculations have no meaning.
(2) As in the previous uses of the χ^2 distribution, the accompanying illustrations follow the "usual" shape – even though that shape is not correct for df = 1 and df = 2.
(3) As in the previous sections, a subscript (the df) may be used to identify which χ^2 distribution to use in the tables.
(4) Exact P-values are found as in previous sections using the χ^2cdf command on the TI-83/84$^+$ calculator.

7. original claim: the four categories are equally likely
   H_o: $p_1 = p_2 = p_3 = p_4 = 0.25$
   H_1: at least one $p_i \neq 0.25$
   $\alpha = 0.05$ and df = 3
   C.V. $\chi^2 = \chi^2_\alpha = \chi^2_{0.05} = 7.815$
   calculations:

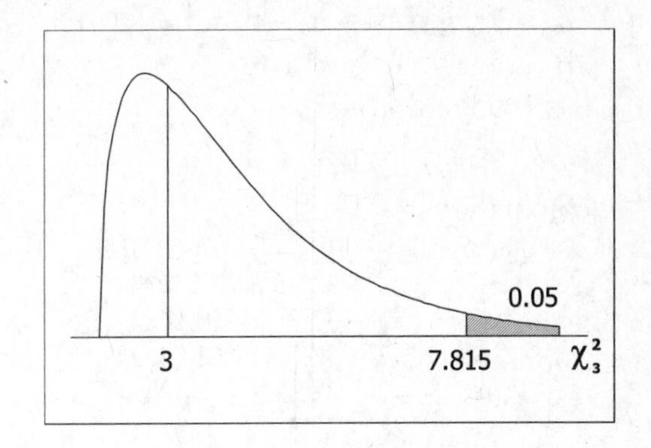

outcome	O	E	$(O-E)^2/E$
0-24	61	25	51.840
25-49	17	25	2.560
50-74	10	25	9.000
75-99	12	25	6.760
	100	100	70.160

   $\chi^2 = \Sigma[(O - E)^2/E] = 70.160$
   P-value = χ^2cdf (70.160,99,3) = 3.9E-15 ≈ 0
   conclusion:
   Reject Ho; there is sufficient evidence to reject the claim that $p_1 = p_2 = p_3 = p_4 = 0.25$. There is sufficient evidence to reject the claim that the four categories are equally likely. Yes; informally speaking, the results appear to support the statement that the frequency for the first category is disproportionately high. [See the following NOTE.]

NOTE: Formally, the goodness-of-fit test can only consider two cases: either the observed frequencies fit the expected frequencies (which must be the null hypothesis), or the observed frequencies do not fit the expected frequencies (which must be the alternative hypothesis). The test cannot address specific variations of the alternative hypothesis – i.e., it cannot address specific manners in which the observed frequencies do not fit the expected frequencies. When the null hypothesis of a goodness-of-fit is rejected, the conclusion can be only that at least one of the observed frequencies (or proportions) is not as expected – and the test cannot identify in which one(s) of the categories that discrepancy occurs.

9. original claim: the four categories are equally likely
   H_o: $p_1 = p_2 = p_3 = p_4 = 0.25$
   H_1: at least one $p_i \neq 0.25$
   $\alpha = 0.05$ and df = 3
   C.V. $\chi^2 = \chi^2_\alpha = \chi^2_{0.05} = 7.815$
   calculations:

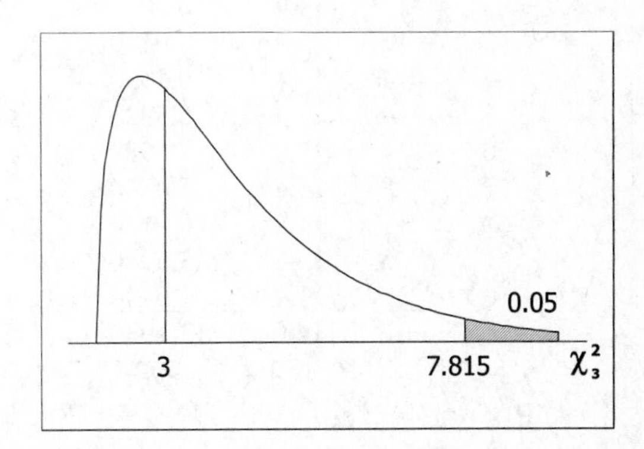

outcome	O	E	$(O-E)^2/E$
0-24	33	25	2.560
25-49	16	25	3.240
50-74	23	25	0.160
75-99	28	25	0.360
	100	100	6.320

   $\chi^2 = \Sigma[(O - E)^2/E] = 6.320$
   P-value = χ^2cdf (6.320,99,3) = 0.0970
   conclusion:
   Do not reject Ho; there is not sufficient evidence to reject the claim that $p_i = 0.25$ for each category. There is not sufficient evidence to reject the claim that the four categories are equally likely. No; the results appear to support the statement that the frequency for the first category is disproportionately high. [See the NOTE following Exercise 7.]

11. original claim: the outcomes are not equally likely

H_o: $p_1 = p_2 = p_3 = p_4 = p_5 = p_6 = 1/6$
H_1: at least one $p_i \neq 1/6$
$\alpha = 0.05$ and df = 5
C.V. $\chi^2 = \chi^2_\alpha = \chi^2_{0.05} = 11.071$
calculations:

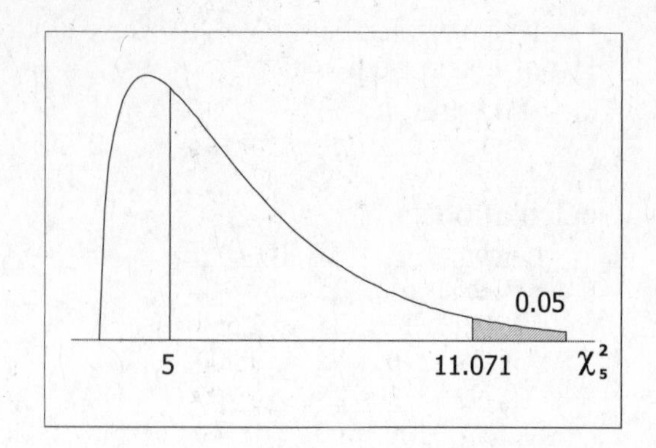

outcome	O	E	(O-E)²/E
1	27	33.33	1.2033
2	31	33.33	0.1633
3	42	33.33	2.2533
4	40	33.33	1.3333
5	28	33.33	0.8533
6	32	33.33	0.0533
	200	200.00	5.8600

$\chi^2 = \Sigma[(O - E)^2/E] = 5.860$
P-value = χ^2cdf (5.86,99,5) = 0.3201

conclusion:

Do not reject Ho; there is not sufficient evidence to support the claim that $p_i \neq 1/6$ for at least one outcome. There is not sufficient evidence to support the claim that the outcomes of the loaded die are not equally likely. No; it does not appear that the loaded die behaves differently than a fair die.

13. original claim: the likelihood of winning is the same for all post positions

H_o: $p_1 = p_2 = p_3 = \ldots = p_{10} = 1/10$
H_1: at least one $p_i \neq 1/10$
$\alpha = 0.05$ and df = 9
C.V. $\chi^2 = \chi^2_\alpha = \chi^2_{0.05} = 16.919$
calculations:

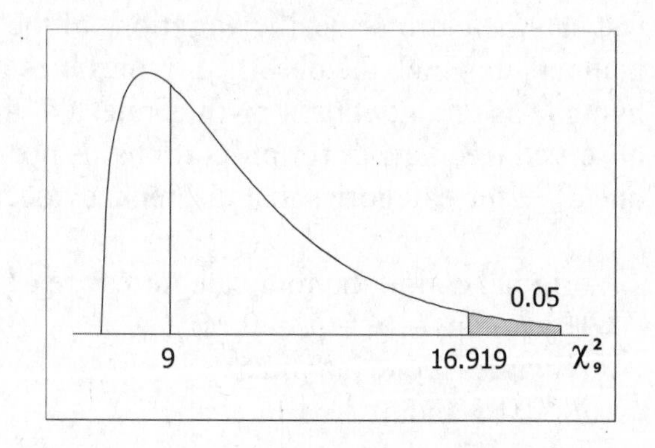

position	O	E	(O-E)²/E
1	19	11.4	5.0667
2	14	11.4	0.5930
3	11	11.4	0.0140
4	14	11.4	0.5930
5	14	11.4	0.5930
6	7	11.4	1.6982
7	8	11.4	1.0140
8	11	11.4	0.0140
9	5	11.4	3.5930
10	11	11.4	0.0140
	114	114.0	13.1930

$\chi^2 = \Sigma[(O - E)^2/E] = 13.193$
P-value = χ^2cdf (13.193,99,9) = 0.1541

conclusion:

Do not reject Ho; there is not sufficient evidence to reject the claim that $p_i = 1/10$ for each position. There is not sufficient evidence to reject the claim that the likelihood of winning is the same for all post positions. Based on these results, post position is not a significant consideration when betting on the Kentucky Derby.

15. original claim: UFO sightings occur in the different months with equal frequency

H_o: $p_{Jan} = p_{Feb} = p_{Mar} = \ldots = p_{Dec} = 1/12$
H_1: at least one $p_i \neq 1/12$
$\alpha = 0.05$ and df = 11
C.V. $\chi^2 = \chi^2_\alpha = \chi^2_{0.05} = 19.675$

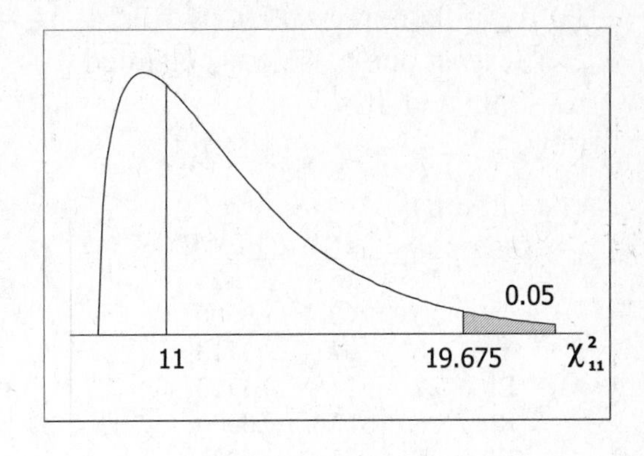

calculations:

month	O	E	$(O-E)^2/E$
Jan	1239	1506.08	47.3636
Feb	1111	1506.08	103.6402
Mar	1428	1506.08	4.0483
Apr	1276	1506.08	35.1497
May	1101	1506.08	108.4159
Jun	1225	1506.08	52.4591
Jul	2233	1506.08	350.8490
Aug	2012	1506.08	169.9452
Sep	1680	1506.08	20.0832
Oct	1994	1506.08	158.0674
Nov	1648	1506.08	13.3727
Dec	1125	1506.08	96.4253
	18073	18073.00	1159.8196

$\chi^2 = \Sigma[(O - E)^2/E] = 1159.820$
P-value = χ^2cdf (1159.820,9999,11) ≈ 0

conclusion:

Reject Ho; there is sufficient evidence to reject the claim that $p_i = 1/12$ for each month. There is sufficient evidence to reject the claim that UFO sightings occur in the different months with equal frequency. July and August likely have the highest frequency because those are summer months when more people are more likely to be outdoors.

17. original claim: the observed frequencies fit the expected proportions

H_o: $p_1 = 9/16$, $p_2 = 3/16$, $p_3 = 3/16$, $p_4 = 1/16$
H_1: at least one p_i is not as claimed
$\alpha = 0.05$ and df = 3
C.V. $\chi^2 = \chi^2_\alpha = \chi^2_{0.05} = 7.815$

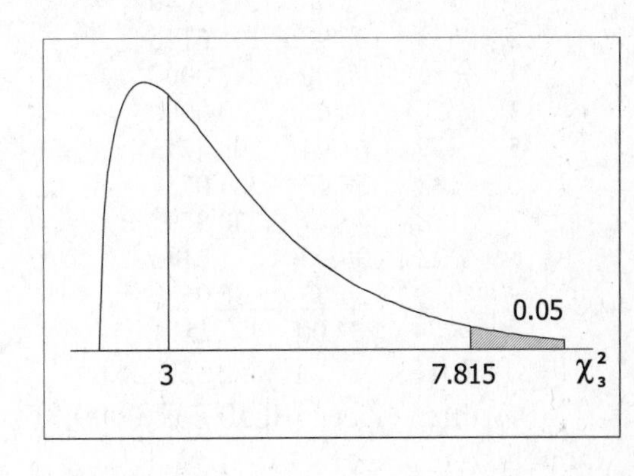

calculations:

category	O	E	$(O-E)^2/E$
1	59	45	4.3556
2	15	15	0.0000
3	2	15	11.2667
4	4	5	0.2000
	80	80	15.8222

$\chi^2 = \Sigma[(O - E)^2/E] = 15.822$
P-value = χ^2cdf (15.822,99,3) = 0.0012

conclusion:

Reject Ho; there is sufficient evidence to reject the claim that the proportions are as claimed. There is sufficient evidence to reject the claim that the observed frequencies fit the proportions that were expected according to the principles of genetics.

19. original claim: the color distribution is as stated by Mars

H_o: $p_G = 0.16$, $p_O = 0.20$, $p_Y = 0.14$, $p_{Bl} = 0.24$, $p_R = 0.13$, $p_{Br} = 0.13$

H_1: at least one p_i is not as claimed

$\alpha = 0.05$ and df = 5

C.V. $\chi^2 = \chi^2_\alpha = \chi^2_{0.05} = 11.071$

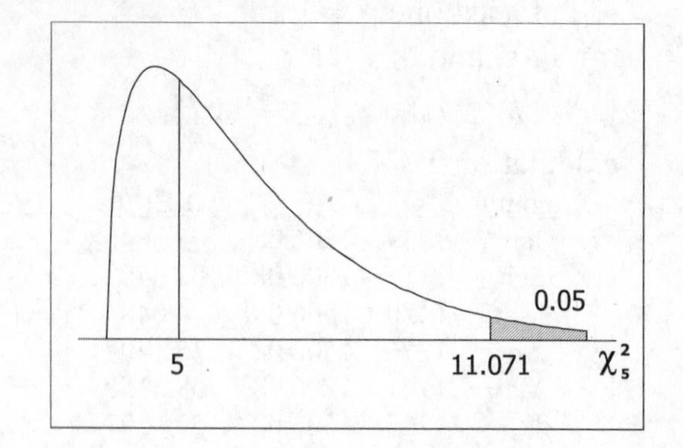

calculations:

color	O	E	$(O-E)^2/E$
G	19	16	0.5625
O	25	20	1.2500
Y	8	14	2.5714
Bl	27	24	0.3750
R	13	13	0.0000
Br	8	13	1.9231
	100	100	6.6820

$\chi^2 = \Sigma[(O-E)^2/E] = 6.682$

P-value = χ^2cdf (6.682,99,5) = 0.2454

conclusion:

Do not reject Ho; there is not sufficient evidence to reject the claim that the proportions are as stated. There is not sufficient evidence to reject the claim that the color distribution is as stated by Mars.

21. original claim: the leading digits follow Benford's law.

H_o: $p_1 = 0.301$, $p_2 = 0.176$, $p_3 = 0.125$, $p_4 = 0.097, \ldots, p_9 = 0.046$

H_1: at least one p_i is not as claimed

$\alpha = 0.01$ and df = 8

C.V. $\chi^2 = \chi^2_\alpha = \chi^2_{0.01} = 20.090$

0.01
8
20.090
χ^2_8

calculations:

digit	O	E	$(O-E)^2/E$
1	0	235.98	235.9840
2	15	137.98	101.6146
3	0	98.00	98.0000
3	76	76.05	0.0000
5	479	61.94	2808.4213
6	183	52.53	324.0737
7	8	45.47	30.8795
8	23	39.98	7.2143
9	0	36.06	36.0640
	784	784.00	3650.2514

$\chi^2 = \Sigma[(O-E)^2/E] = 3650.251$

P-value = χ^2cdf (3650.251,9999,8) ≈ 0

conclusion:

Reject Ho; there is sufficient evidence to reject the claim that the observed frequencies fit the stated proportions. There is sufficient evidence to reject the claim that there is goodness-of-fit of the leading digits with Benford's law. Yes; it appears that the checks are the result of fraud.

23. original claim: the leading digits follow Benford's law.
H_o: $p_1 = 0.301$, $p_2 = 0.176$, $p_3 = 0.125$, $p_4 = 0.097,\ldots, p_9 = 0.046$
H_1: at least one p_i is not as claimed
$\alpha = 0.01$ and df = 8
C.V. $\chi^2 = \chi^2_\alpha = \chi^2_{0.01} = 20.090$
calculations:

digit	O	E	$(O-E)^2/E$
1	52	63.82	2.1865
2	40	37.31	0.1936
3	23	26.50	0.4623
3	20	20.56	0.0155
5	21	16.75	1.0795
6	9	14.20	1.9066
7	8	12.30	1.5009
8	9	10.81	0.3037
9	30	9.75	42.0408
	212	212.00	49.6894

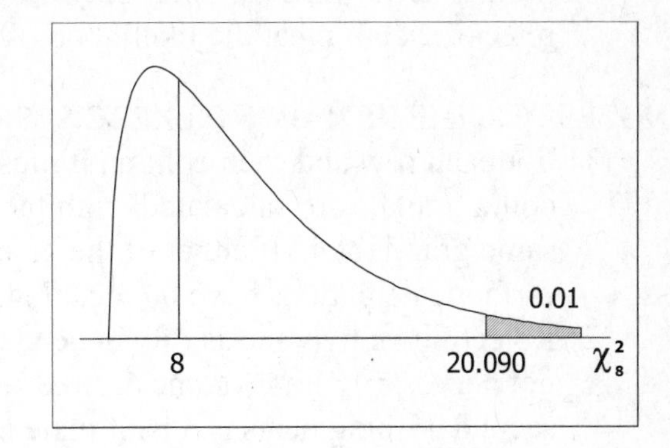

$\chi^2 = \Sigma[(O - E)^2/E] = 49.689$
P-value = χ^2cdf (49.689,99,8) = 4.9E-8 = 0.00000005
conclusion:
Reject Ho; there is sufficient evidence to reject the claim that the observed frequencies fit the stated proportions. There is sufficient evidence to reject the claim that there is goodness-of-fit of the leading digits with Benford's law. No; it appears the campaign contributions are not legitimate.

25. An outlier in one of the categories changes the total number of observations and the expected value for each of the other categories. In Example 1 the calculated $\chi^2 = \Sigma[(O - E)^2/E]$ statistic changes from $\Sigma[(O-8)^2/8] = 11.250$ to $\Sigma[(O-14.3)^2/14.3] = 247.280$, and the effect of the outlier is dramatic.

11-3 Contingency Tables

1. To move from scientific notation to standard notation, the E-11 indicates that the decimal point should be moved 11 places to the left, which places 10 zeros between the decimal point and the first significant digit, so that 1.73251E-11 = 0.0000000000173. Because the P-value is so low, reject the claim that getting paralytic polio is independent of receiving the Salk vaccine. The vaccine appears to be effective.

3. The P-value is the probability of getting results as extreme as or more extreme than the ones obtained whenever the null hypothesis is true. In this context, it is the probability of achieving a success rate at least as effective as the one observed be chance alone – i.e., if the vaccine has no effect, and getting paralytic polio truly is independent of receiving the vaccine.

5. H_o: winning team (home/visitor) and the sport played are independent
H_1: winning team (home/visitor) and the sport played are related
$\alpha = 0.05$ and df = 3
C.V. $\chi^2 = \chi^2_\alpha = \chi^2_{0.05} = 7.815$
Calculations:
$\chi^2 = \Sigma[(O - E)^2/E] = 4.737$ [TI-83/84$^+$]
P-value = χ^2cdf (4.737,99,3) = 0.1921 [TI-83/84$^+$]

conclusion:

Do not reject Ho; there is not sufficient evidence to reject the claim that the winning team (home/visitor) and the sport played are independent. There is not sufficient evidence to reject the claim that the likelihood of the home team winning is independent of the sport.

NOTES FOR THE REMAINING EXERCISES:

(1) For each row and each column it must be true that $\Sigma O = \Sigma E$. After the marginal row and column totals are calculated, both the row totals and column totals must sum to produce the same grand total. If either of the preceding is not true, then an error has been made and further calculations have no meaning.

(2) Rejecting an hypothesis of independence to conclude that two variables are related does not necessarily imply some desired result (e.g., "the drug is effective"). Making such secondary judgments is a legitimate informal procedure, but it requires checking to make sure the frequencies in the table occur in the appropriate direction.

(3) All contingency table analyses in this manual use the following conventions.
- The E values for each cell are given in parentheses below the O values.
- The addends used to calculate the χ^2 test statistic follow the arrangement of the cells in the contingency table – making it easier to monitor the large number of intermediate steps involved and helping to prevent errors caused by missing or double-counting cells.
- As before, P-values are obtained using the χ^2cdf command on the TI-83/84$^+$ calculator.
- The accompanying illustrations follow the "usual" chi-square shape, even though that shape is not correct for df=1 or df=2.

7. H_o: challenge success and gender are independent
H_1: challenge success and gender are related
$\alpha = 0.05$ and df = 1
C.V. $\chi^2 = \chi^2_\alpha = \chi^2_{0.05} = 3.841$
calculations:

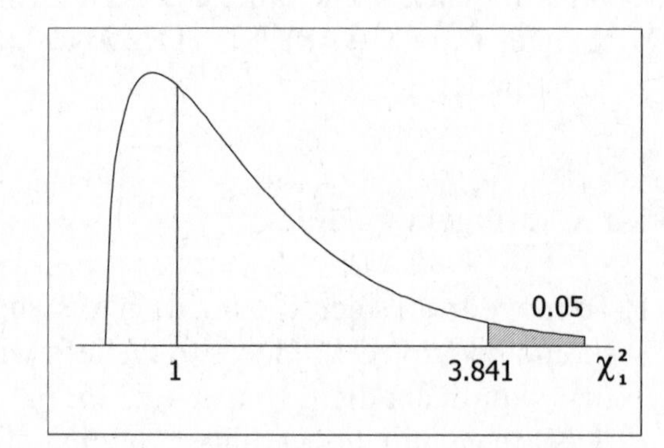

		SUCCESSFUL? yes	no	
GENDER	M	201 (190.59)	288 (298.41)	489
	F	126 (136.41)	224 (213.59)	350
		327	512	839

$\chi^2 = \Sigma[(O - E)^2/E]$
$= 0.569 + 0.363$
$0.795 + 0.598 = 2.235$

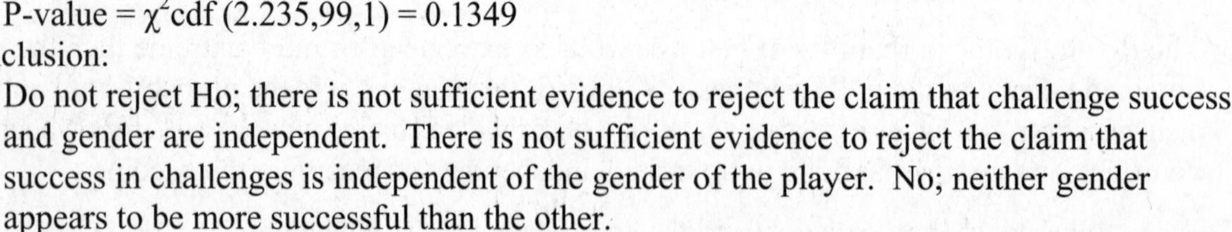

P-value = χ^2cdf (2.235,99,1) = 0.1349

conclusion:

Do not reject Ho; there is not sufficient evidence to reject the claim that challenge success and gender are independent. There is not sufficient evidence to reject the claim that success in challenges is independent of the gender of the player. No; neither gender appears to be more successful than the other.

9. H_o: subject truthfulness and test result are independent
H_1: subject truthfulness and test result are related
$\alpha = 0.05$ and df = 1
C.V. $\chi^2 = \chi^2_\alpha = \chi^2_{0.05} = 3.841$

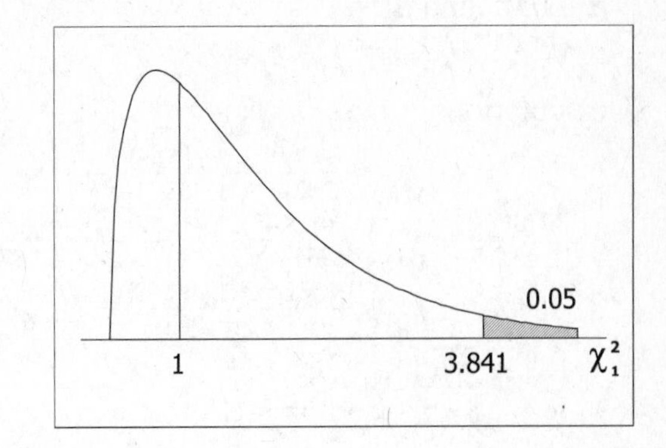

calculations:

		SUBJECT LIE?		
		no	yes	
TEST	lie	15 (190.59)	42 (298.41)	57
	truth	32 (136.41)	9 (213.59)	41
		47	51	98

$\chi^2 = \Sigma[(O - E)^2/E]$
$= 5.567 + 5.131$
$7.740 + 7.133 = 25.571$

P-value = χ^2cdf (25.571,99,1) = 4.3E-7 = 0.0000004

conclusion:

Reject Ho; there is sufficient evidence to reject the claim that subject truthfulness and test results are independent. There is sufficient evidence to reject the claim that the test result is independent of whether the subject was telling the truth. Yes; the results suggest that polygraphs are effective in distinguishing between truth and lies.

11. H_o: type of filling and adverse health conditions are independent
H_1: type of filling and adverse health conditions are related
$\alpha = 0.05$ and df = 1
C.V. $\chi^2 = \chi^2_\alpha = \chi^2_{0.05} = 3.841$

calculations:

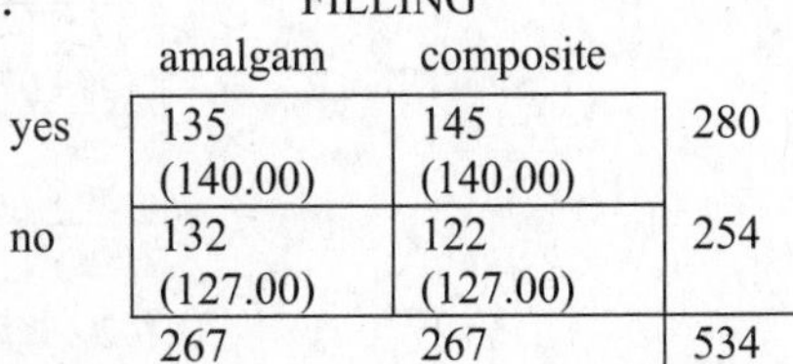

		FILLING		
		amalgam	composite	
ADVERSE?	yes	135 (140.00)	145 (140.00)	280
	no	132 (127.00)	122 (127.00)	254
		267	267	534

0.05
1
3.841
χ^2_1

$\chi^2 = \Sigma[(O - E)^2/E]$
$= 0.179 + 0.179$
$0.197 + 0.197 = 0.751$

P-value = χ^2cdf (0.751,99,1) = 0.3862

conclusion:

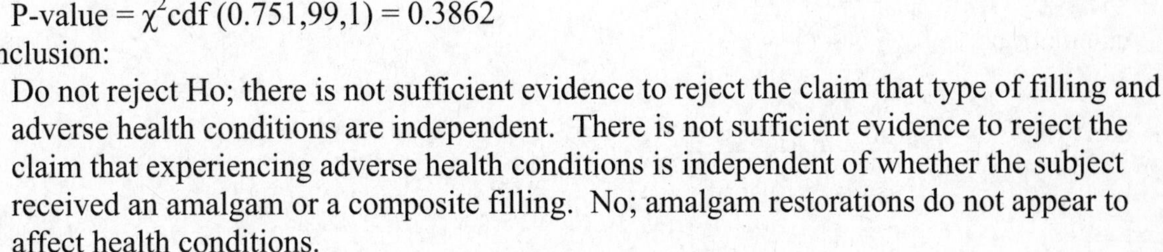

Do not reject Ho; there is not sufficient evidence to reject the claim that type of filling and adverse health conditions are independent. There is not sufficient evidence to reject the claim that experiencing adverse health conditions is independent of whether the subject received an amalgam or a composite filling. No; amalgam restorations do not appear to affect health conditions.

13. H_o: plea and sentence are independent
H_1: plea and sentence are related
$\alpha = 0.05$ and df = 1
C.V. $\chi^2 = \chi^2_\alpha = \chi^2_{0.05} = 3.841$
calculations:

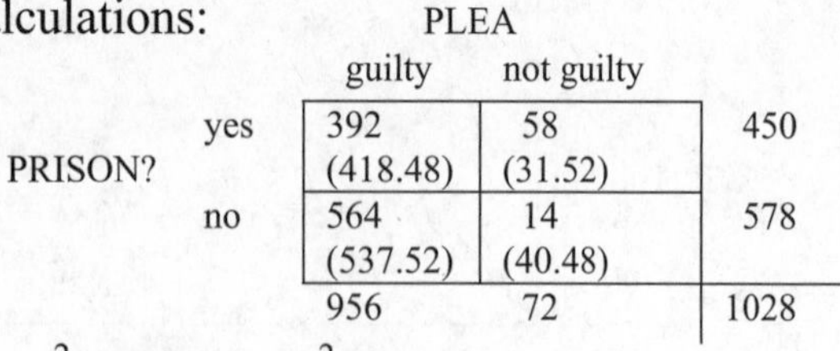

		PLEA		
		guilty	not guilty	
PRISON?	yes	392 (418.48)	58 (31.52)	450
	no	564 (537.52)	14 (40.48)	578
		956	72	1028

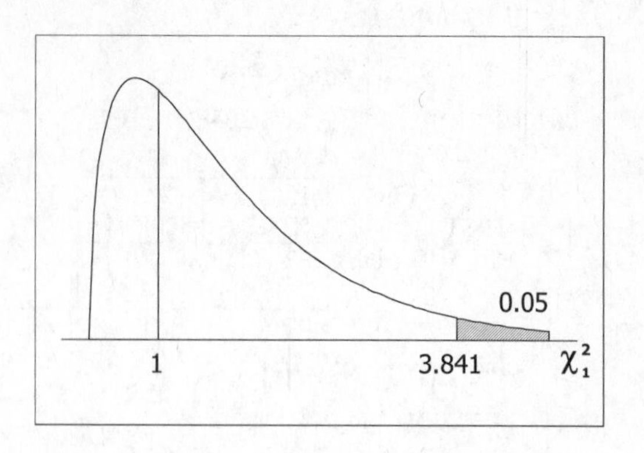

$\chi^2 = \Sigma[(O - E)^2/E]$
$= 1.676 + 22.252$
$1.305 + 17.324 = 42.557$
P-value = χ^2cdf (42.557,99,1) = 6.9E-11 ≈ 0
conclusion:
Reject Ho; there is sufficient evidence to reject the claim that plea and sentence are independent. There is sufficient evidence to reject the claim that whether or not a person is sent to prison is independent of the plea. It appears that those entering a guilty pleas are less likely to be sent to prison. Yes; assuming those who are really guilty will be convicted with a trial, these results suggest that a guilty plea should be encouraged for guilty persons.
NOTE: The study could be misleading because it appears to compare those who plead guilty only to those who are convicted in trials. Apparently those who plead not guilty and are found not guilty (and therefore receive no prison sentence) are not included – and perhaps a guilty person who has a chance for acquittal should plead not guilty. If 50 guilty people with not guilty pleas were acquitted, as shown at the right, including them in the table in the no prison category changes the above conclusion since $\chi^2 = \Sigma[(O - E)^2/E] = 0.125 + 0.982 + 0.090 + 0.704 = 1.901$.

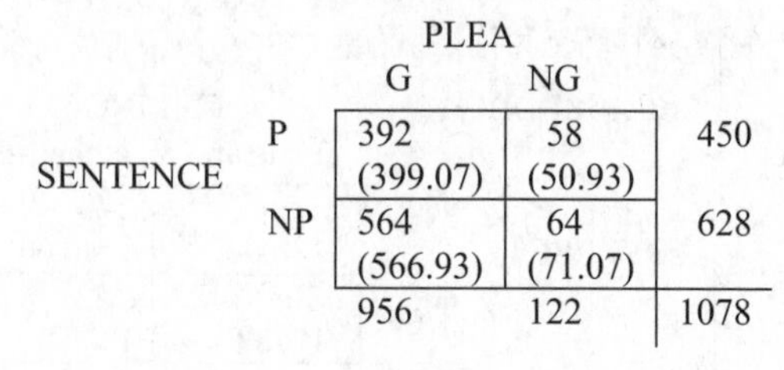

		PLEA		
		G	NG	
SENTENCE	P	392 (399.07)	58 (50.93)	450
	NP	564 (566.93)	64 (71.07)	628
		956	122	1078

15. H_o: success and type of treatment are independent
H_1: success and type of treatment are related
$\alpha = 0.01$ and df = 1
C.V. $\chi^2 = \chi^2_\alpha = \chi^2_{0.01} = 6.635$
calculations:

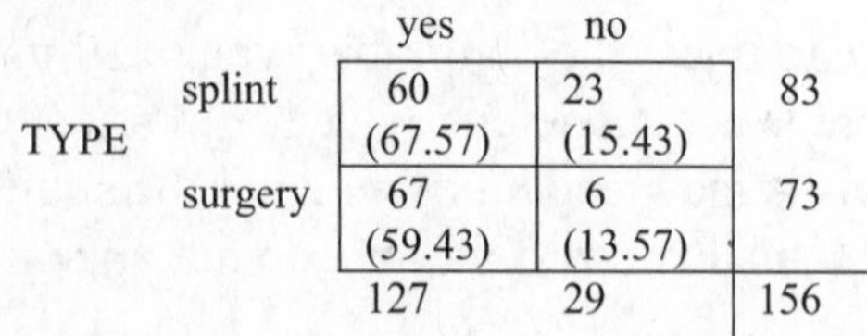

		SUCCESS?		
		yes	no	
TYPE	splint	60 (67.57)	23 (15.43)	83
	surgery	67 (59.43)	6 (13.57)	73
		127	29	156

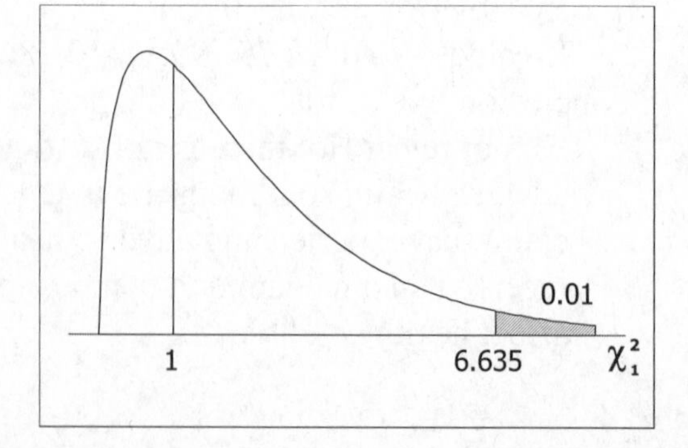

$\chi^2 = \Sigma[(O - E)^2/E]$
$= 0.848 + 3.714$
$0.964 + 4.223 = 9.750$
P-value = χ^2cdf (9.750,99,1) = 0.0018
conclusion:
Reject Ho; there is sufficient evidence to reject the claim that success and type of treatment are independent. There is sufficient evidence to reject the claim that successfully treating carpal tunnel syndrome is independent of the type of treatment (splint vs. surgery) used. The results suggest that the surgery treatment is associated with a higher rate of success.

17. H_0: gender and opinion are independent
H_1: gender and opinion are related
$\alpha = 0.05$ and df = 2
C.V. $\chi^2 = \chi^2_\alpha = \chi^2_{0.05} = 5.991$
calculations:

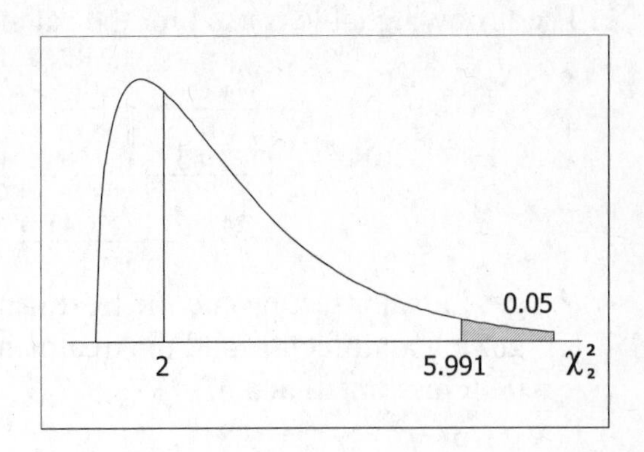

		OPINION			
		human	natural	other	
GENDER	M	314 (307.34)	146 (152.19)	44 (44.47)	504
	F	308 (314.66)	162 (155.81)	46 (45.53)	516
		622	308	90	1020

$\chi^2 = \Sigma[(O - E)^2/E]$
$= 0.144 + 0.252 + 0.005$
$0.141 + 0.246 + 0.005 = 0.792$
P-value = χ^2cdf (0.792,99,2) = 0.6730
conclusion:
Do not reject Ho; there is not sufficient evidence to reject the claim that gender and opinion are independent. There is not sufficient evidence to reject the claim that the gender of the respondent is independent of that respondent's opinion about the cause of global warming. Men and women appear to generally agree.

19. H_0: reaction and treatment are independent
H_1: reaction and treatment are related
$\alpha = 0.01$ and df = 2
C.V. $\chi^2 = \chi^2_\alpha = \chi^2_{0.01} = 9.210$
calculations:

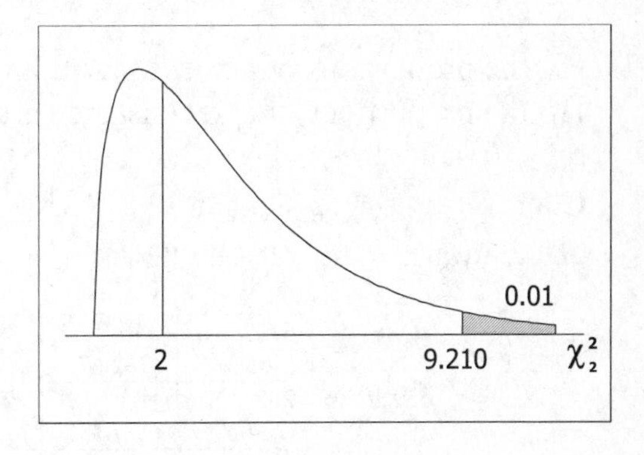

		TREATMENT			
		placebo	C-1332	C-1998	
REACT?	Y	344 (284.12)	89 (143.72)	8 (13.16)	441
	N	13621 (1421.88)	774 (719.28)	71 (65.84)	2207
		1706	863	79	2648

$\chi^2 = \Sigma[(O - E)^2/E]$
$= 12.621 + 20.837 + 2.021$
$2.522 + 4.164 + 0.404 = 42.568$
P-value = χ^2cdf (42.568,99,2) = 5.7E-10 ≈ 0
conclusion:
Reject Ho; there is sufficient evidence to reject the claim that reaction and treatment are independent. There is sufficient evidence to reject the claim that having an adverse effect on the digestive system is not independent of the treatment group. Yes; since the observed frequencies in the REACT?-yes row of the two Campral columns are less than the expected frequencies in those cells, Campral appears to be associated with reduced adverse effects on the digestive system.

21. The following table is used for the calculations in exercise #21.

		TREATMENT				
		placebo	A-10	A-40	A-80	
INFECTION?	Y	27 (27.08)	89 (86.56)	8 (7.92)	7 (9.43)	131
	N	243 (242.92)	774 (776.44)	71 (71.08)	87 (84.57)	1175
		270	863	79	94	1306

H_o: getting an infection and the treatment used are independent
H_1: getting an infection and the treatment used are related
$\alpha = 0.05$ [assumed] and df = 3
C.V. $\chi^2 = \chi^2_\alpha = \chi^2_{0.05} = 7.815$
calculations:

$$\chi^2 = \Sigma[(O - E)^2/E]$$
$$= 0.000 + 0.069 + 0.001 + 0.626$$
$$0.000 + 0.008 + 0.000 + 0.070$$
$$= 0.773$$

P-value = χ^2cdf (0.773,99,3)
= 0.8559

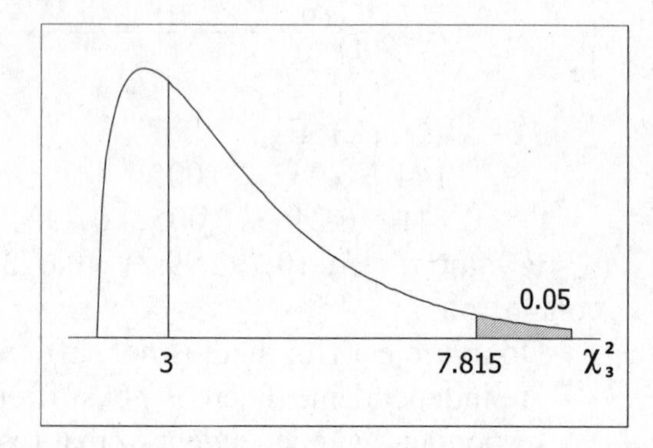

conclusion:
Do not reject Ho; there is not sufficient evidence to reject the claim that getting an infection and the treatment used are independent. There is not sufficient evidence to reject the claim that getting an infection is independent of the atorvastatin treatment used. No; the drug does not seem to have an effect on infection rates.

23. H_o: the proportions of agree/disagree are the same for male and female interviewers
H_1: the proportions of agree/disagree are not the same for male and female interviewers
$\alpha = 0.01$ and df = 1
C.V. $\chi^2 = \chi^2_\alpha = \chi^2_{0.01} = 6.635$
calculations:

		INTERVIEWER		
		male	female	
REPLY	agree	512 (565.33)	336 (282.67)	848
	disagree	288 (234.67)	64 (117.33)	352
		800	400	1200

$$\chi^2 = \Sigma[(O - E)^2/E]$$
$$= 5.051 + 10.063$$
$$12.121 + 24.242 = 51.458$$

P-value = χ^2cdf (51.458,99,1) = 7.3E-13 ≈ 0

conclusion:
Reject Ho; there is sufficient evidence to reject the claim that the proportions of agree/disagree are the same for male and female interviewers. There is sufficient evidence to reject the claim that women given the proportions of agree/disagree responses to male interviewers as they do to female interviewers. Yes; it appears that the gender of the interviewer affects the responses of women.

25. The test from exercise 7 as a difference between two proportions is as follows.
Let the males be group 1.
$\hat{p}_1 = x_1/n_1 = 201/489 = 0.411$ $\hat{p}_1 - \hat{p}_2 = 0.411 - 0.360 = 0.051$
$\hat{p}_2 = x_2/n_2 = 126/350 = 0.360$
$\bar{p} = (201+126)/(489+350)$
$= 327/839 = 0.390$

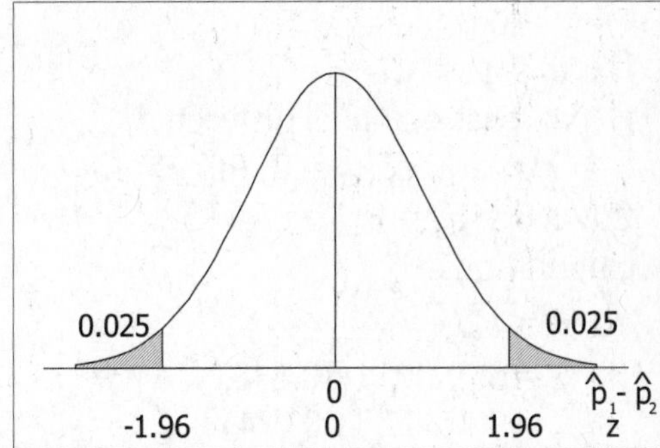

original claim: $p_1 - p_2 \neq 0$
H_o: $p_1 - p_2 = 0$
H_1: $p_1 - p_2 \neq 0$
$\alpha = 0.05$
C.V. $z = \pm z_{\alpha/2} = \pm z_{0.025} = \pm 1.96$
calculations:
$$z_{\hat{p}_1-\hat{p}_2} = (\hat{p}_1 - \hat{p}_2 - \mu_{\hat{p}_1-\hat{p}_2})/\sigma_{\hat{p}_1-\hat{p}_2}$$
$$= (0.051 - 0)/\sqrt{(0.390)(0.610)/489 + (0.390)(0.610)/350} = 0.051/0.03415 = 1.4948$$
P-value = $2\cdot P(z>1.4948) = 2\cdot\text{normalcdf}(1.4948,99) = 0.1350$
conclusion:

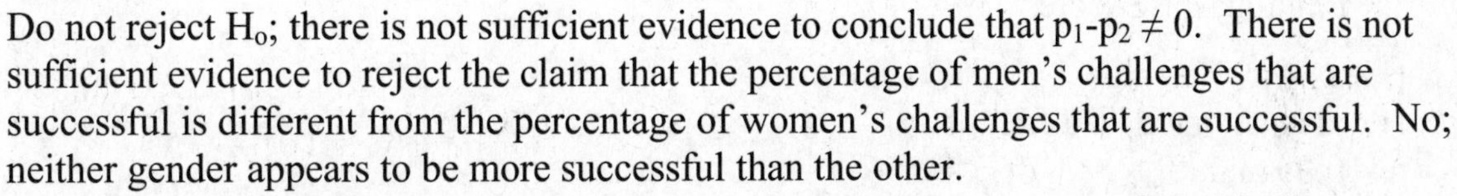

Do not reject H_o; there is not sufficient evidence to conclude that $p_1 - p_2 \neq 0$. There is not sufficient evidence to reject the claim that the percentage of men's challenges that are successful is different from the percentage of women's challenges that are successful. No; neither gender appears to be more successful than the other.

The test as a test of independence is given in Exercise 7. The tests are mathematically equivalent, the relationship being that $(z)^2 = \chi^2$. In particular,
calculated statistic: $(z)^2 = (1.4918)^2 = 2.235 = \chi^2$
critical region: $(z)^2 = (\pm 1.96)^2 = 3.842 = \chi^2$ [$\chi^2 = 3.841$, there is a slight round-off error]
P-value: 0.1350 = 0.1349 [there is a slight round-off error]

11-4 Analysis of Variance

1. a. One-way analysis of variance is appropriate for these data because they represent three or more populations categorized by a single characteristic that distinguishes the populations from each other. The distinguishing characteristic in this case is epoch.
   b. One-way analysis of variance tests the equality of two or more population means by analyzing sample variances. It finds a difference in the population means if the variance between the sample means is larger than can be expected considering the variance within the samples.

3. We should reject the hypothesis that the three epochs have the same mean skull breadth. There is sufficient evidence to conclude that at least one of the means is different from the others.

NOTE: When testing the hypothesis that three or more groups have the same mean, the test statistic is F, where F is the ratio is the ratio of the variance between the groups to the variance within the groups as defined in the text. The following conventions are used in this manual regarding the F test.

- If the desired df does not appear in Table A-7, the closest entry is used. For a desired entry exactly halfway between two tabled values, the conservative approach of using the smaller df is used – and 120 is used for all df larger than 120. The F distribution always "bunches up" around 1.0 regardless of the df values.

- Since the F value depends on two degrees of freedom, the df for "between groups" (numerator) and "within groups" (denominator) may be used with the F as a superscript and a subscript respectively to clarify which F distribution is being used.
- This manual generally uses the generic notation $F = s_B^2/s_p^2$.

5. H_o: $\mu_1 = \mu_2 = \mu_3$
   H_1: at least one μ_i is different
   $\alpha = 0.05$ and $df_{num} = 2$, $df_{den} = 33$
   C.V. $F = F_\alpha = F_{0.05} = 3.3158$
   calculations:
   $F = s_B^2/s_p^2$
   $= 669.0011/70.6481$
   $= 9.4695$ [TI-83/84+]
   P-value $= P(F_{33}^2 > 9.4695) = 0.0006$ [TI-83/84+]
   conclusion:
   Reject Ho; there is sufficient evidence to reject the claim that $\mu_1 = \mu_2 = \mu_3$. There is sufficient evidence to reject the claim that the three books have the same mean Flesch Reading Ease score.

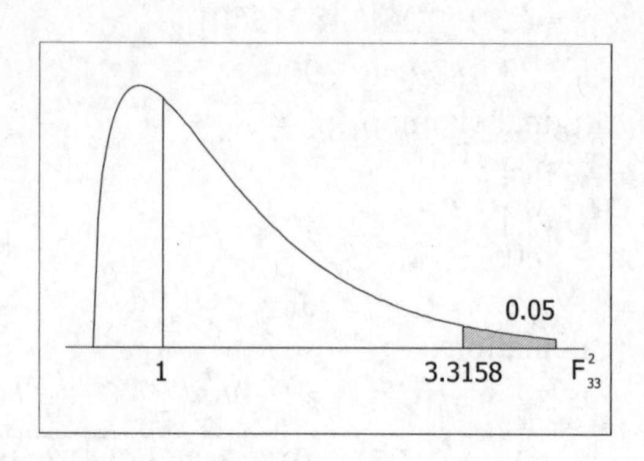

7. H_o: $\mu_1 = \mu_2 = \mu_3 = \mu_4$
   H_1: at least one μ_i is different
   $\alpha = 0.05$ and $df_{num} = 3$, $df_{den} = 156$
   C.V. $F = F_\alpha = F_{0.05} = 2.6626$ [Excel]
   calculations:
   $F = s_B^2/s_p^2$
   $= 11.99995/34.08497$
   $= 0.35206$ [Excel]
   P-value $= P(F_{156}^3 > 0.3521) = 0.7877$ [Excel]
   conclusion:
   Do not reject Ho; there is not sufficient evidence to reject the claim that $\mu_1 = \mu_2 = \mu_3 = \mu_4$. There is not sufficient evidence to reject the claim that the mean weight loss is the same for all four diets. Given the four mean losses that range from 2.1 lbs to 3.3 lbs, it appears that one year of following the diet does not result in a weight loss worth the effort. The effort may be justified, however, when other potential benefits are considered – viz., maintaining one's weight (i.e., not gaining weight), enjoying a healthy lifestyle, etc.

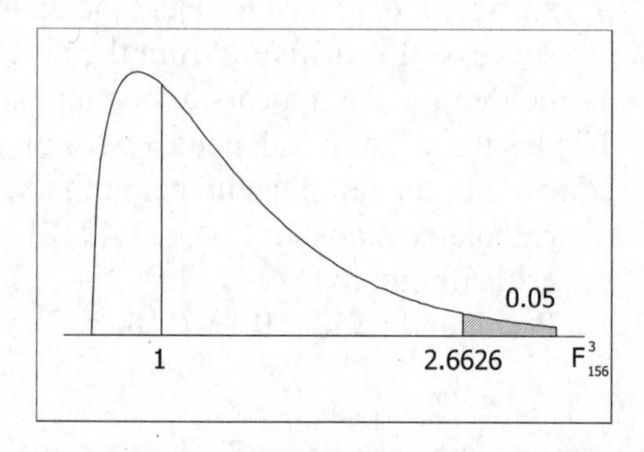

9. H_o: $\mu_1 = \mu_2 = \mu_3$
   H_1: at least one μ_i is different
   $\alpha = 0.05$ and $df_{num} = 2$, $df_{den} = 32$
   C.V. $F = F_\alpha = F_{0.05} = 3.3158$
   calculations:
   $F = s_B^2/s_p^2$
   $= 0.42400.933/7981.129$
   $= 5.313$ [SPSS]
   P-value $= P(F_{32}^2 > 5.313) = 0.010$ [SPSS]
   conclusion:
   Reject Ho; there is sufficient evidence to reject the claim that $\mu_1 = \mu_2 = \mu_3$. There is sufficient evidence to reject the claim that PG, PG-13 and R movies have the same gross amount.

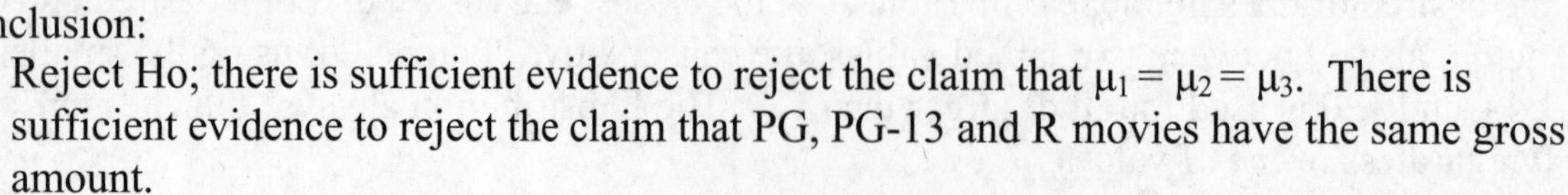

NOTE FOR THE REMAINING EXERCISES IN THIS SECTION: This section is calculation-oriented. Do not get so involved with the formulas that you miss the concepts. This manual arranges the calculations to promote both computational efficiency and understanding of the underlying principles. The following notation is used.

k = the number of groups
n_i = the number of scores in group i (where i = 1,2,…,k)
$\bar{x}_i$ = the mean of group i
s_i^2 = the variance of group i
$\bar{\bar{x}}$ = the overall mean of all the scores in all the groups
$= (\Sigma n_i \bar{x}_i)/\Sigma n_i$ = the (weighted) mean of the group means
$= \Sigma \bar{x}_i / k$ = simplified form when each group has equal size n
s_B^2 = the variance between the groups
$= \Sigma n_i(\bar{x}_i - \bar{\bar{x}})^2/(k-1)$
$= n\,\Sigma(\bar{x}_i - \bar{\bar{x}})^2/(k-1) = ns_{\bar{x}}^2$ = simplified form when each group has equal size n
s_p^2 = the variance within the groups
$= (\Sigma df_i s_i^2)/\Sigma df_i$
$= \Sigma s_i^2 / k$ = simplified form when each group has equal size n
numerator df = k-1
denominator df = Σdf_i
= k(n-1) = simplified form when each group has equal size n
$F = s_B^2/s_p^2$ = (variance between groups)/(variance within groups)
P-value = Fcdf(F, 99, numerator df, denominator df) from the TI-83/84$^+$ calculator

As a crude check against errors, and to help get a feeling for the problem, always verify that the following "overall" values are realistic in that
$\bar{\bar{x}}$ is a value between the lowest and the highest of the $\bar{x}_i$ values.
s_p^2 is a value between the lowest and the highest of the s_i^2 values.

11. Since each group has equal size n=10, use the simplified form of the calculations. The following preliminary values are identified.

	small	medium	large
n	10	10	10
Σx	4315	3906	3858
Σx^2	1918833	1620816	1764878
$\bar{x}$	431.5	390.6	385.8
s^2	6323.389	10570.267	30717.956

$k = 3$ $\qquad \bar{\bar{x}} = \Sigma \bar{x}_i/k$
$n = 10$ $\qquad = 402.633$
$s_{\bar{x}}^2 = \Sigma(\bar{x}_i - \bar{\bar{x}})^2/(k-1)$ $\qquad s_p^2 = \Sigma s_i^2/k$
$= 630.723$ $\qquad = 15870.537$

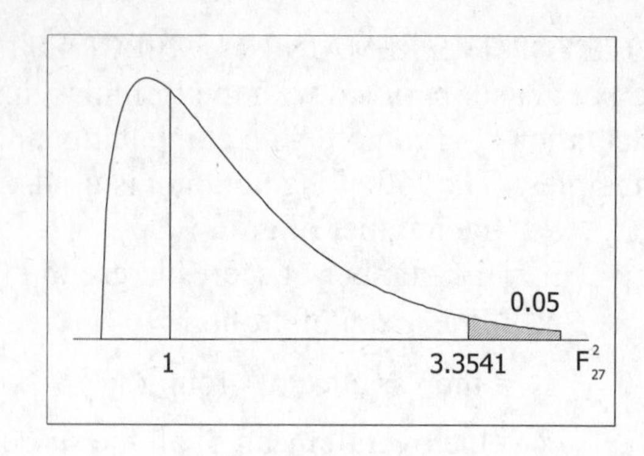

H_o: $\mu_1 = \mu_2 = \mu_3$
H_1: at least one μ_i is different
$\alpha = 0.05$ and $df_{num} = 2$, $df_{den} = 27$
C.V. $F = F_\alpha = F_{0.05} = 3.3541$
calculations:

$F = ns_{\bar{x}}^2/s_p^2$
$= 10(630.723)/15870.537$
$= 0.3974$

P-value = Fcdf(0.3974,99,2,27) = 0.6759

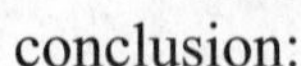
conclusion:

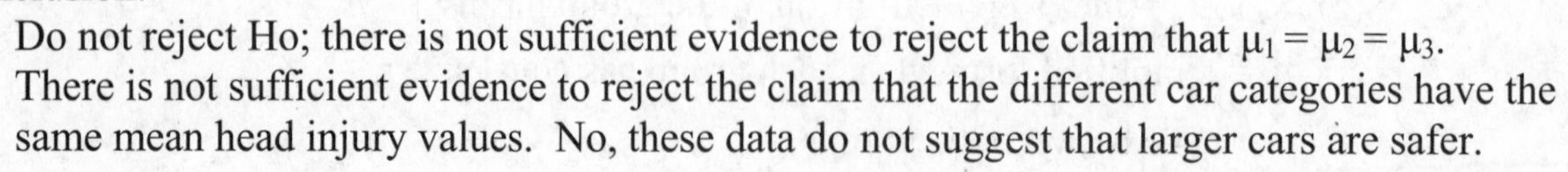
Do not reject Ho; there is not sufficient evidence to reject the claim that $\mu_1 = \mu_2 = \mu_3$. There is not sufficient evidence to reject the claim that the different car categories have the same mean head injury values. No, these data do not suggest that larger cars are safer.

NOTE: Since subtracting the same value from each will not affect the differences between the group means or any variances in the problem, that strategy may be used to reduce the mathematical magnitude of each of the following problems. In Exercise 13, for example, one could subtract 3 minutes from each time and deal only with the excess seconds over 3 minutes as follows.
mile 1: 15 24 23 22 21
mile 2: 19 22 21 17 19
mile 3: 34 31 29 31 29

13. Since each group has equal size n=5, use the simplified form of the calculations. The following preliminary values are identified (all measurements in seconds).

	mile 1	mile 2	mile 3
n	5	5	5
Σx	1005	998	1054
Σx^2	202055	199216	222200
$\bar{x}$	201.0	199.6	210.8
s^2	12.50	3.80	4.20

k = 3 $\qquad \bar{\bar{x}} = \Sigma\bar{x}_i/k$
n = 5 $\qquad = 203.8$
$s_{\bar{x}}^2 = \Sigma(\bar{x}_i - \bar{\bar{x}})^2/(k-1)$ $\qquad s_p^2 = \Sigma s_i^2/k$
$= 37.24$ $\qquad = 6.8333$

H_o: $\mu_1 = \mu_2 = \mu_3$
H_1: at least one μ_i is different
$\alpha = 0.05$ and $df_{num} = 2$, $df_{den} = 12$
C.V. $F = F_\alpha = F_{0.05} = 3.8853$
calculations:

0.05
1
3.8853
F^2_{12}

$F = ns_{\bar{x}}^2/s_p^2$
$= 5(37.24)/6.8333$

$= 27.2488$

P-value = Fcdf(27.2488,99,2,12) = 3.45E-5 = 0.00003

conclusion:

Reject Ho; there is sufficient evidence to reject the claim that $\mu_1 = \mu_2 = \mu_3$. There is sufficient evidence to reject the claim that it takes the same time to ride each of the miles. Yes; the data suggest that there may be a hill on mile 3.

15. The Minitab output is as follows

Level	N	Mean	StDev
king	25	1.2560	0.2329
menthol	25	0.8720	0.2424
filter	25	0.9160	0.2478

Source	DF	SS	MS	F	P
Factor	2	2.2083	1.1041	18.99	0.000
Error	72	4.1856	0.0581		
Total	74	6.3939			

H_o: $\mu_1 = \mu_2 = \mu_3$
H_1: at least one μ_i is different
$\alpha = 0.05$ and $df_{num} = 2$, $df_{den} = 72$
C.V. $F = F_\alpha = F_{0.05} = 3.1504$
calculations:
$F = s_B^2/s_p^2$
$= 1.1041/0.0581$
$= 18.99$ [Minitab]
P-value = $P(F_{72}^2 > 18.99) = 0.000$ [Minitab]

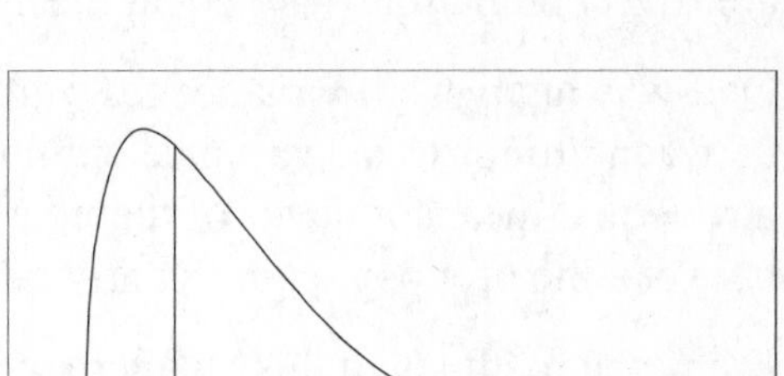

conclusion:
Reject Ho; there is sufficient evidence to reject the claim that $\mu_1 = \mu_2 = \mu_3$. There is sufficient evidence to reject the claim that the three types of cigarettes yield the mean amount of nicotine. Yes; given that only the king size cigarettes are not filtered, it appears (although not formally justifiable by the ANOVA) that the filters do make a difference.

Statistical Literacy and Critical Thinking

1. The numbers in the table are the frequency counts that summarize the data, and they are not the actual data. The actual data are the (145+23+78+4001+322+1786) = 6355 "nausea" or "no nausea" responses.

2. This is a two-way table because each piece of data is categorized in two dimensions (i.e., according to two different variables): according to which treatment it represents (Celebrex, Ibuprofen, placebo), and according to the response (nausea, no nausea).

3. No. In general, statistics deals with mathematical values and associations – not with cause and effect. The proof of causality requires physical evidence, not just statistical evidence.

4. Since two of the three samples are from the same power source, the three sets of sample data are not independent. Since one-way analysis of variance requires independent sets of sample data, the one-way analysis of variance should not be used. Even if the three data sets came from three independent power sources, if all the measurements were taken at precisely the same time the data would consist of matched triples and the one-way analysis of variance (since it ignores the time characteristic) would not be the optimal technique to use.

Chapter Quick Quiz

1. H_o: $p_{Sun} = p_{Mon} = p_{Tue} = p_{Wed} = p_{Thu} = p_{Fri} = p_{Sat} = 1/7$
   H_1: at least one $p_i \neq 1/7$

2. The observed frequency for Sunday is 40. The expected frequency for Sunday is $(\Sigma f)/7 = (40+24+25+28+29+32+38)/7 = 216/7 = 30.857$.

3. The critical value is the value in the chi-square distribution with 6 degrees of freedom that has 0.05 in the upper tail: $\chi_\alpha^2 = \chi_{0.05}^2 = 12.592$.

4. Since 0.2840 > 0.05, we fail to reject the null hypothesis. There is not sufficient evidence to reject that claim that fatal DWI crashes occur equally on the different days of the week.

5. The one-way analysis of variance is used to test the null hypothesis that three or more samples are from populations with equal means.

6. One-way analysis of variance tests are right tailed. The test statistic is the ratio of the variance between groups to the variance within groups, and we reject the hypothesis that the groups have equal means in favor of them having means that are not all equal only if the variance between the groups is significantly larger than the variance between groups.

7. Using the methods of this chapter (section 11-3),
   H_0: response and job position are independent
   H_1: response and job position are related
   NOTE: There is another possibility. Using the methods of section 9-2,
   H_0: $p_1 - p_2 = 0$
   H_1: $p_1 - p_2 \neq 0$
   where p_1 = the proportion of workers who say it is unethical to monitor employee e-mail
   p_2 = the proportion of bosses who say it is unethical to monitor employee e-mail

8. For the test of independence in exercise 7, the critical value is the value in the chi-square distribution with 1 degree of freedom that has 0.05 in the upper tail: $\chi^2_\alpha = \chi^2_{0.05} = 3.841$.

9. Since 0.0302 < 0.05, we reject the null hypothesis. There is sufficient evident to reject the claim that a person's response and job position are independent. There is sufficient evidence to conclude that a person's opinion about monitoring employee e-mail is related to whether that person is a worker or a boss.

10. Since 0.0302 > 0.01, we fail to reject the null hypothesis. There is not sufficient evidence to reject the claim that a person's response and job status are independent. There is not sufficient evidence to reject the claim that person's opinion about monitoring employee e-mail is independent of whether that person is a worker or a boss.

Review Exercises

1. original claim: experiencing nausea is independent of the treatment
   H_o: nausea and treatment are independent
   H_1: nausea and treatment are related
   $\alpha = 0.05$ and df = 2
   C.V. $\chi^2 = \chi^2_\alpha = \chi^2_{0.05} = 5.991$
   calculations:

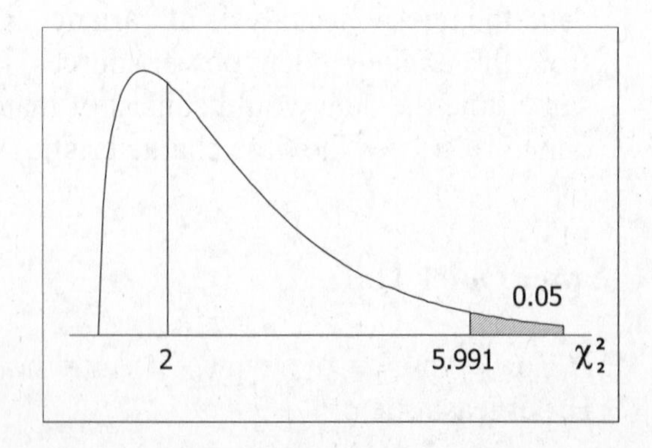

		TREATMENT			
		celebrex	ibupro	placebo	
NAUSEA?	Y	145 (160.45)	23 (13.35)	78 (72.15)	246
	N	4001 (3985.51)	322 (331.65)	1786 (1791.85)	6109
		4146	345	1864	6355

$\chi^2 = \Sigma[(O - E)^2/E]$
$= 1.495 + 6.966 + 0.474$
$0.060 + 0.281 + 0.019 = 9.294$
P-value = χ^2cdf (9.294,99,2) = 0.0096

conclusion:

Reject Ho; there is sufficient evidence to reject the claim that nausea and treatment are independent. There is sufficient evidence to reject the claim that a subject's experiencing nausea is independent of the treatment received. No; the adverse reaction of nausea does not appear to be about the same for the different treatments.

2. original claim: deaths by lightning occur on the different days with equal frequency

H_o: $p_{Sun} = p_{Mon} = p_{Tue} = \ldots = p_{Sat} = 1/7$

H_1: at least one $p_i \neq 1/7$

$\alpha = 0.01$ and df = 6

C.V. $\chi^2 = \chi^2_\alpha = \chi^2_{0.01} = 16.812$

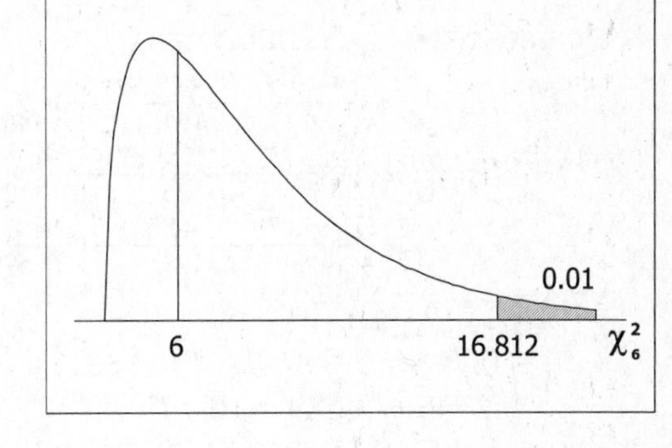

calculations:

day	O	E	$(O-E)^2/E$
Sun	574	462.57	26.8420
Mon	445	462.57	0.6675
Tue	429	462.57	2.4365
Wed	473	462.57	0.2351
Thu	428	462.57	2.5838
Fri	422	462.57	3.5585
Sat	467	462.57	0.0424
	3238	3238.00	36.3657

$\chi^2 = \Sigma[(O - E)^2/E] = 36.366$

P-value = χ^2cdf (36.366,99,6) = 2.34E-6 = 0.000002

conclusion:

Reject Ho; there is sufficient evidence to reject the claim that $p_i = 1/7$ for each day. There is sufficient evidence to reject the claim that lightning deaths occur on the different days with equal frequency. The disproportionate number of deaths on Sunday is due to the fact that more people are involved in outdoor activities on Sundays than on the other days of the week.

3. original claim: the distribution of participants fits the population distribution

H_o: $p_W = 0.757$, $p_H = 0.091$, $p_B = 0.108$, $p_{AP} = 0.038$, $p_{NA} = 0.007$

H_1: at least one p_i is not as claimed

$\alpha = 0.01$ and df = 4

C.V. $\chi^2 = \chi^2_\alpha = \chi^2_{0.05} = 9.488$

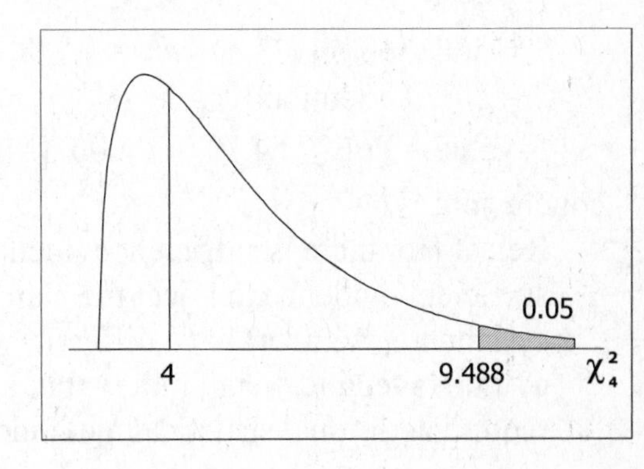

calculations:

ethnic	O	E	$(O-E)^2/E$
W	644	569.26	9.8117
H	23	68.43	30.1623
B	69	81.22	1.8375
AP	14	28.58	7.4349
NA	2	5.26	2.0239
	752	752.75	51.2703

$\chi^2 = \Sigma[(O - E)^2/E] = 51.270$

P-value = χ^2cdf (51.270,99,4) =1.96E-10 ≈ 0

conclusion:

Reject Ho; there is sufficient evidence to reject the claim that the observed frequencies fit the stated proportions. There is sufficient evidence to reject the claim that the distribution of clinical trial participants fits the population distribution. If there is not proportionate representation in clinical trials, then the sample is not representative of the general population and the results should not be applied to the general population.

NOTE: The E's sum to 752.75 because the given probabilities sum to 1.001. In Exercise 20 of section 11-3, a similar problem gives p_W as 0.756 and the given probabilities sum to 1. Using that value gives a calculated $\chi^2 = 51.482$ and the difference is trivial.

4. original claim: continuing to smoke is independent of the treatment received
   H_o: smoking and treatment are independent
   H_1: smoking and treatment are related
   $\alpha = 0.05$ and df = 1
   C.V. $\chi^2 = \chi^2_\alpha = \chi^2_{0.05} = 3.841$
   calculations:

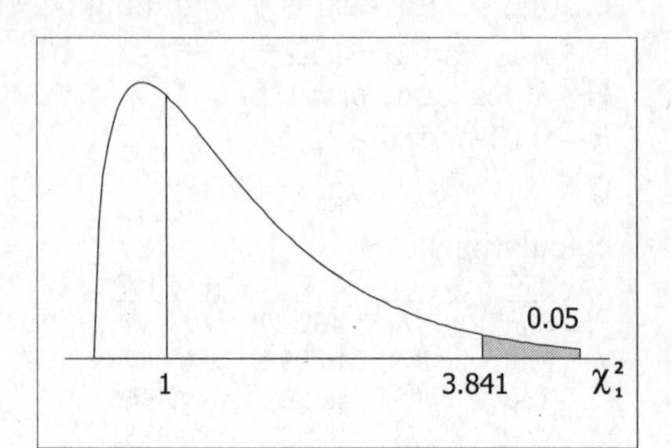

		TREATMENT		
		Bu-Hy	placebo	
SMOKING?	yes	299 (314.33)	167 (151.67)	466
	no	101 (85.67)	26 (41.33)	127
		400	193	593

   $\chi^2 = \Sigma[(O - E)^2/E]$
   $= 0.748 + 1.550$
   $2.745 + 5.689 = 10.732$
   P-value = χ^2cdf (10.732,99,1) = 0.0011

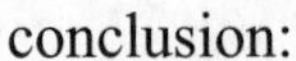

   conclusion:
   Reject Ho; there is sufficient evidence to reject the claim that smoking and treatment are independent. There is sufficient evidence to reject the claim that whether or not a person continues smoking is independent of the treatment received vaccinated. Yes; since the "Bu-Hy/no" cell contains fewer than expected subjects, the vaccine appears to be effective.

5. H_o: $\mu_1 = \mu_2 = \mu_3$
   H_1: at least one μ_i is different
   $\alpha = 0.05$ and $df_{num} = 2$, $df_{den} = 29$
   C.V. $F = F_\alpha = F_{0.05} = 3.3277$
   calculations:
   $F = s_B^2/s_p^2$
   $= 3083363/56373$
   $= 54.70$ [Minitab]
   P-value = $P(F^2_{29} > 54.70) = 0.000$ [Minitab]
   conclusion:
   Reject Ho; there is sufficient evidence to reject the claim that $\mu_1 = \mu_2 = \mu_3$. There is sufficient evidence to reject the claim that the different car categories (4 cylinder, 6 cylinder, 8 cylinder) have different weights. Yes, it does appear that cars with more cylinders tend to weigh more – but a formal conclusion about such a relationship would require additional statistical methodology.

6. These are precisely the values from Data Set 4. The Minitab output is as follows.

```
Level        N    Mean  StDev          Source  DF      SS     MS     F      P
king        25  15.720  0.936          Factor   2    12.1    6.0  0.50  0.608
menthol     25  14.960  4.168          Error   72   868.0   12.1
non-menth25     14.800  4.233          Total   74   880.1
```

H_o: $\mu_1 = \mu_2 = \mu_3$
H_1: at least one μ_i is different
$\alpha = 0.05$ and $df_{num} = 2$, $df_{den} = 72$
C.V. $F = F_\alpha = F_{0.05} = 3.1504$
calculations:

$F = s_B^2/s_p^2$
$= 6.0/12.1$
$= 0.50$ [Minitab]

P-value $= P(F^2_{72} > 0.50) = 0.608$ [Minitab]

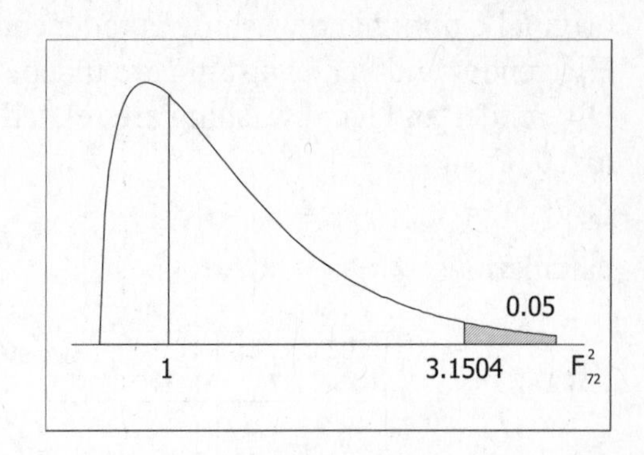

conclusion:

Do not reject Ho; there is not sufficient evidence to reject the claim that $\mu_1 = \mu_2 = \mu_3$. There is not sufficient evidence to reject the claim that the three types of cigarettes yield the same mean amount of carbon monoxide. No; it appears that filters do not make a difference in the amount of carbon monoxide – but a formal test using only two groups (filtered and non-filtered) would provide a better test for answering that question.

Cumulative Review Exercises

1. a. This is an observational study, since the researched merely watched the subjects without applying a treatment or otherwise introducing a factor that would modify the results.
   b. The given numbers are discrete, since they must be whole numbers.
   c. The given numbers are statistics, since they describe the samples and not the populations.
   d. If the subjects could tell that they were being observed, they might modify their behavior. Since the study was sponsored by an organization with a vested interest in the results, it is possible that those interests may have (consciously or subconsciously) biased in some way the conducting of the study.

2. Let the males be group 1.

$\hat{p}_1 = x_1/n_1 = 2023/3065 = 0.660$ $\qquad$ $\hat{p}_1 - \hat{p}_2 = 0.660 - 0.880 = -0.220$
$\hat{p}_2 = x_2/n_2 = 2650/3011 = 0.880$
$\bar{p} = (2023+2650)/(3065+3011)$
$= 4673/6076 = 0.769$

original claim: $p_1 - p_2 \neq 0$
H_o: $p_1 - p_2 = 0$
H_1: $p_1 - p_2 \neq 0$
$\alpha = 0.05$
C.V. $z = \pm z_{\alpha/2} = \pm z_{0.025} = \pm 1.96$
calculations:

$$z_{\hat{p}_1-\hat{p}_2} = (\hat{p}_1 - \hat{p}_2 - \mu_{\hat{p}_1-\hat{p}_2})/\sigma_{\hat{p}_1-\hat{p}_2}$$
$$= (-0.220 - 0)/\sqrt{(0.769)(0.231)/3065 + (0.769)(0.231)/3011} = -0.220/0.01081 = -20.35$$

P-value $= 2 \cdot P(z < -20.35) = 2 \cdot (0.0001) = 0.0002$

conclusion:

Reject H_o; there is sufficient evidence to reject the claim that $p_1 - p_2 = 0$ and conclude that $p_1 - p_2 \neq 0$ (in fact, that $p_1 - p_2 < 0$). There is sufficient evidence to reject the claim that the proportion of men who wash their hands is equal to the proportion of women who wash their hands Yes; there is a significant difference.

3. original claim: hand washing is independent of gender

H_o: gender and hand washing are independent

H_1: gender and hand washing are related

$\alpha = 0.05$ and df = 1

C.V. $\chi^2 = \chi^2_\alpha = \chi^2_{0.05} = 3.841$

calculations:

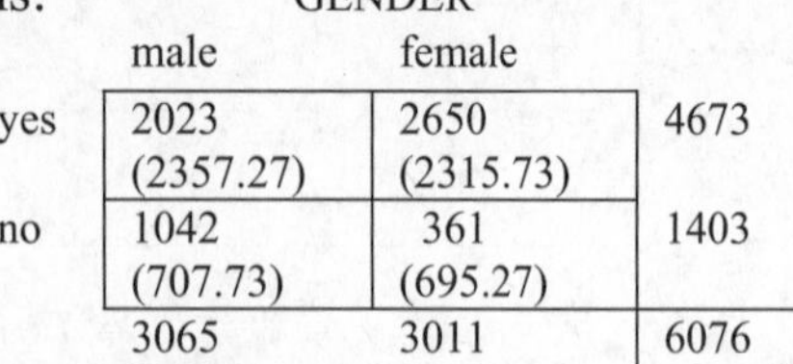

		GENDER		
		male	female	
WASH?	yes	2023 (2357.27)	2650 (2315.73)	4673
	no	1042 (707.73)	361 (695.27)	1403
		3065	3011	6076

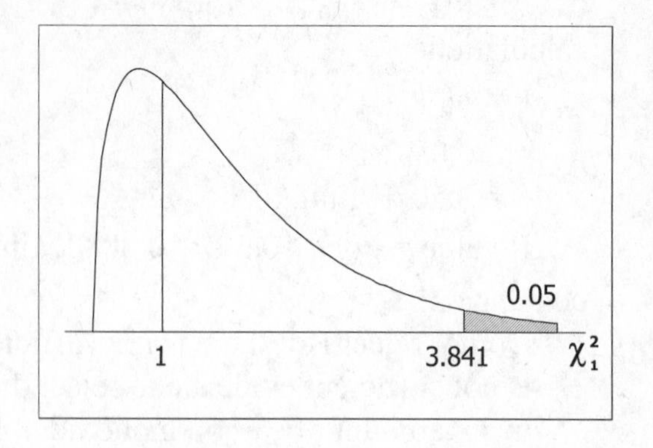

$\chi^2 = \Sigma[(O - E)^2/E]$

$= 47.400 + 48.250$

$157.875 + 160.706 = 414.230$

P-value = χ^2cdf (414.230,999,1) ≈ 0

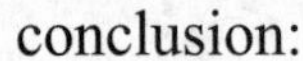

conclusion:

Reject Ho; there is sufficient evidence to reject the claim that gender and hand washing are independent. There is sufficient evidence to reject the claim that hand washing is independent of gender. Since the "male/yes" cell contains fewer than expected subjects, it appears that the proportion of males who wash their hands is lower than the proportion of females who do so.

The following summary statistics apply to exercises 4-6.

$n = 6$
$\Sigma x = 427$
$\Sigma y = 422$
$\Sigma xy = 30021$
$\Sigma x^2 = 30439$
$\Sigma y^2 = 29696$

$n(\Sigma xy) - (\Sigma x)(\Sigma y) = 6(30021) - (427)(422) = -68$
$n(\Sigma x^2) - (\Sigma x)^2 = 6(30439) - (427)^2 = 305$
$n(\Sigma y^2) - (\Sigma y)^2 = 6(29696) - (422)^2 = 92$

4. first round

$\bar{x} = (\Sigma x)/n = (427)/6 = 71.2$
$\tilde{x} = (71+72)/2 = 71.5$ (use ordered scores)
$R = 75 - 67 = 8$ (use ordered scores)
$s^2 = [n(\Sigma x^2) - (\Sigma x)^2]/[n(n-1)]$
$= [305]/[6(5)]$
$= 305/30 = 10.167$
$s = \sqrt{10.167} = 3.189$, rounded to 3.2

fourth round

$\bar{x} = (\Sigma x)/n = (422)/6 = 70.3$
$\tilde{x} = (69+70)/2 = 69.5$ (use ordered scores)
$R = 73 - 69 = 4$ (use ordered scores)
$s^2 = [n(\Sigma x^2) - (\Sigma x)^2]/[n(n-1)]$
$= [92]/[6(5)]$
$= 92/30 = 3.067$
$s = \sqrt{3.067} = 1.751$, rounded to 1.8

The fourth round scores are slightly lower and closer together.

5. $r = [n(\Sigma xy) - (\Sigma x)(\Sigma y)]/[\sqrt{n(\Sigma x^2) - (\Sigma x)^2}\sqrt{n(\Sigma y^2) - (\Sigma y)^2}]$
$= -68/[\sqrt{305}\sqrt{92}]$
$= -0.406$

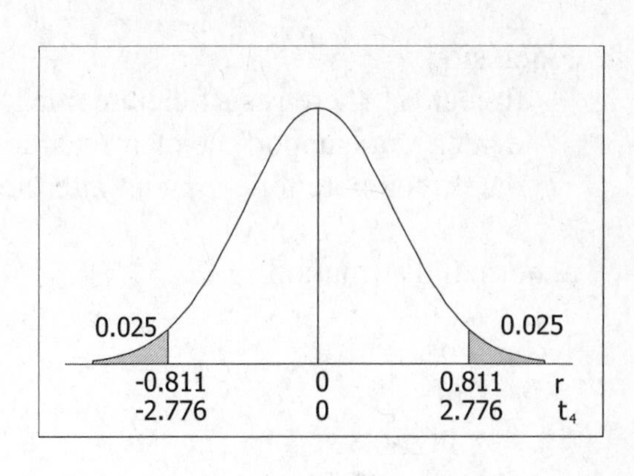

original claim: there is a linear correlation
H_o: $\rho = 0$
H_1: $\rho \neq 0$
$\alpha = 0.05$ and df = 4
C.V. $t = \pm t_{\alpha/2} = \pm t_{0.025} = \pm 2.776$ [or $r = \pm 0.811$]
calculations:

$t_r = (r - \mu_r)/s_r$
$= (-0.406 - 0)/\sqrt{[1 - (-0.406)^2]/4}$
$= -0.406/0.4569$
$= -0.888$

P-value = 2·tcdf(-99,-0.888,4) = 0.4245

conclusion:

Do not reject H_o; there is not sufficient evidence to conclude that $\rho \neq 0$. There is not sufficient evidence to support the claim of a linear correlation between the first round scores and the fourth round scores.

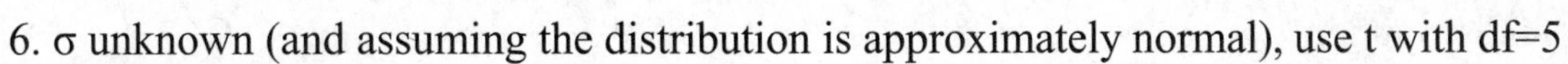

6. σ unknown (and assuming the distribution is approximately normal), use t with df=5
$\alpha = 0.05$, $t_{df,\alpha/2} = t_{5,0.025} = 2.571$

$\bar{x} \pm t_{\alpha/2} \cdot s/\sqrt{n}$

$71.167 \pm 2.571(3.1885)/\sqrt{6}$
71.167 ± 3.347
$67.8 < \mu < 74.5$
We have 95% confidence that the limits of 67.8 and 74.5 contain the true mean first round golf score of all golfers in the tournament.

7. $\alpha = 0.05$, $z_{\alpha/2} = z_{0.025} = 1.96$ and $\hat{p} = x/n = 132/150 = 0.88$

$\hat{p} \pm z_{\alpha/2}\sqrt{\hat{p}\hat{q}/n}$

$0.8800 \pm 1.96\sqrt{(0.88)(0.12)/150}$
0.8800 ± 0.0520
$0.828 < p < 0.932$
To increase the chances of being hired, it appears that an applicant would be wise to send a thank-you note after every job interview for a position that he desired.

8. original claim: $p > 0.75$

$\hat{p} = x/n = 132/150 = 0.880$

Ho: $p = 0.75$
H_1: $p > 0.75$
$\alpha = 0.01$
C.V. $z = z_\alpha = z_{0.01} = 2.326$
calculations:

$z_{\hat{p}} = (\hat{p} - \mu_{\hat{p}})/\sigma_{\hat{p}}$
$= (0.880 - 0.750)/\sqrt{(0.750)(0.250)/150}$
$= 0.130/0.03536 = 3.68$

P-value = P(z>3.68) = 1 – 0.9999 = 0.0001

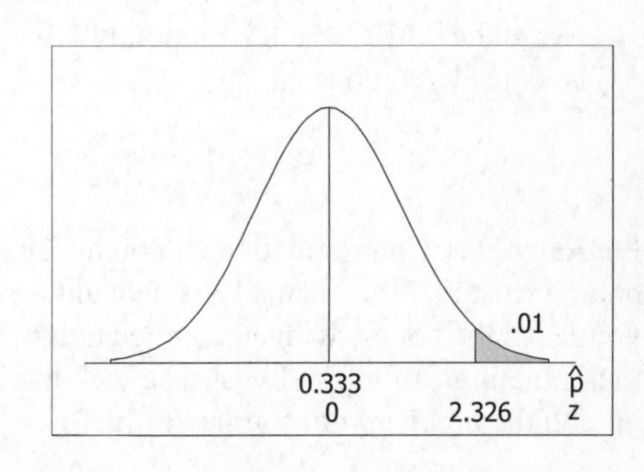

conclusion:
Reject H_o; there is sufficient evidence to conclude that $p > 0.75$. There is sufficient evidence to support the claim that more than 75% of all senior executives believe that a thank-you note following an interview increases the chances of an applicant being hired.

9. a. normal distribution
$\mu = 42.6$
$\sigma = 2.9$
$P(x>45)$
$= P(z>0.83)$
$= 1 - 0.7967$
$= 0.2033$

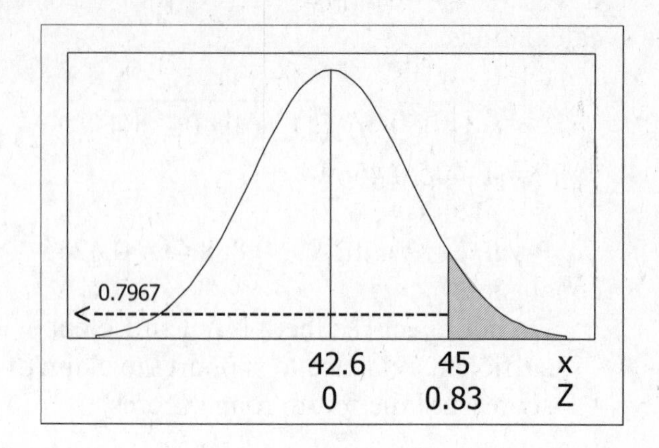

b. normal distribution,
since the original distribution is so
$\mu_{\bar{x}} = \mu = 42.6$
$\sigma_{\bar{x}} = \sigma/\sqrt{n} = 2.9/\sqrt{16} = 0.725$
$P(\bar{x}>45)$
$= P(z>3.31)$
$= 1 - 0.9995$
$= 0.0005$

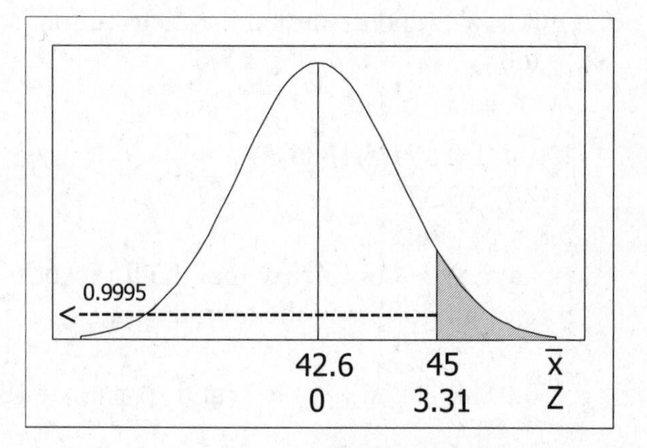

c. The result from (a) is more meaningful, since the cockpit will be occupied by one person at a time.

10. Since repeated random selections are independent,
$P(T_1 \text{ and } T_2 \text{ and } T_3 \text{ and } T_4 \text{ and } T_5) = P(T_1)\cdot P(T_2)\cdot P(T_3)\cdot P(T_4)\cdot P(T_5)$
$= (0.925)(0.925)(0.925)(0.925)(0.925)$
$= (0.925)^5 = 0.6772$
No; since $0.6772 > 0.05$, randomly selecting five women that are all over 5 feet tall would not be considered unusual.

FINAL NOTE: Congratulations! You have completed statistics – the course that everybody likes to hate. I trust that this manual has helped to make the course a little more understandable – and that you leave the course with an appreciation of broad principles, and not merely memories of manipulating formulas. I wish you well in your continued studies, and that you achieve your full potential wherever your journey of life may lead.